Complex Systems

Complex Systems

Theory and Applications

Editors

G. Rzveski
The Open University, UK

C.A. Brebbia
Wessex Institute, UK

WIT*PRESS* **Southampton, Boston**

Editors:

G. Rzveski
The Open University, UK

C.A. Brebbia
Wessex Institute, UK

Published by

WIT Press
Ashurst Lodge, Ashurst, Southampton, SO40 7AA, UK
Tel: 44 (0) 238 029 3223; Fax: 44 (0) 238 029 2853
E-Mail: witpress@witpress.com
http://www.witpress.com

For USA, Canada and Mexico

Computational Mechanics International Inc
25 Bridge Street, Billerica, MA 01821, USA
Tel: 978 667 5841; Fax: 978 667 7582
E-Mail: infousa@witpress.com
http://www.witpress.com

British Library Cataloguing-in-Publication Data

A Catalogue record for this book is available
from the British Library.

Library of Congress Catalog Card Number: 2016917269

ISBN: 978-1-78466-235-6
eISBN: 978-1-78466-236-3

The text of the papers in this volume were set individually by the authors or under their supervision. Only minor corrections to the text may have been carried out by the publisher.

Printed in Great Britain by Lightning Source, UK.

The material contained herein is reprinted from a special edition of the International Journal of Design & Nature and Ecodynamics, Volume 11, Number 4, published by WIT Press.

Preface

This Volume comprises a series of papers dealing with new approaches for resolving complex systems, a topic that has numerous applications in modern society.

The papers demonstrate how complex issues, which cannot be formulated using conventional mathematical or software models, can be resolved.

Complex systems occur in an infinite variety of problems, not only in the realm of physical sciences and engineering, but encompassing fields as diverse as the economy, environment, humanities, social and political sciences. This book demonstrates how complex problems can be solved in different disciplines.

The Editors hope that this work will promote further applications and interaction between researchers working in different fields, as the solution of complex systems nearly always require interdisciplinary approaches.

George Rzevski and Carlos A. Brebbia
Editors
The New Forest, UK
2017

Contents

HARNESSING THE POWER OF SELF-ORGANISATION

G. RZEVSKI
The Open University, Milton Keynes & Multi-Agent Technology Ltd, UK.

ABSTRACT
Complexity of our social, political, economic and technological environment has increased recently to unprecedented levels creating frequent disruptive events (and occasional extreme events) that affect our lives and work. As a contribution to the knowledge on how to cope with complexity, the paper reports research on self-organisation. A practical approach to designing self-organising systems, based on author's substantial experience, is outlined.
Key words: agent autonomy, business processes, complex social systems, emergent creativity, emergent intelligence, emergent leadership, self-organisation.

1 INTRODUCTION

There exists ample evidence that social, political, economic and technological environments in which we live and work are *Complex*; and that complexity is relentlessly increasing [1]. This condition puts a considerable stress on our traditional institutions and organisations, which were developed to operate under conditions of a stable and predictable, rather than complex, environment.

It has been shown that for organisations expected to operate under conditions of complexity, the best way of coping with complexity is to develop a capacity for *Self-Organisation* [2], i.e. the ability to autonomously adjust own connectivity and behaviour in order to eliminate, or at least reduce, consequences of unpredictable external or internal *Disruptive Events*.

And since only complex systems can exhibit self-organisation, it follows that we shall have in the near future to design complexity into our institutions and organisations to enable them to become adaptive in order to survive and prosper in complex environments.

2 COMPLEX SYSTEMS REVISITED

2.1 Definition

Complex Systems are open systems consisting of large numbers of diverse, partially autonomous and richly interacting components, known as Agents, which are not centrally controlled. Global behaviour of complex systems emerges from the interactions of agents and is unpredictable but not random. Complex systems are characterised by the following seven properties: connectivity, autonomy, emergence, nonequilibrium, nonlinearity, self-organisation and co-evolution [3].

2.1.1 Connectivity

Agents are interconnected. Complexity of the system increases with the number of links that connect agents to each other. The strengths of agent links also affects system complexity; the weaker the

DOI: 10.2495/DNE-V11-N4-483-494

links, the easier is to break them and form new ones, which increases system complexity. Adjusting agent connectivity is an effective method for tuning complexity. Complex systems often consist of regions of high connectivity (and high complexity) interconnected by low-connectivity (and low complexity) links, as exemplified by clustering of activities in the human brain.

2.1.2 Autonomy

Agents have certain freedom of behaviour (autonomy), which is always limited by norms, rules, regulations, and/or laws (natural or man-made). The increase in autonomy of agents increases complexity and if all constraints on agent behaviour are removed the system switches from complex to random behaviour. Inversely, if autonomy of agents is reduced (by tightening of laws and/or regulations), the system complexity will decrease, and in the extreme, the system will become deterministic. Agents are not centrally controlled.

2.1.3 Emergence

Behaviour of complex systems emerges from the interactions of agents and is not predictable and yet it is not random. Uncertainty about the outcome of agent interactions is always between 0 and 1. Emergence, in general, denotes a property of a system that is evident in the system as a whole but it is not present in any of its components.

2.1.4 Nonequilibrium

Complex systems are subjected to perpetual change experienced as a succession of discrete disruptive events and/or as a slow, almost imperceptible drift into failure. Frequency of disruptive events varies with complexity. In systems of high complexity disruptive events occur so frequently that the system has no time to return to stable equilibrium before the next disruption occurs. When complexity levels are very high the system is said to be at the edge of chaos because the uncertainty of behaviour is close to 1.

2.1.5 Nonlinearity

Relations between agents are nonlinear. Nonlinearity may amplify a small, insignificant disruptive event and cause a catastrophic outcome (an extreme event), the property called butterfly effect. The butterfly effect increases with complexity. In complex systems outcomes are, as a rule, consequences of numerous interacting causes, and therefore, the cause-effect analysis is inappropriate.

2.1.6 Self-organisation

Complex systems have a propensity to react to disruptive events by autonomously self-organizing with the aim of eliminating or, at least, reducing consequences of the disruption. This property is called *Adaptation*. Self-organisation may be also caused by a propensity to improve own performance, the property called *Creativity* or *Innovation*. To initiate and perform adaptive and creative activities the system must be *Intelligent*. Intelligence, adaptation and creativity are emergent properties exclusive to complex systems; their levels increase with complexity. The Artificial Intelligence (AI) found in complex adaptive software is normally referred to as *Emergent Intelligence*. It emerges from agent interaction and it is greater than the sum of agent intelligence.

2.1.7 Co-evolution

With time, complex systems change as their environments change and, in turn, they affect their environments. Co-evolution is irreversible.

2.2 Further readings

Literature on Complexity is very extensive. Nevertheless it is best to start with real pioneers and founders of the Science of Complexity, Prigogine [4, 5], Kaufman [6] and Holland [7].

3 SELF-ORGANISING SOFTWARE

3.1 Introduction

Conventional computer programs (software) are deterministic systems with strictly predictable behaviour determined by instructions specified and written by the designers and coders.

Can we design software to be complex?

And if we could, why would anyone prefer complex software with unpredictable behaviour more than deterministic computer programs, which behave as instructed?

The answer to the first question is: yes, we can. During the last 20 years the author and his team have designed and implemented a large number of complex adaptive computer programs [1].

The answer to the second question is counterintuitive and not widely known or accepted. *Complex adaptive software is capable of self-organising in a way that eliminates consequences of disruptive events and therefore it is perfectly suited for modelling (and/or managing) business, social, economic, political, geopolitical and many other complex systems.*

Conventional software is not suitable for operating in a complex environment because it cannot cope with unpredictable disruptive inputs.

3.2 Architecture

Complex adaptive software is an open system. It interacts with its environment, which is also complex and it consists of three key components: *Ontology*, *Virtual World* and *Connections to the Environment*, as depicted in Fig. 1.

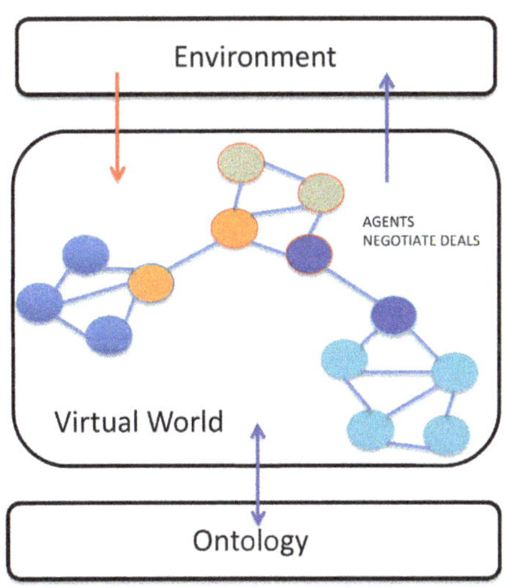

Figure 1: Architecture of a complex adaptive software system.

Ontology contains conceptual domain knowledge organised as a network of Agent Classes and their Relations. Agent Classes are defined by their attributes, which include patterns of behaviour. Virtual World is a network of Agent Instances and their relations. Agent instances are defined by the values of their attributes. Complex software being open, is always connected to other systems – its environment.

3.3 How complex software works

Complex software components are computational objects known as Software Agents.

The key feature of complex software is that its components, software agents, accomplish given tasks by exchanging messages with each other rather than by computation.

Typically, agents negotiate deals, they match supply to demand, allocate resources to orders, determine which word will have which meaning, which record will belong to which cluster, etc.

Whenever a disruptive event occurs agents autonomously and rapidly re-negotiate affected deals and thus eliminate consequences of the disruption.

The re-negotiation of deals amounts to self-organisation.

3.3.1 Connectivity

Which agents are designed to communicate with each other is determined by Agent Class Relations in Ontology. When two agents agree on a deal, they form a Connection, as depicted in Fig. 1 above.

These connections are not permanent and agents can cancel or modify existing connections and/or establish new ones, if necessary. By varying connection strength and the number of connections between software agents we can tune software complexity.

3.3.2 Autonomy

Agents' autonomy is constrained by the set of Standard Messages that each Agent Class is permitted to exchange with other agents and which is specified in Ontology. Before acting agents consult Ontology where they select from the set of Standard Messages a message that is appropriate for their task in hand. If no message is fully appropriate, agents select a message that appears to be most appropriate. In other words, if domain knowledge stored in Ontology is incomplete, agents will try to overcome this limitation by informed guessing. By varying the degree of autonomy of software agents we can tune software complexity. However, the autonomy of software agents is bound to be restrictive and therefore it is far more effective to tune software complexity by varying agent connectivity.

3.3.3 Emergence

The global behaviour of complex software emerges from agent negotiation but the task is never fully accomplished as long as disruptive events continue to occur. Complex software keeps adapting to its environment throughout its life.

3.3.4 Nonequilibrium

The perpetual occurrence of disruptive events does not allow software behaviour to stabilise. After every disruption software moves to a different state.

3.3.5 Nonlinearity

Nonlinearity of connections between agents may cause the propagation and amplification of instabilities. Insignificant disruptions may cause extreme Events, such as destruction of a carefully built agent network. We can vary the resistance to extreme events by modifying properties of connections with a view to restricting propagation of instabilities, e.g. oscillations.

3.3.6 Self-organisation

Software self-organises by autonomously changing connections between agents whenever such change helps to accomplish a task. Typically, software self-organises in order to eliminate consequences of disruptive events. Here is an example.

Let assume that the software task is to schedule the delivery of cargo to the International Space Station (ISS) in real time, and that after a considerable activity of agents they have agreed on a schedule that meets all requirements. If then a message is received that one of the items scheduled to be delivered by the next spacecraft is no longer required, the message will activate the relevant Item Agent, which will send a message to the relevant Container Agent that the cargo capacity, which was occupied by the cargo item that is no longer required, can be used for delivering an alternative item. Container Agent will immediately start negotiations with agents representing cargo items on the waiting list to select which of them will be scheduled for delivery by the next spacecraft. The decision is typically reached in a few milliseconds.

Self-organisation aimed at eliminating consequences of the occurrence of an unpredictable disruptive external event enables software to adapt to changes in its environment.

Adaptation results may not be optimal if time available for adaptation (negotiation between agents how to adapt plus all the actions required to implement the required changes) is too short to search for the optimal solution. In such cases agents will accept, for the time being, the first feasible solution and would attempt to improve this solution during intervals of time without disruptions, if possible.

Self-organisation aimed at correcting nonoptimal adaptation decisions can be interpret as a creative or innovative activity, which enables software to improve it's own performance.

3.3.7 Co-evolution

In time, the pattern of disruptive events generated by the complex software environment will change and, inevitably, because of perpetual adaptation, the changes in the environment will cause changes in software. The process is reciprocal, the changed software will cause changes in its environment.

4 SELF-ORGANISING BUSINESS

4.1 Introduction

Business is all about the allocation of available resources to demands – people to tasks, projects or departments; money to departments, projects, teams or individuals; machine tools to manufacturing processes; transportation facilities to cargos; storage capacities to parts and components; etc. A typical business has a massive amount of resources: human, financial, physical and knowledge resources (imbedded in data) and it is not surprising therefore that Enterprise Resource Planning (ERP) systems are a backbone of every enterprise.

ERP systems and various planners and schedulers, which work in batch mode, do the allocation effectively under stable market conditions. They typically require 8 h to calculate the optimal allocation schedule and then the business is expected to work according to the schedule for the next 8 h. Any changes in demand and the availability of resources during this 16-h period are ignored.

When the environment in which business operates is seriously complex (as is the Internet based global market) and characterised by the frequent occurrence of unpredictable disruptive events, the allocation of resources to demands must be done in real time, which implies the capability of the business processes to self-organise.

For real-time scheduling we have to have systems capable of performing autonomously the following activities:

1. Rapid detection of a disruptive event
2. Immediate identification of demands and resources that will be affected by the disruption, and
3. Autonomous re-scheduling the affected demands and resources with the aim to eliminate, or at least reduce, consequences of the disruptive event before the next one occurs.

Note that autonomous re-scheduling of resources to demands amounts to self-organisation.

Only complex systems are capable of self-organisation, and therefore, businesses supplying products or services to the Internet-based global market should have complex resource allocation systems [8–10].

When complexity of the environment is reasonably low, the real-time allocation of resources can be done by human operators (dispatchers) but, with complexity on the rise, the periods of time between consecutive disruptive events is becoming too short to allow people to gather relevant information, make decisions and implement actions.

4.4 Domain knowledge

In business systems the domain knowledge (knowledge how a business should work) is contained in business constitutions, rules and regulations; business process specifications; operational manuals; skills of managers and employees; national and international laws; and norms of business behaviour.

Business would be deterministic if knowledge how the business should work were precise and explicit and neither managers nor employees were allowed any initiative. In the past many corporations almost succeeded in achieving such a goal by strict Command and Control management approach, although a fully deterministic (predictable) business never existed.

Modern approaches to business management increase complexity of businesses by

- Wider distribution of decision-making powers
- Drastic reduction of levels of managerial hierarchies
- Replacement of individual problem solving by teamwork and consultation at all levels of management.

And the increase in complexity enables businesses to self-organise under conditions of frequent disruptions

4.5 Disruptive events

Let's look closer why do we need self-organisation in business. To answer this question we have to look into complexity of the contemporary global market. The biggest issue here is market dynamics resulting in the unpredictability of demand and supply.

Under conditions of frequent unpredictable demand failures (nonarrival of the expected order, the arrival of an unexpected order, modification or cancellation of an accepted order) and resource failures (failures of resources allocated to demands) it is very difficult to conduct a business unless we have a mechanism in place ensuring that the business will self-organise *rapidly* in reaction to any demand/resource failure and eliminate the consequences.

4.6 Conflict resolution

In addition to demand/resource failures, in complex environments businesses often experience conflicting demands. A typical conflict arises when two or more different demands each requests the same time slot to perform two or more different activities. An alternative conflict arises when two or more different demands request to perform a joint activity each proposing a different time slot.

4.7 Adaptive business

A business is Adaptive if it is capable of achieving specified results under conditions of frequent disruptive events and conflicting demands.

Conditions for adaptability are certain degree of business complexity, which enables self-organisation, and complex adaptive resource allocation systems, which ensure that self-organisation is sufficiently rapid to eliminate consequences of a disruption before the next one occurs.

The same conditions apply for a business to become *Resilient* to electronic attacks.

4.8 Drifting into failure

When a business operates successfully and manages to increase its value over a long period of time (measured in years), a subtle change in behaviour slowly emerges: increased tolerance of reduced thoroughness, mistakes, laziness and even of small-scale fraud. And then everyone seems surprised when the accumulation of small incidents reaches the Tipping point and the business is hit by an extreme event – loss of a significant client or even insolvency.

Perhaps the best example of a drift into failure of the global scale is the widespread cheating with subprime loans among financial institutions during a long period of global economic growth, which led to the global financial crises of 2008/2009.

A self-organising business can eliminate the drift into failure by tightening up control over behaviour of agents during the long periods of prosperity and relaxing controls when the going is tough in order to enable emergent creativity.

4.9 How complex adaptive business works

A diagram how a Real World of business is managed by a Virtual World of software agents engaged in the real-time allocation of resources to demands, is shown if Fig. 2.

5 SELF-ORGANISING SOCIO-POLITICAL SYSTEMS

5.1 Introduction

For the purposes of this paper socio-political systems are communities, towns, regions, nations and unions of nations. The largest socio-political system of them all, the geo-political constellation will be considered separately.

It is assumed here that the goal of a socio-political system is, or it should be, to provide for its members an environment conducive to satisfactory existence, as defined by members themselves, directly or through their representatives.

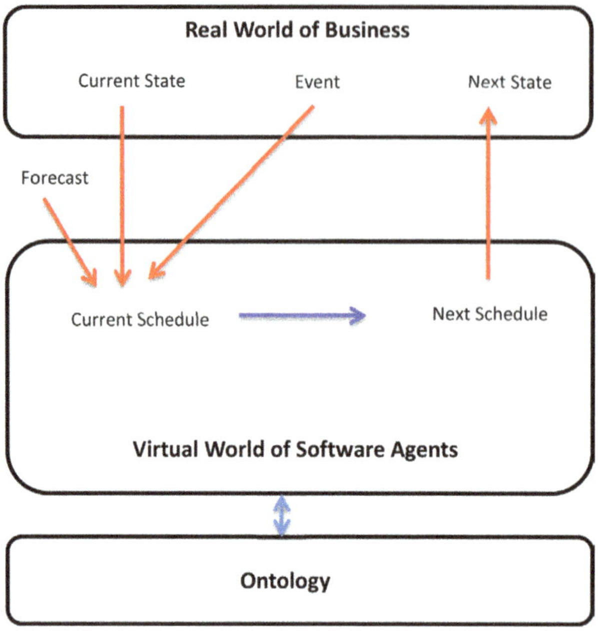

Figure 2: Virtual world manages in real time the real world of business.

In rigidly structured socio-political systems where members (social agents) are centrally controlled (dictatorship) self-organisation is not allowed. Dictators make the assumption that society is deterministic and it will behave as instructed. Attempts to self-organise are nevertheless almost always made at least by some social agents (see section on dissident behaviour).

Whenever constituent agents are given certain freedom to make autonomous decisions (democracy), they will make use of this freedom to self-organise with a view to eliminating consequences of any disruption interfering with their work or life. In other words, by self-organising a complex socio-political system may achieve its goals in spite of disruptions.

5.2 Domain knowledge

In socio-political systems the domain knowledge (knowledge how the system works) is contained in constitutions, statutes, rules and regulations but also in precedencies, rituals and unwritten traditions (culture). And, of course, a considerable part of it is locked in member's memories, very often in a tacit form, and applied without a need for the formal representation.

It is important to realise that knowledge on how a social system works is never complete and without contradictions; and therefore, it cannot be literally applied, it is necessary to interpret it, which can be done only in the presence of considerable intelligence.

5.3 Disruptive events

Socio-political systems experience a variety of disruptive events with increasing frequency, although the frequency is nowhere as high as in business, where, the Internet-based global economy is in a real turmoil.

Examples of disruptive events are (1) changes in legislations concerning employment, pensions, holidays, working hours, minimal wage, etc., (2) changes in prosperity caused by the loss of employment, increase/decrease of remuneration, (3) changes in living costs, (4) disruptive actions of constituent agents aimed against each other (internal attacks), (5) disruptive actions generated by other social systems (external attacks).

Social systems as a whole, or parts of social systems, self-organise aiming to reduce consequences of disruptive events or to maximise own advantages.

5.4 Emergent intelligence

The key difference between social complex systems and biological, physical or chemical ones is in the degree of *intelligence* of constituent agents. For the purposes of this paper let us define intelligence as 'the capability to formulate and achieve goals under conditions of uncertainty'. Intelligence subsumes propensity for self-expression and the ability to learn, investigate and communicate. This feature of social systems is very important because intelligence provides agents with the ability to exercise choices. However, social agents may or may not be allowed (empowered) to use their intelligence, depending on the social order.

A social system comprising intelligent agents that are given the appropriate autonomy exhibits *Emergent Intelligence*, which is far greater than the sum total of constituent agent intelligence. The 'appropriate' autonomy here is the degree of freedom given to individual agents to make decisions, which yields the best global behaviour of the system.

The appropriate agent autonomy depends on intelligence of agents (not all agents have the same intelligence) and on complexity of the environment in which the social system is embedded. In that respect each system, at any given point in time, is different. We can ascertain only that the appropriate autonomy is always greater than none and smaller than total.

5.5 Emergent creativity

To survive and prosper in the complex world there is a need to perpetually review goals and invent new ways of achieving agreed aims and objectives. To satisfy this need agents should not just react to disruptive events, they should be *creative* – able to anticipate trends and generate new opportunities. Creative agents can be appointed (research & development staff) or allowed to emerge when creativity is required (emergent creativity). Experience shows that the latter approach is more promising. Any social agent appears to be able to exhibit certain degree of creativity when circumstances demand. It is worth remembering that in complex systems emergent creativity is greater than the sum of agent creativities. A good reason to encourage complexity.

5.6 Emergent leadership

To achieve difficult goals the available intelligence and skills may not be sufficient. A *Leader* is an agent that is capable of motivating, mobilising and advising other agents to undertake difficult tasks and, in particular, tasks that are critical for the achievement of system goals. Leaders can be appointed or they can be allowed to emerge at the time when leadership is required (emergent leadership). The idea of emergent leadership is rather new and untested but much more aligned to the complex thinking than the current practice of appointing leaders.

5.7 Social order

The freedom of exercising choices in social systems is never complete. Autonomy of social agents, and thus their freedom of choice, is limited by social conventions and norms, by ethical standards, by rules and regulations imposed by social system statutes and by national and international laws enforceable by punishment, which can be severe (expulsion from a school, club, business; deportation from a country), or very severe (imprisonment, capital punishment). The purpose of limiting agent autonomy has always been to eliminate or restrict the unpredictability of the emergent behaviour of social systems, in other words, to ensure that the systems behave as nearly as possible as intended by system creators. However the effort to control the system by controlling constituent agents is often self-defeating.

The whole idea that it is possible to control a social system by excessively restricting autonomy of constituent agents should be carefully re-examined. The notion is fully valid only if the system is closed, or if its environment is stable and without disruptive events. Such situations do not exist naturally in the real world but are occasionally artificially imposed (Berlin Wall). When these conditions are not satisfied, i.e. when the system is open and its environment is complex (ever changing in unpredictable manner) attempts to control a social system by imposing excessive restrictions on agent autonomy are counterproductive. They prevent agents to react positively to disruptive events and thus stifle self-organisation, which in time leads to system disintegration (centrally planned economies).

Even more importantly, when perception of the desirable autonomy of agents by those attempting to control a social system and by constituent agents themselves, differ significantly, each agent tends to formulate a private (nondeclared) set of goals, which may not be compatible with the publicly declared social goals. Activities aimed at achieving nondeclared goals are often conducted in a covertly manner resulting in 'deviant' behaviour (infidelity, lying, theft, fraud and murder) and/or in organizing resistance aimed at changing official goals (rebellions and revolutions).

In large socio-political systems, such as nations, a greater autonomy is often given to a small number of elected agents, each representing a group of individual agents, rather than to individual agents themselves (democracy). This arrangement is a compromise, which may be satisfactory in a transient period from a dictatorship to fully participative democracy, where decision-making is prerogative of all society stakeholders.

5.8 Dissident behaviour

In social systems with strict centralised control there is a tendency of groups of agents to form an unofficial (dissident) system, which exhibits all features of complexity, including emergence and self-organisation. Complexity of a dissident group of agents ensures its long-term viability as a parallel social system, as experienced by disciplinarian businesses and totalitarian political regimes.

6 SELF-ORGANISING GEOPOLITICAL SYSTEM

The geopolitical system is the most complex social system encompassing all seven or so billions of people who live on our planet. At present, until we find our footing in this new field of research, it is wise to assume that the smallest component of the geopolitical system (its constituent Agent) is a nation or a large international movement. Then we could build models of geopolitical constellations using complex adaptive software and exercise them to gain understanding how adaptive and resilient they are.

A quick analysis shows that although we never had a global dictatorship, we had the geopolitical constellation of very low complexity in the 20th century after the World War 2, which exhibited equilibrium behaviour due to the existence of two powerful political blocks facing each other armed with nuclear weapons. We experienced a geopolitical system consisting of two antagonistic agents, both powerful because both possessed weapons capable of destroying life on the whole planet, and yet powerless, because they could not use these weapons without causing self-destruction.

With the exception of this period in the 20th century, the Cold War, throughout the recent history the geopolitical system behaved as a genuine complex system. Whenever a nation attacked another nation, the attacked side managed to self-organise and to successfully defend itself. According to my analysis, in recent history, no nation that started a war managed to win it. Consider the Napoleon's disastrous attempt to conquer Russia. As superior Napoleon's army approached Moscow, highly centralised state of Russia abandoned the defence of its capital and withdrew its army allowing Napoleon to enter into the city. But Moscow was empty, its *citizens decided* to abandon it and when Napoleon entered Moscow they set fires to their own homes and expelled the enemy without suffering any human losses. A perfect example of self-organisation in self-defence. Napoleon was then chased by partisans and regular army right back to Paris. Hitler's defeat at Stalingrad followed a similar pattern.

Current geopolitical conflicts are even more complex. Mobile phones and the Internet empower the side, which is attacked to rapidly self-organise into a formidable defence network. No recent military intervention by one geopolitical agent against another achieved its goals because of resilience created by social complexity.

Technology-driven complexity of the 21st century world calls for new geopolitical behaviour - conflict resolution rather than aggressive intervention.

ACKNOWLEDGEMENT

Research results reported in this paper are experimental. The author acknowledges contributions to his research by his team members who helped to design and code large-scale complex adaptive software systems for commercial clients, which were used to experimentally gain insight into Complexity. In particular, the author is grateful to his long-term research and business partner, Professor Petr Skobelev for his valuable contributions to topics discussed in this paper.

CONCLUSIONS

Under conditions of complexity, which currently prevail in our technological, business, social, economic and geopolitical environments, the power of self-organisation is evident. Since the trend is for complexity to increase, we shall have to learn to adjust our way of working and living to accommodate complexity rather than ignore it. And the best way is to re-design our organisations and institutions to make them self-organising.

REFERENCES

[1] Rzevski, G. & Skobelev, P., *Managing Complexity*, WIT Press: Southampton, Boston, 2014. ISBN 978-1-84564-936-4.

[2] Rzevski, G., A practical methodology for managing complexity. *Emergence: Complexity & Organization – An International Transdisciplinary Journal of Complex Social Systems*, **13**(1–2), pp. 38–56, 2011, ISSN 1521-3250.

[3] Rzevski, G., Complexity as the defining feature of the 21st century. *International Journal of Design & Nature and Ecodynamics*, **10**(3), pp. 191–198, 2015, ISSN: 1755-7437.

[4] Prigogine, I., *Is Future Given?* World Scientific Publishing Co., 2003. ISBN 981-238-508-8.

[5] Prigogine, I., *The End of Certainty: Time, Chaos and the New Laws of Nature,* Free Press, 1997, ISBN 0-684-83705-6.

[6] Kaufman, S., *At Home in the Universe: The Search for the Laws of Self-Organization and Complexity*, Oxford Press, 1995. ISBN 0-19-511130-3.
[7] Holland, J., *Emergence: from Chaos to Order*, Oxford University Press, 1998. ISBN 0-19-850409-8.
[8] Rzevski, G. & Skobelev, P., Emergent intelligence in large scale multi-agent systems. *International Journal of Education and Information Technology*, **1**(2), pp. 64–71, 2007.
[9] Rzevski, G., Using complexity science framework and multi-agent technology in design. In *Embracing Complexity in Design*, eds. K. Alexiou, J. Johnson, T. Zamenopoulos, Routledge, ISBN 978-0-415-49700-8, pp. 61–72, 2010.
[10] Rzevski, G., Using tools of complexity science to diagnose the current financial crisis. *Optoelectronics, Instrumentation and Data processing*, **46**(2), 2010, ISSN 8756-6990.

DISCOVERIES ON DEMAND

BORIS STILMAN
University of Colorado Denver & STILMAN Advanced Strategies, Denver, USA.

ABSTRACT

This paper is the next logical step in discovering the algorithm for inventing new algorithms, the Algorithm of Discovery. This research is within the scope of investigation of the structure of the Primary Language of the human brain as introduced by J. von Neumann in 1957. In this paper, we revise our hypothesis that the Primary Language is a collection of algorithms crucial for survival and development of humanity including two major components, the algorithm optimizing warfighting, Linguistic Geometry (LG), and the Algorithm of Discovery. We suggest that the Primary Language is the Language of Visual Streams ("mental movies") that empower directly both of those components as well as all the human symbolic languages and sciences. The Algorithm of Discovery is based on multiple thought experiments, which manifest themselves via visual streams. Those streams can run concurrently and exchange information between each other. The streams may initiate additional thought experiments, program them, and execute them in due course. The streams may operate employing the algorithm of "a child playing a construction set" that includes a construction model, a construction set and the Ghost. The streams are focused employing several types of reasoning including proximity and mosaic reasoning. In this paper we introduce a concise discovery program generated as a result of execution of the Algorithm of Discovery. A discovery program consists of a recorded series of thought experiments in a way that every experiment either builds new theme of visual streams or utilizes themes built previously. A series of themes of streams is the key for making discoveries.
Keywords: algorithm of discovery, artificial intelligence, linguistic geometry, mosaic reasoning, primary language, proximity reasoning, visual stream.

1 THE PRIMARY LANGUAGE

In 1957, J. von Neumann hypothesized existence of the Primary Language of the human brain [1]. He suggested that the external language (including multiplicity of natural languages as well as mathematics and computer science) that we use in communicating with each other may be quite different from the internal language used for computation by the human brain. He argued that we are still totally unaware of the nature of the Primary Language for mental calculation. He writes: "It is only proper to realize that [human] language is largely an historical accident. The basic human languages are traditionally transmitted to us in various forms, but their very multiplicity proves that there is nothing absolute and necessary about them. Just as languages like Greek or Sanskrit are historical facts and not absolute logical necessities, it is only reasonable to assume that logics and mathematics are similarly historical, accidental forms of expression. They may have essential variants, i.e., they may exist in other forms than the ones to which we are accustomed. The language here involved may well correspond to a short code in the sense described earlier, rather than to a complete code [in modern terms, he means high-level vs. low-level programming languages]: when we talk mathematics, we may be discussing a secondary language, built on the primary language truly used by the central nervous system." Over the last 50 years since J. von Neumann hypothesized existence of the Primary Language the nature of this language has been unknown.

DOI: 10.2495/DNE-V11-N4-495-507

Our original hypothesis was that the Primary Language is a collection of major algorithms crucial for survival and development of humanity, the underlying "invisible" foundation of all the modern languages and sciences [2]. We suggested that those components include Linguistic Geometry (LG) [3–16, 17–22], the algorithm of efficient warfighting, and the Algorithm of Discovery. This hypothesis was further developed in [2, 23–29]. Extensive investigation of the mental visual streams as a foundation of the Algorithm of Discovery and its application to rediscovering LG led us to observe powerful universal nature of those streams. We realized that our original hypothesis about the Primary Language should be revised. It appears that the Primary Language is the Language of Visual Streams (Section 2). In such capacity, it certainly serves as the language utilized directly by both algorithms, LG and the Algorithm of Discovery. Moreover, it is the underlying foundation of all the sciences and human languages as it should have been according to J. von Neumann [1].

Following J. von Neumann, we call it a language although it is not a language in mathematical sense, that is, it is not a set of strings of symbols. The Primary Language is the engine with the ability to generate visual streams and focus them in a desired direction. Visual streams serve as the driving force behind the Algorithm of Discovery for constructing new algorithms and, essentially, making discoveries. Visual streams empower LG for generating winning strategies for opposing games.

We suggest that visual streams drive generation for all the human symbolic languages including sound, pictorial, and written languages. This generation happens as real-time narration to running visual streams. Essentially, when talking, drawing, and writing, we describe what we "see" in visual stream using symbolic "summaries". For this purpose a stream can play real or imaginary world scenarios in real time or faster than real time. It can also branch into other streams, morph itself, zoom in and out, etc. Though these activities are controlled to support narration, the stream (or streams) has a lot of freedom of direction. It is likely that this freedom is not just a random choice but influenced by external features such as emotions. In addition, a visual stream is the key for translation from one human language to another. Essentially, every phrase or write-up in natural language is accompanied by a respective visual stream which is initiated after being retrieved from the mental archive. Then translation of this phrase to a different language is a narration of the visual stream employing this language. The Language of Visual Streams certainly supports all the semantics which is missing in conventional translation algorithms.

With respect to the Algorithm of Discovery the streams control is substantially different from the one utilized for generation and translation of languages because its' purpose is not simply a narration. In this case, the streams serve as construction tools of the discovery program whose execution leads to the discovery (Sections 6). It is even more interesting, following the J. von Neumann suggestion, to consider how the visual streams engine empowers all the sciences. Indeed, in our investigation of the Algorithm of Discovery, we applied it to rediscovering various components of LG, 3D structure of DNA, and the Theory of Special Relativity [2, 23–29]. In all cases, those re-discoveries happened first in the form of visual streams that eventually were transformed (recorded) into the conventional symbolic algorithms. It is reasonable to suggest that those visual streams introduced in [2, 23–29] is the "primary language" of LG, 3D structure of DNA, and the Theory of Special Relativity and, likely, all other sciences. The role of the Language of Visual Streams in empowering symbolic languages and sciences will be further investigated in our future research. Additional details are also provided in Section 3 in the description of the symbolic shell.

We will provide more details about visual streams in the following sections. The purpose of this paper is twofold, to make the next step in discovering the Algorithm on Discovery and to better understand the nature and usage of the visual streams as components of the Primary Language.

2 TOWARDS ORDINARY DISCOVERIES

What if discoveries are produced routinely as an output of computer programs? What a leap this would mean for humanity? Approaching this time of making discoveries on demand is the purpose of our efforts.

In a number of papers, we have been developing a hypothesis that there is a universal Algorithm of Discovery driving all the innovations and, certainly, the advances in all sciences [2, 23–29]. All the human discoveries from mastering fire more than a million years ago to understanding the structure of our Solar System to inventing airplane to revealing the structure of DNA to mastering nuclear power utilized this algorithm. The Algorithm of Discovery should be a major ancient item "recorded" in the Primary Language due to its key role in the development of humanity. This line of research involved investigating past discoveries and experiences of construction of various new algorithms, beginning from those, which we were personally involved in [2–36, 38–49].

Thought experiments allow us, by pure reflection, to draw conclusions about the laws of nature [30]. For example, Galileo before even starting dropping stones from the Tower in Pisa used pure imaginative reasoning to conclude that two bodies of different masses fall at the same speed. The Albert Einstein's thought experiments that inspired his ideas of the special and general relativity are known even better [31, 32]. The efficiency and the very possibility of thought experiments show that our mind incorporates animated models of the reality, e.g., laws of physics, mathematics, human activities, etc. Scientists managed to decode some of the human mental images by visualizing their traces on the cortex [30, 33]. It was shown that when we imagine a shape "in the mind's eye", the activity in the visual areas of the brain sketches the contours of the imagined object; thus, mental images have the analogical nature. It appears that we simulate the laws of nature by physically reflecting the reality in our brain. The human species and even animals would have had difficulty to survive without even minimal "understanding" of the laws of environment. Over the course of evolution and during development of every organism, our nervous system learns to comprehend its environment, that is, to "literally take it into ourselves" in the form of mental images, which is a small scale reproduction of the laws of nature. Neuropsychologists discovered that "we carry within ourselves a universe of mental objects whose laws imitate those of physics and geometry" [30]. In [2], we suggested that we also carry the laws of the major human relations including the laws of optimal warfighting. The laws of nature and human relations manifest themselves in many different ways. However, the clearest manifestation is in perception and in action. For example, we can say that the sensorimotor system of the human brain "understands kinematics" when it anticipates the trajectories of objects. It is really fascinating that these same "laws continue to be applicable in the absence of any action or perception when we merely imagine a moving object or a trajectory on a map" [30]. This observation, of course, covers actions of all kinds of objects, natural and artificial. Scientists have shown that the time needed to rotate or explore these mental images follows a linear function of the angle or distance traveled as if we really traveled with a constant speed. They concluded that "mental trajectory imitates that of a physical object" [30].

Our hypothesis is that the Algorithm of Discovery is based not on formal logic but on the visual streams, that is, mental imaginary movies which run in our brain. (By the way, LG is based on those streams as well [2, 23–26].) This is how it may work. Within the brain, the visual streams run consciously and subconsciously and may switch places from time to time (in relation to conscious/subconscious use). We may run several visual streams concurrently, morph them, and even use logic for such morphing, although this use is auxiliary. Then, we mentally tag some of the objects, actions, and events shown in the movie and create the so-called symbolic shell around the main visual stream.

This shell eventually becomes a standard symbolic algorithm that can be communicated to others employing familiar language, logic, mathematics, etc.

We named this approach "visual reasoning" [26]. According to our original hypothesis of visual streams we assumed that the "visual" component (including pattern recognition) is, in general, pretty sophisticated, while the "reasoning" component is relatively simple. We also assumed that the full scale mental visibility (with pattern recognition) is rarely used in discoveries, while the limited visibility can be simulated with a reasonable effort. The "reasoning" component is certainly within the scope of the modern software development. Our assumption of the need for "standard" visibility was based on the following. When we introduced mental visual streams we assumed that we "mentally" observe events happening within those streams, that is, we see what is going on. And, for all of us observing or seeing something means recognizing it. Thus, as part of our original hypothesis, we assumed that pattern recognition should be an inherent part of those visual streams. In other words, we assumed that in order to simulate and animate the natural or artificial reality in the form of visual streams we have to understand it, and, certainly, recognize it in detail as a necessary component of this understanding. We also assumed that the same understanding and recognition should happen dynamically within the visual streams. In our revised hypothesis, we state that utilization of visual streams does not require pattern recognition at all. We utilize our brain's innate ability to simulate outside world, that is, to "photocopy" the nature or invent and mentally construct artificial reality (which is, essentially, a modified nature) [34]. This ability goes further because we can animate simulation in the form of visual streams. In case of simulating (observing) the nature we periodically verify simulation, simply, "pixel by pixel", employing information coming through our sensors, images, sounds, etc., by-passing conscious understanding of this information. In case of artificial reality verification is not needed. In all cases simulation of reality within visual streams happens by construction, i.e., via internal subconscious understanding of the simulated reality and, consequently, without pattern recognition. Specifically, due to the fact that this simulation is driven by the internal engine (visual stream) the "mechanics" of this engine is used to understand the simulated reality. Roughly, we do not have to look at this simulation from outside, even mentally, (and employ pattern recognition). The visual stream can analyze it from inside because the stream *is* the one who created it. For example, to realize similarity of two different visual streams we should recall that those streams are driven by the specific algorithms each, and these algorithms are available to the streams engine. Thus, to realize similarity of the streams we have to realize similarity between the algorithms driving them which in its turn can be done by their comparison (or mutual analysis). We conclude that visual reasoning is essentially a mental construction based reasoning which does not involve pattern recognition.

Our approach to discovering the Algorithm of Discovery is analogous to an attempt to understand the algorithm of a program while watching its execution. In addition to this Algorithm, we are trying to discover the instruction set of the "computer" running this program, i.e., the means of the human brain to run it. In our investigation we assume that this program simulates sequences of events while its interface includes color movies demonstrating visually those events. With multiple published introspections of great scientists we can recreate clips from various movies, that is, their imaginary thought experiments. What really helps is the assumption that all those movies were "demonstrated" by the programs running essentially the same algorithm, the Algorithm of Discovery. With our own developments in LG, we have additional power of asking what-if questions via morphing our own movies and getting answers by watching those morphed movies until the very end. Unfortunately, we do not have this power with the discoveries of other scientists. However, we could utilize their published and oral introspections.

3 THE ALGORITHM OF DISCOVERY

In this section, we briefly summarize the results introduced in [2, 23–29]. The Algorithm of Discovery operates as a series of thought experiments, which are performed via imaginary animated movies (plays), named visual streams. These streams may or may not reflect the reality. In addition, those streams serve as the only interface of the Algorithm of Discovery. This interface is constructive, that is, visual streams could be morphed in the desired direction.

The input to the Algorithm is also a visual stream, which includes several visual instances of the object whose structure has to be understood or whose algorithm of construction has to be developed. Sometimes, the object is dynamic, that is, its structure is changing in time. Then the input visual stream includes this visual dynamics. As a rule, neither the structure of the object nor the details of the dynamics are present in the stream. It simply replicates (mimics) the natural or imaginary phenomenon. The task of the Algorithm of Discovery is to understand its structure including dynamics and/or develop an algorithm for reconstructing this object including its changes in time. This understanding happens in several stages. Importantly, it always ends up with the process of actual reconstruction of the object employing the construction set developed by the Algorithm on the previous stages. If the Algorithm investigates a natural real life object this imaginary reconstruction may be totally unrelated to the construction (replication) utilized by the nature. Usually, this reconstruction process is artificially developed by the Algorithm of Discovery with the only purpose to reveal the structure of the object. However, if the algorithm of natural replication is the goal of discovery, then the Algorithm of Discovery will employ a set of different visual streams (or different themes) to reveal the relevant components utilized by the nature [28, 35, 36].

All the visual streams are divided into classes, Observation, Construction, and Validation. They usually follow each other but may be nested hierarchically, with several levels of depth.

The visual streams operate in a very simple fashion similar to a child construction set. The Construction stream utilizes a construction set and a mental visual prototype, a model to be referenced during construction. This is similar to a list of models pictured in a manual (or a visual guide) enclosed to every commercial construction set. It appears that all the thought experiments in LG related to construction investigated so far, utilized those manuals. Imagine a child playing a construction set. He needs a manual to construct an object by looking constantly at its picture included in this manual. This model comes from the Observation stream as its output. It is not necessarily a real world model. It is not even a model from the problem statement. It is created by the Observation stream out of various multiple instances of the real world objects by abstraction, specifically, by "erasing the particulars". A final version of the object constructed by the Construction stream should be validated by the Validation stream.

The Algorithm of Discovery initiates the Observation stream, which must carefully examine the object. It has to morph the input visual stream and run it several times to observe (mentally) various instances of the object from several directions. Often, for understanding the object, it has to observe the whole class of objects considered analogous. If the object is dynamic (a process) it has to be observed in action. For this purpose, the Observation stream runs the process under different conditions to observe it in different situations. The purpose of all those observations is erasing the particulars to reveal the general relations behind them. A good example of multiple observations of processes is related to the thought experiments with various objects with respect to the inertial reference frames when discovering the Theory of Special Relativity. This includes experiments with uniformly moving ships, trains, experiments with ether as well as experiments for catching a beam of light [29, 31, 32]. Once the relations have been revealed, a construction set and a visual model have to be constructed by the Observation stream. Both are still visual, i.e., specific, – not abstract. However, they should visually represent an abstract concept, usually, a class of objects or processes, whose structure is being investigated. For construction, the Observation stream utilizes the Construction stream with auxiliary purpose (which differs from its prime purpose – see below). Note that the model construction is different from the subsequent reconstruction of the object intended to reveal its structure. This model may differ substantially from the real object or class of objects that are investigated. Its purpose is to serve as a manual to be used for references during reconstruction. Various discoveries may involve a series of models.

When the model and the construction set are ready, the Algorithm of Discovery initiates the Construction stream with its prime purpose. This purpose is to construct the object (or stage the process) by selecting appropriate construction parts of the set and putting them together. If an object has a sequential nature the construction also takes place sequentially, by repetition of similar steps. If multiple models have been produced the final object construction can also be considered as a model construction. At some point of construction, the parts are tagged symbolically and, in the end, visual reasoning with symbolic tagging turns into a conventional symbolic algorithm to be verified by the subsequent Validation stream.

Models and construction sets may vary significantly for different problems. Construction of the model begins from creation of the construction set and the relations between its components. Both items should be visually convenient for construction. The Algorithm of Discovery may utilize a different model for the same object if the purpose of development is different. Such a different model is produced by a different visual stream.

In many cases the Algorithm of Discovery employs "a slave" to visually perform simple tasks for all types of visual streams. This slave may be employed by the Construction stream to move around, to observe construction parts, and put them together. More precisely, imagine a child playing a simplistic construction set. To avoid offending children, I had named this personality a Ghost. This Ghost has very limited skills, knowledge and, even, limited visibility. The Observation stream may utilize the Ghost to familiarize itself with the optional construction set, to investigate its properties. Next, the Construction stream may use the Ghost to perform the actual construction employing those properties. Eventually, the Validation stream may use the Ghost to verify visually, if properties of the constructed object match those revealed by the Observation stream. In all cases, the Ghost is guided by the Algorithm of Discovery or, more precisely, by the respective visual streams.

As was already discussed, the initial visual model is usually guided by a very specific prototype, where the Observation stream has actually erased the particulars. However, this specificity does not reduce generality in any way. This sounds like a paradox. Essentially, every component of this model carries an abstract class of components behind it. This way visual reasoning about the model drives reasoning about abstract classes, which is turned eventually into the standard formal reasoning. This happens as follows. A visual model drives construction of the formal symbolic model so that the key items in a visual model have tags representing the respective formal model. At first, the formal model is incomplete. At some stage, a running visual stream is accompanied by a comprehensive formal symbolic shell. Running a shell means doing formal derivation, proof, etc. synchronized with a respective visual stream. While the shell and the stream are not completely synchronized, the visual stream drives execution of the shell, not the other way around. For example, a formal proof is driven by animated events within the respective visual stream. The visual streams, usually, run the creation of the visual model, the construction set and the final construction of the object several times. During those runs as a result of persistent tagging the symbolic shell appears. Multiple runs utilize the same visual components but during initial runs the synchronization of the stream and the shell is not tight. Further on, synchronization is tightened by morphing the visual model and/or adjusting symbolic derivation if they initially mismatch. Eventually, the stream and the shell switch their roles. In the end, it appears that the stream becomes the animated set of illustrations, a movie, driven by the running symbolic shell. For example, during the final runs (and only then), the visual streams, presented in [2, 23–28], are driven by the constraints of the abstract board game, the abstract set theory and/or the productions of the controlled grammars. At this point, the visual stream and the symbolic shell can be completely separated, and the visual stream can be dropped and even forgotten.

While the symbolic shell mechanism has been introduced several years ago and demonstrated in several papers, only now we realized that it provides the link between the conventional symbolic science and languages and the same knowledge "recorded" directly in the Primary Language, i.e., the Language of Visual Streams (Section 1). Existence of "other forms [of knowledge] than the ones to which we are accustomed" was suggested by J. von Neumann, [1] and Section 1. Persistent tagging leading to creation of the symbolic shell links visual streams to the conventional science and languages. However, as it stands now, this is a one-directional link. A reverse link, i.e., initiation of the visual stream by the natural language or scientific statement happens by invoking this stream from the archive (or the data base) with subsequent morphing for a better match. Note that switching from the visual stream to symbolic shell creates symbolic knowledge (even formal knowledge in case of science). However, this switch comes at a cost of depriving this knowledge from the "creative force". Interestingly, that many years ago we realized that development of the set of sophisticated mathematical tools of LG, the Hierarchy of Formal Languages [4–15], did not increase the inventive power of this approach. Unfortunately, at that time it looked like a particular observation.

There are several means to focus visual streams in the desired direction. They include Proximity and Mosaic reasoning.

4 PROXIMITY REASONING

Proximity reasoning as a type of visual reasoning is utilized due to the need for approaching optimum for many discoveries. It is likely that all the technological inventions and discoveries of the laws of nature include "optimal construction" or, at least, have optimization components [36]. Thus, various construction steps performed by the Algorithm of Discovery require optimization, which, certainly, makes construction more difficult. As the appearance of this Algorithm is lost in millennia, for its main purpose, it could not certainly utilize any differential calculus even for the problems, where it would be most convenient. For the same reason, it could not utilize any approximations based on the notion of a limit of function. Those components of differential calculus could certainly serve as auxiliary tools. In that sense, in order to reveal the main optimization components of this Algorithm, the most interesting problems to be investigated should lack continuity compelling the Algorithm of Discovery to employ explicitly those components. Based on several case studies [27], we suggested that this optimization is performed by the visual stream that operates imaginary movement approaching a location (or area) in the appropriate imaginary space. Having such space and means, the Algorithm employs an agent to catch sight of this location, pave the way, and approach it. Contrary to the function-based approach, which is static by its nature, the Algorithm operates with dynamic processes, the visual streams. Some of those streams approach optimum (in a small number of steps); other streams show dynamically wrong directions that do not lead to the optimum and prevent the Algorithm from pursuing those directions. Both types of streams represent proximity reasoning. We suggested that proximity reasoning plays a special role for the Algorithm of Discovery as the main means for optimization.

Proximity reasoning is a type of visual reasoning. This implies that the Algorithm should reason about the space where distances are "analogous" to the 3D Euclidian distances. Roughly, when we approach something, the distance must be visually reduced, and this should happen gradually. The space for proximity reasoning should provide means to evaluate visually if the animated images representing various abstract objects approach each other or specific locations [27]. Construction of those spaces is the key component of the Algorithm of Discovery. There could be situations when an object approaching the desired location (of an expected optimal state) hits an obstacle, i.e., another object, but the time constraints do not permit to remove this obstacle. This way, the desired location and optimal state appear to be unreachable [10, 11]. In such cases, the Algorithm may consider a total retreat to one of the initial states where time constraints are conducive for removal of the obstacle. After removal, the Algorithm may attempt to return objects to the originally desired locations which may lead to reaching the optimum.

5 MOSAIC REASONING

The name of mosaic reasoning was introduced due to the analogy of the Construction stream operation with assembling a mosaic picture of small colorful tiles [28]. Another, maybe, even more transparent analogy is known as a jigsaw puzzle when a picture is drawn on a sheet of paper and then this paper is cut into small pieces, mixed up, to be assembled later into the original picture. As Sir G. Thompson [37] pointed "… the progress of science is a little like making a jigsaw puzzle. One makes collections of pieces which certainly fit together, though at first it is not clear where each group should come in the picture as a whole, and if at first one makes a mistake in placing it, this can be corrected later without dismantling the whole group". Both analogies, the pictorial mosaic and the jigsaw puzzle, represent well the key feature of the Algorithm of Discovery construction set. However, we prefer the former because the jigsaw puzzle looks more like an assignment in reassembling a construct, a picture, which has already been created and, certainly, well known. In that sense, a tile mosaic is created from scratch, including choosing or even creating necessary tiles. In addition, a jigsaw puzzle is reassembled out of pieces based on random cuts. On the contrary, in pictorial mosaic, in many cases, every tile should have unique properties; it should be shaped and colored to match its neighbors precisely. A similar specificity is related to a group of adjacent tiles, the aggregate.

Returning to the Algorithm of Discovery, for many discoveries, the components of the construction set should be developed with absolute precision, in the way that every part should be placed to its unique position matching its neighbors. We will use the same name, the tiles, for those construction parts. If precision is violated the final mosaic will be ruined and the discovery will not happen. Though a group of tiles, an aggregate, may be configured properly, its correct placement in the mosaic may be unclear and requires further investigation. Moreover, a tile itself may have complex structure which may require tailoring after placement in the mosaic. In some cases, a tile is a network of rigid nodes with soft, stretchable links.

Mosaic reasoning may lead to the observation, construction, and validation steps of the Algorithm of Discovery operating with tiles and aggregates of tiles. Overall, mosaic reasoning requires tedious analysis of the proper tiles and their matching rules. Investigation of the matching rules is the essential task of the Observation stream. Multiplicity of those rules and their specificity with respect to the classes of construction tiles make the actual construction very complex. Selecting a wrong tile, wrong tailoring, choosing a wrong place, or incompatible neighbors may ruin the whole mosaic. The matching rules are the necessary constraints that control the right placement of the tiles. Missing one of them, usually, leads to the wrong outcome because the Algorithm of Discovery is pointed in the wrong direction.

Some of the matching rules impact mosaic locally while other rules provide global constraints. The global matching rules include the requirement of the top-down analysis and construction, the global complementarity rule, certain statistical rules, the transformation rules, etc. For many if not all natural objects and processes, their structure is not reducible to a combination of the components. Large groups of tiles, i.e., large aggregates, may obey the rules which are hardly reducible to the rules guiding placement of singular tiles. This matching rule must be understood globally first, implemented in the mosaic skeleton construction, and, only then, reduced to the placement of the specific tiles. The rule of the global complementarity means that placement of one aggregate may determine precisely the adjacent aggregate. The global statistical rules related to the whole mosaic may reflect the relationship between major structural components, the large aggregates. If understood and taken into account by the Observation stream, they may focus the Construction stream and lead to a quick discovery. Yet another class of global matching rules is called transformation rules. This is an algorithm for reconstructing an aggregate out of another aggregate and placing this aggregate in the proper location. Applied sequentially, such a rule permits to turn an aggregate, the so-called generator, into the set of adjacent aggregates. This way the whole mosaic could be constructed. Interestingly, this type of construction may be utilized by the Algorithm of Discovery as a convenient procedure to reveal the structure of an object, while the nature may use a totally different algorithm for generating the same object.

The local matching rules include the local complementarity rule, the interchangeability rule, etc. The local complementarity means, roughly, that a protrusion on one tile corresponds to the cavity on the complementary adjacent tile. The local complementarity often manifests itself in the requirement of various kinds of symmetry within the pairs of matching construction tiles. The whole class of the local matching rules is based on interchangeability. In simple terms, if two aggregates that include several tiles are not identical but interchangeable, their internal structure may be unimportant. There are several levels of interchangeability. Two aggregates could be essentially the same, i.e., their skeletons coincide. Importantly, those skeletons must include nodes which serve as the attaching points of the aggregates to the rest of the mosaic. The notion of an internal skeleton depends on the problem domain and is specific for different types of mosaic. Another lower level of interchangeability of the aggregates does not require their skeletons to coincide. The only requirement is that the attaching points of those aggregates are identical. In all cases interchangeability means that the stream can take one aggregate off the mosaic and replace it with another. This will certainly change the picture but the whole structure will stand. We named those aggregates plug-ins.

Besides mosaic structural components that include tiles, aggregates, global and local matching rules, there is an unstructured component that we named a mosaic environment. Such environment may impact the structure of tiles, aggregates, application of matching rules, and the whole mosaic while being relatively unstructured itself.

A transparent example of application of mosaic reasoning to the discovery of the 3D structure of the DNA molecule is considered in [28].

6 PROGRAMMING DISCOVERIES

The process of discovery is a sequence of thought experiments. This sequence could be considered as "program of discovery" as follows. This program may include subprograms. The actual creators of those programs and subprograms are visual streams. A stream may schedule other streams by creating a program with experiment "calls". It may schedule a thought experiment to be executed immediately. After completion of this experiment based on its outcome, the stream can schedule and invoke another thought experiment, etc. We suggest that such sequence of executed experiments recorded visually (as sequence of movies) is a program of discovery. However, this is not a program in conventional sense.

The program of discovery may be executed several times, each time with different outcome. This might happen due to the different input received by each of the experiments included in the program, beginning from the first one. Moreover, this program may change during execution. Indeed, the output of an experiment usually includes request for the next experiment. Different output may include request for the experiment that differs from the one recorded previously.

Besides experiments created, executed, and recorded sequentially, one after another, a stream can create a sequence of thought experiments to be executed in the future. These experiments could, in their turn, initiate new visual streams (with new experiments). The purpose, the nature, and the general outcome of those future experiments should be known to the stream created this sequence. This sequence is different from the list of procedure calls in conventional procedural (or imperative) programming. The algorithms of those "procedures", i.e., the algorithms to be utilized by the respective thought experiments are generally unknown. The experiments are not programmed – they are staged. The actual algorithm should be developed as a result of execution of such experiment. In a sense, this is analogous to the notion of declarative programming when a function is invoked by a problem statement while the function's body does not include an algorithm for solving this problem. The ability of a visual stream to schedule a sequence of thought experiments (with delayed execution) permits to create a nested top-down structure of visual streams with several levels of depth. Though, we suspect that the actual depth of the nested programmed experiments never exceeds two or three.

A discovery, i.e., a development of the final algorithm for the object construction is based usually on constructing a series of themes of streams. Each of those themes may, in its turn, be based on multiple experiments and may result from multiple uses of the Observation, Construction, and Validation streams. Interestingly, those themes may represent the same object, though, be totally different. The purpose of these themes is to look at the object from different prospective to reveal different properties. The themes do not appear at once. Experiments with one theme may lead to demonstrating the need for the next one. After being invoked several themes could be utilized simultaneously. The themes construction is based on the wide use of the principle of erasing the particulars. For each theme, some of the particulars of an object under investigation are erased while other particulars are emphasized.

All the thought experiments of a discovery program could be broken into several groups. It appears that a group of experiments that operate with the same theme are related to each other while they differ from those groups that operate with another theme. Once the theme has been invoked it can usually schedule experiments within itself to be executed in the future. As a rule, after being invoked the theme is used in several experiments. Then another theme could be invoked. At that time, previous theme is turned into the idle mode, which means that the stream freezes to pass control to the stream of a new theme. The experiments with different themes cannot be scheduled until those themes have actually been invoked. The problem is that until that moment the stream is not aware of their existence and even the need for them, unless this theme has already been invoked, used, and currently preserved in an idle mode. Consequently, scheduling of experiments with delayed execution is possible with the active and idle visual themes, only.

The programming system of the Algorithm of Discovery reminds remotely a conventional compiler. The output of this compiler is analogous to the executable code, i.e., the program of discovery. Different inputs of this compiler (different "source code") produce different outputs, i.e., different programs of discovery. Almost all of them, except for one or two, are incorrect, and do not lead to the discovery. These incorrect inputs reflect our current knowledge, e.g., incorrect experimental data, incorrect theoretical principles, etc. The failure of making a discovery, verified by the Validation stream, usually causes the Algorithm of Discovery to rerun several parts of the discovery program with different inputs. The program does not have to be rerun from the very beginning. Different parts of the program may be restarted with new inputs. As we already mentioned, different inputs to the program of discovery may change this program completely. These partial reruns with new inputs lead to the effect that, eventually, the discovery program looks like a small tree with branches at different levels. Usually, only one branch leads to discovery. This tree is a recording of the history of making a discovery including failures. With correct input the program of discovery will produce and reproduce the right branch of the tree and, this way, will produce and reproduce the discovery as many times as it will be executed.

It appears that the most sophisticated steps of execution of the Algorithm of Discovery, i.e., generation and execution of the discovery program, are those related to invoking new themes of visual streams. While reasoning within the active and idle themes requires just logic, activating a new theme requires constructing it from scratch or retrieving it from the archive of the past themes and morphing it to match current needs.

Components of the Algorithm of Discovery described in this paper (as well as those presented in [2, 23–29]) will be hypothetical until they will be verified by software implementations. A number of implementations have been initiated at the University of Colorado Denver. More advanced implementations and experiments are planned at STILMAN Advanced Strategies.

REFERENCES

[1] Von Neumann, J., *The Computer and the Brain*, Yale University Press, 1958.

[2] Stilman, B., Thought experiments in linguistic geometry. *Proc. of the 3d Int. Conf on Advanced Cognitive Technologies and Applications – COGNITIVE'2011*, Rome, Italy, pp. 77–83, 2011.

[3] Stilman, B., *Formation of the Set of Trajectory Bundles, Appendix 1 to the book: On the Cybernetic Goal of Games*, by Botvinnik, M.M., Soviet Radio, Moscow (in Russian), pp. 70–77, 1975.

[4] Stilman, B., Ierarhia formalnikh grammatik dla reshenia prebornikh zadach (Hierachy of Formal Grammars for Solving Search Problems), Tech. Report, VNIIE, Moscow (in Russian), p. 105, 1976.

[5] Stilman, B., A formal language for hierarchical systems control. *International Journal Languages of Design*, **1**(4), pp. 333–356, 1993.

[6] Stilman, B., A linguistic approach to geometric reasoning. International *Journal of Computers & Mathematics with Applications*, **26**(7), pp. 29–58, 1993.
http://dx.doi.org/10.1016/0898-1221(93)90049-2

[7] Stilman, B., Network languages for complex systems. *International Journal of Computers & Mathematics with Application*, **26**(8), pp. 51–80, 1993.
http://dx.doi.org/10.1016/0898-1221(93)90331-O

[8] Stilman, B., Linguistic geometry for control systems design. *International Journal of Computers and their Applications*, **1**(2), pp. 89–110, 1994.

[9] Stilman, B., Translations of network languages. *International Journal of Computers & Mathematics with Application*, **27**(2), pp. 65–98, 1994.
http://dx.doi.org/10.1016/0898-1221(94)90037-X

[10] Stilman, B., Deep search in linguistic geometry, symp. on linguistic geometry and semantic control. *Proc. of the First World Congress on Intelligent Manufacturing: Processes and Systems*, Mayaguez, Puerto Rico, pp. 868–879, 1995.

[11] Stilman, B., A linguistic geometry for 3d strategic planning. *Proc. of the 1995 Goddard Conference on Space Applications of Artificial Intelligence and Emerging Information Technologies, NASA Goddard Space Flight Center*, Greenbelt, MD, USA, pp. 279–295, 1995.

[12] Stilman, B., Linguistic geometry tools generate optimal solutions. *Proc. of the 4th Int. Conference on Conceptual Structures - ICCS'96*, Sydney, Australia, pp. 75–99, 1996.
http://dx.doi.org/10.1007/3-540-61534-2_5

[13] Stilman, B., Managing search complexity in linguistic geometry. IEEE Transaction on Systems, **27**(6), pp. 978–998, 1997.
http://dx.doi.org/10.1109/3477.650058

[14] Stilman, B., Network languages for concurrent multi-agent systems. *International Journal of Computers & Mathematics with Application*, **34**(1), pp. 103–136, 1997. http://dx.doi.org/10.1016/S0898-1221(97)00102-8
[15] Stilman, B., *Linguistic Geometry: From Search to Construction*, Kluwer Academic Publishers (now Springer), p. 416, 2000.
[16] Stilman, B., Linguistic geometry and evolution of intelligence. *ISAST Transaction on Computers and Intelligent Systems*, **3**(2), pp. 23–37, 2011.
[17] Stilman, B., Yakhnis, V. & Umanskiy, O., Winning strategies for robotic wars: defense applications of linguistic geometry. *Artificial Life and Robotics*, **4**(3), 2000. http://dx.doi.org/10.1007/BF02481336
[18] Stilman, B., Yakhnis, V. & Umanskiy, O., Knowledge acquisition and strategy generation with lg wargaming tools. *International Journal of Computer Intelligence and Applications*, **2**(4), pp. 385–409, 2002. http://dx.doi.org/10.1142/S1469026802000713
[19] Stilman, B., Yakhnis, V. & Umanskiy, O., Chapter 3.3. Strategies in large scale problems. In *Linguistic Geometry: From Search to Construction* , pp. 251–285, 2007.
[20] Stilman, B., Yakhnis, V. & Umanskiy, O., Linguistic geometry: the age of maturity. *Journal of Advanced Computational Intelligence and Intelligent Informatics*, **14**(6), pp. 684–699, 2010.
[21] Stilman, B., Yakhnis, V. & Umanskiy, O., Revisiting history with linguistic geometry. *ISAST Transaction on Computers and Intelligent Systems*, **2**(2), pp. 22–38, 2010.
[22] Stilman, B., Yakhnis, V. & Umanskiy, O., The primary language of ancient battles. *International Journal of Machine Learning and Cybernetics*, **2**(3), pp. 157–176, 2011. http://dx.doi.org/10.1007/s13042-011-0029-9
[23] Stilman, B., Discovering the discovery of linguistic geometry. *International Journal of Machine Learning and Cybernetics*, **4**(6), pp. 575–594, 2012. http://dx.doi.org/10.1007/s13042-012-0114-8
[24] Stilman, B., Discovering the discovery of the no-search approach. *International Journal of Machine Learning and Cybernetics*, p.27, 2012. http://dx.doi.org/10.1007/s13042-012-0127-3
[25] Stilman, B., Discovering the discovery of the hierarchy of formal languages. *International Journal of Machine Learning and Cybernetics Springer*, p. 25, 2012. http://dx.doi.org/10.1007/s13042-012-0146-0
[26] Stilman, B., Visual reasoning for discoveries. *International Journal of Machine Learning and Cybernetics, Springer*, p. 23, 2013. http://dx.doi.org/10.1007/s13042-013-0189-x
[27] Stilman, B., Proximity reasoning for discoveries. *International Journal of Machine Learning and Cybernetics, Springer*, p. 31, 2014.
[28] http://dx.doi.org/10.1007/s13042-014-0249-x
[29] Stilman, B., Mosaic reasoning for discoveries. *Journal of Artificial Intelligence and Soft Computing Research*, **3**(3), pp. 147–173, 2013. http://dx.doi.org/10.2478/jaiscr-2014-0011
[30] Stilman, B., The primary language of battles and discoveries. *International Journal of Applied Mathematics and Fundamental Informatics*, **1**, pp. 160–166, 2014.
[31] Deheaene, S., A few steps toward a science of mental life. *Mind, Brain and Education*, **1**(1), pp. 28–47, 2007. http://dx.doi.org/10.1111/j.1751-228X.2007.00003.x

[32] Einstein, A., *On the Electrodynamics of Moving Bodies*, Annalen der Physik, 17, 891, 1905 (in German).
http://dx.doi.org/10.1002/andp.19053221004
[33] Miller, A.L., *Insights of Genius: Imagery and Creativity in Science and Art*, Copernicus, an imprint of Springer-Verlag, 1996.
http://dx.doi.org/10.1007/978-1-4612-2388-7
[34] Kosslyn, A., Thompson, W., Kim, I. & Alpert, N., Representations of mental images in primary visual cortex. *Nature*, **378**, pp. 496–498, 1995.
http://dx.doi.org/10.1038/378496a0
[35] Iacoboni, M., *Mirroring People, The New Science of How We Connect with Others, Farrar, Straus and Giroux*, New York, 2008.
[36] Watson, J.D., *The Double Helix: A Personal Account of the Discovery of the Structure of DNA, Atheneum*, New York, 1968. [Scribner Classics Edition, NewYork, 1996.]
[37] Watson, J.D. & Crick, F.H.C., A structure for deoxyribose nucleic acid. *Nature*, **171**, pp. 737–738, 1953.
http://dx.doi.org/10.1038/171737a0
[38] Thomson, G., *The Inspiration of Science*, Oxford University Press, London, 1961.
[39] Astbury, W.T., X-Ray studies of nucleic acids. *Symposia of the Society of Experimental Biology*, **1**, pp. 66–76, 1947.
[40] Botvinnik, M., *Chess, Computers, and Long-Range Planning*, Springer-Verlag, New York, 1970.
http://dx.doi.org/10.1007/978-1-4684-6245-6
[41] Botvinnik, M., Blok-skema algorithma igry v shahmaty (in Russian: A Flow-Chart of the Algorithm for Playing Chess), Sovetskoe Radio, 1972.
[42] Botvinnik, M., *Computers in Chess: Solving Inexact Search Problems*, Springer-Verlag, 1984.
http://dx.doi.org/10.1007/978-1-4612-5204-7
[43] Chargaff, E., Structure and function of nucleic acids as cell constituents. *Federation Proceedings*, **10**, pp. 654–659, 1951.
[44] Crick, F.H.C. & Watson, J.D., The complementary structure of deoxyribonucleic acid Proceedings of the Royal Society A, **223**, pp. 80–96, 1954.
http://dx.doi.org/10.1098/rspa.1954.0101
[45] Foppl, O., *Introduction to Maxwell's Theory of Electricity*, B.G. Teubner: Leipzig, 1894 (in German).
[46] Galilei, G., Dialogue Concerning the Two Chief World Systems, 1632 (tranl. by Stillman Drake).
[47] Furberg, S., On the structure of nucleic acids. *Acta Chemica Scandinavica*, **6**, pp. 634–640, 1952.
http://dx.doi.org/10.3891/acta.chem.scand.06-0634
[48] Kott, A. & McEneaney, W. (eds), *Adversarial Reasoning: Computational Approaches to Reading the Opponent's Mind*, Chapman & Hall/CRC, 2007.
[49] Pauling, L. & Corey, R.B., A proposed structure for the nucleic acids. *Nature*, **171**, pp. 346–359, 1953.
http://dx.doi.org/10.1038/171346a0
[50] Watson, J.D. & Crick, F.H.C., The structure of DNA. *Cold Spring Harbor Symposia on Quantitative Biology*, **18**, pp. 123–131, 1953.
http://dx.doi.org/10.1101/SQB.1953.018.01.020

PHENOMENA OF FUNCTIONAL DIFFERENTIATION AND FRACTAL FUNCTIONALITY

IRINA TROFIMOVA
Department of Psychiatry and Behavioral Neuroscience, McMaster University, Hamilton, Canada.

ABSTRACT

The phenomena of vertical and horizontal emergence are analysed in terms of functional differentiation (FD), the concept of fractal functionality (FF), the concept of the zone of proximate development and an application to iterative map techniques. These theoretical components are used to trace an evolution of structures at various levels of organization and to derive several universal evolutionary principles. The implications of these principles are: (1) "building blocks" of natural systems are performed uniquely and only once, they emerge, change and disappear, and therefore cannot be considered as Lego-like "bricks" for these systems; (2) "building blocks" develop not prior to, but simultaneously with the emergence of a macro-system, to which they are associated, and may continue to change even after the macro-system is established; (3) the existence of functional groups "on the diagonal" affects the subdivisions of horizontal distributions; (4) soft associations of elements to functional systems of several levels of complexity speaks against a vertical division of complexity levels and against a horizontal division of building blocks.

Keywords: diagonal evolution, emergence, functional differentiation FD-3 operators, functionality, neurotransmitters, zone of proximate development.

1 EMERGENT NATURE OF LIVING AND SOCIAL SYSTEMS: MOVING AWAY FROM LEGO SCIENCE

The concept of emergence is fundamental to complex systems research, yet its significance is still underestimated in the majority of the sciences. Most natural sciences describe their objects using the "Lego" approach, i.e. as being built up from a collection of smaller, constant fundamental parts. This follows the reductionist philosophy that presumes that the behaviour of natural systems can be derived from the properties of their fundamental components, i.e. from knowing "what kind of Lego blocks this stuff is made of". Scientific analysis describes the interactions between these components mostly in mechanical terms. Thus, physiology is focused on the actions of organ systems within bodies; biology is focused on biomolecules and their interactions, especially those forming genes; physicists are focused on elementary particles and the search for grand theories of everything based on their properties.

Remarkably, the emergent and transient nature of life and social systems has been recognized for several millennia even while ignored by Western science [1–4]. There is abundant evidence that natural systems and phenomena, no matter how large or structurally stable they appear, have limited life spans and throughout their existence continue to change both structurally and functionally and then die. This is readily apparent in the life and social sciences where open, adaptive, developing systems with fungible components are the norms. This is true in physics and chemistry as well although such systems have tended to be ignored for methodological and philosophical reasons. Unfortunately, this has led to the development of mathematical models and methods ideally suited

DOI: 10.2495/DNE-V11-N4-508-521

to reductionist, "Lego" science, and ill suited to the study of natural systems. The wholesale adoption of these methods by biology, economics, social science and psychology, has led to the widespread misapplication of linear methodologies such as factor analysis, leading to stalled progress, misleading and even erroneous results [5, 6]

For example, in evolution theory, many models use a concept of traits/characteristics z_i that are either inherited or not under the process of natural selection, using a fitness function $w_{i.}$, and discuss the covariance of these traits at the individual, group or population levels. Problems arise, however, when we attempt to partition the evolving systems and their traits into the units of analysis. After all, these systems are not isolated but rather are strongly contingent on multiple systems at various levels of a conditional hierarchy of organization. As Cairns-Smith put it, "quite simply, in central biochemistry everything depends on everything" [7] (p. 94). Evolution theories focus primarily on structurally observable properties of biological species, such as phenotype or genotype, but rarely provide across the board comparisons of more fluid features associated with the emergence and maintenance of these species. The partitioning of traits/characteristics of species is also biased by theoretical traditions, often under-estimating characteristics related to within-species, between-species and species-environment interactions. More importantly, these systems are transient, and relying on their presence or that of their traits might lead to wrong results in computer modelling of evolutionary processes. The emergence, divergence and disappearance of biological species through genetic mutations were often attributed to a mechanical re-assembly of previously established genetic structures, while the dynamic nature of DNA and background processes were often not considered important until the mutation took place. However, it is mainly these processes that show up as a mutation.

The concept of emergence is close to the concept of evolution: initially, it was thought that evolved structures depended on their predecessors entirely whereas emergence related to something lacking predecessors. Nowadays evolution theory considers evolution as based on multiple parallel, independent processes that are capable of bringing new emergent features in a sudden fashion [8]. I contend that the concepts of evolution and emergence can be equally applied to all natural phenomena, even very transient phenomena. Indeed evolution may be considered as a special case of emergence.

2 ADDRESSING TRANSIENCE: PERFORMANCES (PRF-S) AS UNITS OF ANALYSIS

Considering the transience of living systems, we have to wonder what holds these systems together as consistent objects thus making them observable. The existence of natural systems is described here as a composition, construction or "performance" via synergetic interactions between multiple dynamical sources. The concept of "*performance*" (denoted by "*prf*") is introduced here for the formal description of a natural system, analogous to features usually attributed to the performance of a play. These features include:

Process-like nature: the performance of a play is a process, a sequence of multiple overlapping events unfolding within a certain time frame. Similarly, natural systems, whether social, biological or physical, are the result of multiple events (or processes, that are also events but spread over time) contributing to their existence. Natural systems also manifest for finite durations and may exhibit a discontinuous existence (emerging several times with some periods of non-existence in-between).

Existent does not mean observed: there is a swarm of components that contribute to a given perf, and the most prominent perf is merely an approximation of the most consistent elements and only those chosen by an observer. For example, the lead up to a play's performance involves multiple rehearsals and many failures – nevertheless these form part of an essential swarm of events required to ensure that the final product is of quality, and so should be considered as part of this perf. Only the final performance is likely to be acknowledged by the audience (observer). Likewise, forest mushrooms have an extensive underground fungal bed providing many semi-ready buds, most of which won't survive to the observation stage. This background network of prfs provides a storage of drafts that new prfs might use as building blocks.

Low rates of being observed: many prfs, even in a semi-ready state, are never observed because their components are not consistent enough to capture the attention of an observer. Some consistent components contributing to prfs ("service components") may still be invisible because only the "artists" have observers who make them visible (anecdotally the ratio of actors to stage crew is roughly 1:10). Thus transience in natural systems may be under-recognized because there is simply a lack of appropriate observers to render the associated prfs visible.

Local utility of consumer/observation bias: Results of an observation of a natural system depend on the nature of the observer. An ant perceives a chair differently than a human. Those aspects of a natural system that are observed depend upon the presence of an observer that reacts to specific consistencies within the swarm of prfs that is the system. Many plays are never observed because the consumer "is not ready to buy it". By analogy, the concept of fitness in evolution theory describes a correspondence between traits, or characteristics of biological prfs and environmental biases towards or against these traits. Whereas globally, evolution and emergence are likely neutral, locally, consistency in the emergence of structures suggests a "lobby" from contributing factors that capitalizes on the properties of these structures. In this sense, we can talk about temporal and local fitness. Vrba and Gould [9] underlined that the natural selection of species encompasses those heritable character-environment interactions that cause differences in rates of birth or death among varying individuals (high fitness), but this doesn't exclude the emergence of less fit characters. Different species use different sensory spectra to examine their environments and generally do not react to the properties of objects that cannot be perceived, thus creating an observation bias. When biologists reason about features of species they might also suffer an observation bias steaming from specific, socio-cultural properties of human nature. For example, it has been noticed that humans of different sex and temperament perceive neutral and abstract concepts differently [4, 6, 10, 11]. In general, the selectivity of observation means, by definition, that most properties are not being perceived. Returning to the play analogy, rehearsals, advertising and administration of a performance are essential to its existence, however, they are not registered by the audience members who only came to see the show. They would appear to be low fitness prfs to the audience yet are vitally important to the performance of the play. Utility is thus local and subject to observer bias.

Global neutrality of emergence: the emergence of features of biological species often has little to do with a better adaptation to their environment, even though those features that facilitate adaptation promote survival. Instead of viewing the properties of a species as always arising from increased fitness and adaptation, several evolution theories pointed to the rather neutral character of evolution [12–15]. This brings the concept of evolution closer to the concept of emergence. Returning to the play metaphor, "consumers", representing the environmental reinforcement of certain traits, are often unresponsive to the majority of a prf's properties and focus only on those aspects that benefit themselves, however, this doesn't mean that "overlooked" traits do not survive in the production if they do not cause harm to a prf's existence.

Mutual agreement principle in multiple choices: it is likely that natural systems, or perfs, are not actively seeking survival, contrary to common belief. It is just that those who have better compatibility with favouring factors (including cooperation and employment strategies) have better chances to survive longer [16]. Systems enter into interactions, regardless of whether the search for required prfs is active or passive, based upon comparisons of their compatibility to one another, in terms of salient features governing their participation in a joint performance. Similar to speed dating or to employee recruitment, the greater the compatibility, the longer the relationship will be maintained, and it will be longest if all prfs performing in an ensemble "agree" to it [18]. This "mutual-agreement" sorting is happening across all levels of complexity and time frames, and thus vertically, horizontally and diagonally.

3 COEVOLUTION AT MULTIPLE LEVELS: IF EVOLUTION GOES DIAGONALLY…

Two distinct forms of emergence are described in the complex systems literature: vertical and horizontal emergence.

"Vertical emergence" is described as a process of composition, a grouping of lower-level "bricks" into some superordinate-system. Fundamental particles form atoms which form molecules, organelles, cells, organs, organisms, families, societies, with causation acting from the bottom up and from the top down (Andersen *et al.* 2000). Hierarchical taxonomies of natural systems are complemented by descriptions of how upper levels of organization regulate the bottom processes, or how the nature of "brick" elements of super-systems determines the functioning of super systems.

"Horizontal" emergence occurs among systems existing within the same phenomenological level. The examples of horizontal emergence are predator-prey and host-parasite dynamics, symbiosis, sex and kin selection [17], collective mode phenomena [18–20] including collective intelligence [21], functional differentiation between cells or organisms [22], and an impact of a spectrum of sociabilities of agents on behaviour of a population [23]. Since Darwin it had been noticed that living systems often co-evolved resulting in correlated changes among their features. Co-evolution was mostly described as applying to systems at the same level of organization (e.g. co-evolution of organisms [23], or even co-evolution of galaxies [24], but not across the levels.

This paper proposes the idea that in evolution emergence may go diagonally. Diagonal evolution, or *dievolution* is the view that observable and structured natural systems developed in evolution "on a diagonal", with several levels of natural selection working simultaneously and changing the structures at these levels accordingly. Suppose that a natural system of interest is said to exist at level i. Then component (lower) levels are denoted as *i-1, i-2, i-3* etc while organizational (higher) levels are denoted *i+1, i+2, i+3* etc. The dievolution concept suggests that specific features of the structures at the level *i* are the results of natural selection occurring reciprocally at a minimum of three levels. Here are the six important features of dievolution that are often overlooked in the literature on evolution and emergence:

1. The dievolution process requires the contribution of (at least) three levels of organization. In other words, a structure at level *i* cannot emerge unless there are processes at the *i*+1 level that consistently select for the properties of the *i*-level structure, and unless there is mutual agreement (correspondence) between the *i*+1 and *i*-1 levels. Consequently, changes at either the *i+1* or *i-1* levels can compromise the compatibility between them, without which the resulting *i*-level structure will start to deteriorate and disassemble. Thus, the factors underlying the mutation or disappearance of biological species are not always observable and always should be analysed in terms of an interaction between lower and upper levels of organization, and not just lower (e.g. accidental gene mutation) or upper (e.g. climate change) levels only. A similar idea about multi-level selection was independently expressed in the Multilevel Selection Theory, MLS [20, 26] that also suggested that higher-level selection can be a significant evolutionary force for lower-level selection. The difference between MLS and dievolution theory is the more strict statement about simultaneous multi-level evolution for at least three levels, with the middle one produced through the interaction with the other two, the lower one becoming more specialized and partitioned while the top is sketched up and starts developing almost simultaneously with the middle one.

2. An increase in the consistency of structures emerging at the i level affects, in turn, selection processes at the i-1 level and creates a dispositional engagement with processes at the i+2 level. Securing emergence at i level starts a process of setting up the i+2 level of organization. This type of dynamics was noticed in developmental psychology by Vigotsky and informed his concept of a *Zone of Proximate Development (ZPD)*. ZPD is observed when children are too young to produce a certain behaviour but, when exposed to suitable degrees of freedom in actions, they choose a ZPD that suits their motivations and needs and tune the development of their skills towards these certain activities. In neuropsychology, it has been noticed that in the development of an action, the choice of neuronal ensembles is determined by the overall task that this action should contribute to, and such neuronal and behavioural integration proceeds with constant feedback adjusting the ensembles and the content of the task itself. In line with Gould's idea [8], the evolution (not development) of natural systems might not come as the gradual growth of a state space; instead it maps distant future limits ahead of time and fills out the intermediate states in a non-sequential manner (Fig. 1).
3. Structures possessing a *diagonal subordination* have more chances to maintain their consistency in step-by-step emergence than structures with either vertical or horizontal subordination alone. *Diagonal subordination* is a contribution of factors from several levels to prf's production in an overlapping manner: distinct prf-s can be produced out of the same components and selecting them levels, and contributing selecting factors act, of course on different prfs. Analogous to a stage crew having part-time jobs in several theatres, the same components of natural systems can be used in different perfs and even provide different functions. Plus there is a multiplicity of time frames in prf-s components: the temporal features (timing of onset, offset, duration) and of the factors contributing to their emergence are often only weakly correlated. By analogy with holography, the more perspectives converge on the same point, the sharper the image, and even in performing a play there are benefits for having several alternatives for leading artists, advertising campaigns, stage crews and so on. Likewise in neurophysiology, it has been found that the same action can be regulated by different neuronal ensembles even though there is some localization

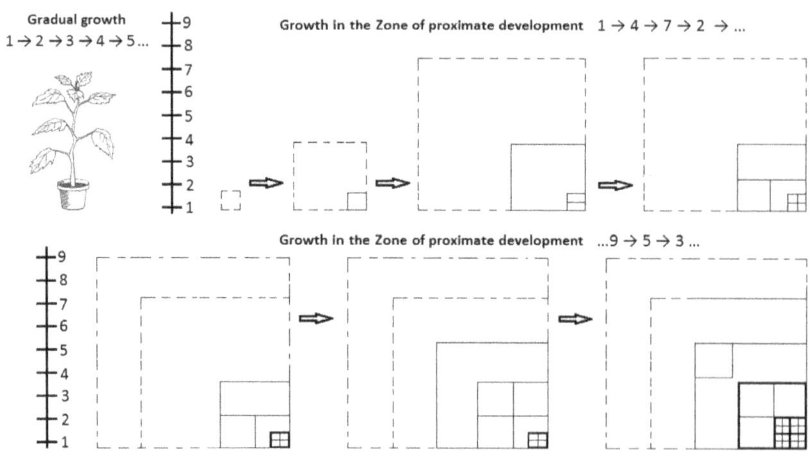

Figure 1: Development of more complex structures is likely not consecutive along the hierarchic levels (1-2-3...) (size of squares) but is a ZPD-like interaction between the upper- and the lower-level (1-4-7-2-...).

at the level of gross anatomy. In sum, in most consistent structures there is no strict association of *i*-1 level structures with only one *i*-level structure as *i*-1 level structures can contribute to several *i*-structures, recursively between levels (i.e. with an overlapping *i*+1 build-up), and these contributions vary in time.

4. Fundamental to the evolution, or emergence of complex structures is the *relative diversity* of sub-structures, in many senses of the word "relative".

   Modern evolution theory recognized that the classical portrayal of evolution as a ladder of progress leading towards better traits for adaptation and fitness is insufficient to explain many phenomena in biology. Instead of a drive to perfection, evolution appears to be associated with a drive for diversification (not necessarily with increasing complexity) [6, 7, 9, 15]. One of the main benefits of having a diversity of systems at say, the *i*-level, is that when selection tendencies from the *i+1*-level are not compatible with the *i-1* components in one *i*-level system, there are alternative *i*-level structures. Otherwise, the *i*-level structures would disappear, compromising existence of other prf-s that relied on it.

   Emerging structures have a greater chance to exist as something consistent when they are produced in bunches. Even if not capable of being reproduced exactly they may still form a family (with shareable components) and a distribution of similar even if not identical configurations has a chance for better fitness in changing environments. *Uniqueness of prf-s also should be noted:* Performances are unique in terms of their components, but should be multiple in numbers. Returning to the play analogy, every single performance of a play and all of the steps leading to each performance are unique in terms of cast and crew, theatre, sets, direction, locale and audience. At the same time *similar prf-s have more chance to be observable and to be used at many levels than a single or multiple diverse prf-s.* Similar performances appear like structures because they lie close to some mean, thus reinforcing it. Diverse performances fail to do so, hence seem less visible, though this doesn't mean that they don't exist. The existence of living systems depends on the consistency of their components which must emerge as a collection of relatively similar and sequentially compatible, but not necessarily identical states.

5. Diagonal subordination and the overlap of inter-level associations (principle 3) and "family production" (principle 4) lead to another important aspect of dievolution: that configurations survive better as ensembles and as contingent cycles of multiple states, rather than as a structure with well-defined and steady properties. A question arises: how many states in such a cycle can an ensemble have?

   "Two types as a minimum" comes to mind, and there is likely more than just matter of a minimum number for multiplicity. As noted above, *i*-level systems have overlapping sets of *i-1*-level components and in theory should compete for these components to maintain their own consistent existence. Competition strategies are one of the extensive areas of evolutionary modelling, however, another overlooked aspect of the *i-1* level overlap in *i*-level structures is the *reciprocity* of rivals that compete for the same *i-1*-level resources. Examples of reciprocal relationships are numerous not only between biological species but also in the neurotransmitter systems: 5-HT vs. NE; GABA vs. glutamate (Glu); prolactin (PRL) vs. Growth Hormone (GH); Neuropeptide Y (NPY) vs. Substance P (SupP); mu- vs. kappa-opioid receptors (MOPr vs. KOPr), and these relationships existed long before the human brain or brain cells in general were formed. The anatomy of the nervous system and the co-localization of neurotransmitter receptors in such reciprocal systems likely are not just the product of the top-level selection (in our case the levels of gross anatomy or of the functional tasks of humans and animals) but the product of an interaction between this top level, *i+1*, and the *i-1* (metabolic/chemical systems) level, leading to the emergence of the structures at the intermediate, *i*-level (such as brain structures).

The reciprocal relationships within these neurotransmitter pairs, when they suppress each other's release, represent the idea of the basic *contingency cycle*, essential for the survival of complex systems [8]. Such cycles are commonly described at a horizontal level of emergence by autocatalytic functions or as prey-predator or host-parasite models, however, these cycles have a better chance of survival if they employ not only horizontal but also diagonal subordination as described above. They have available multiple degrees of freedom and multiple systems and states at several levels, due to the overlap in associated elements, and they can survive contingencies that depend on the status of the *i-1* and the *i+1* level systems better than single-configuration systems.

6. Dievolution likely employs the phenomena of fractal functionality and "cruise controls" described in the next section. Formally, we can introduce the following three operators working recursively between levels $[i-n, i+n]$ when a perf *i* emerges: selection Λ_i, integration $[]_i$ and production O_i. Call them *Functional Differentiation operators, or FD-3.* The O_i denote perf-s that emerge no matter how transient they are, and which don't necessarily work as a selection factor for the lower levels. Only the Λ_i denotes subsets or aspects of these perf processes that are compatible for supporting the emergence of the structures. A most simple prf is: $O_i = ([]_i(\Lambda_{i+1}, O_{i-1})$.

   One important aspects of these operators is that they transform to each other's type on a diagonal between levels in the following manner: novel integrations $[]_i$ at the *i* level become routine establishments that can be used at the *i+1* level if supported by contributing factors, along the ZPD dynamics ($[]_i \leftrightarrow O_{i+n}$) and cruise control creation ($[]_i \leftrightarrow O_{i-n}$). Since the operator $[]_i$ denotes an integration of a prf at the *i* level based on compatibility of offers at the upper (Λ_{i+n}) and lower (Λ_{i-n}) levels of complexity in contribution to the emerging performance, such compatibility might be affected from both sides of the diagonal, with the $\Lambda_{i+n} \leftrightarrow []_i$ for the selection as a contribution from the upper levels. However, it is often an oversight that performance capacities of the lower level work as a selective factor in a integration $\Lambda_{i-1}(O_{i-n}) \leftrightarrow []_i$, and psychological phenomena of "projection through capacities" or "embodiment" are example of such dynamics [6, 10].

4 FUNCTIONAL DIFFERENTIATION (FD) AND FRACTAL FUNCTIONALITY (FF)

4.1 Emerging prf-s as recursions on a diagonal

The various operations that enter into dievolution are not thought of as acting continuously but rather in quazi-discontinuous cycles, *I1, I2, I3* and so on. Since the operators acting in each cycle will vary as a result of the actions of previous cycles, this takes the form of an iterated function system [40]. Such systems generate attracting sets which form persisting dynamical structures akin to speciation. There is a caveat here which stems from the fact that the state spaces upon which these operators act are not, strictly speaking, identical. In fact the phase spaces associated with the different levels are undergoing repeated change, sometimes increasing in dimensionality, sometimes decreasing, with varying partitioning as the upper levels change and along with them their classifying, discriminating and selecting functions, and with varying membership as elements are created and destroyed. Nevertheless if the time frames for the various levels are sufficiently different then it is possible to model this situation at the i-1 level on a fast time scale as an iterated function system and at the i and *i+1* levels on much slower time scales. In this way, there may be time for an attractor to emerge at a lower level for a sufficiently long duration so that there is a transient structure capable of triggering actions from the upper levels onto the lower levels, altering the dynamics and state space, and setting off a new iterated function system for the next cycle. In this manner, one expects to see a very complex form of recursion and fractal structure.

4.2 Cruise controls facilitate construction of consistent prfs but also carry functional specificity of contributing factors: FF

The production of prfs that are similar to each other and consistent is challenging, considering the emergent and transient nature of all of the components of these prfs. One approach is to develop habits. Modern psychology recognises that with repetition and learning, mammals construct their behaviour using more and more automatic elements. The common view of habits is that they are "bricks", the most consistent, well-learned elements of behaviour, whether we consider physical activity, verbal activity or cognition. Once learned, driving becomes automatic, speech becomes fluent, concept formation faster. Previous experience, similar to epigenetic (morphogenetic) fields, seems to determine future performance: previously learned habits are activated more easily than new behaviour.

The more consistent and more observable prfs use pre-made units that are supported by previous developments ("cruise controls"). Similar to cruise control in a car, such pre-made and pre-tried units of performance are only useful in a steady, predictable environment. In spite of the evolution of systems facilitating the emergence of structures in more and more complex and changeable settings, there are multiple "cruise controls" observed at the various levels of complexity. We define a "*cruise control*" as an *epigenetically based repetition of previously integrated performances (prf-s) that run using compatible inter-level elements* $O_i = []_{i-1}(\Lambda_i, O_{i-2})$. These prfs are not being oriented to the upper levels as much as new prfs as they have established support at at least three levels of organization. They are known for their reproductive nature, for example, solitons, genes, memes and habits. These units can be denoted as: $[]_{i-1} = \{O_i([]_i(\Lambda_{i+1}, O_{i-1}))\}_j = \{O_{i-1}O_i\}_j$. A *distribution* of the established prf-s, O_i (more often than a single established O_i) becomes a factor that contributes to selection processes at both lower and upper levels $\{O_{i-n}O_i\}_j . \leftrightarrow \Lambda_{ij-1}$, $\{O_{i+1}O_i\}_j . \leftrightarrow \Lambda_{ij+1}$. Cruise-control type performances (i.e. those whose emergence was facilitated by the epigenetic history of their previous integrations) by definition carry functional specificity of the multi-level factors that contributed to their consistent emergence.

Multiple reports from biological sciences point to the impact of the properties of functional activities shaped by the environment of biological species, including socio-cultural and intellectual activities of humans. If we allow a recursive dynamics between several levels of organization as described above, then it is reasonable to expect *Fractal Functionality (FF)* phenomena: this is when biases in features related to the functional specificity of structures on one level emerge in similar biases at other levels. Indeed, we see similarity in the functional differentiation of elements within living systems related to the functions of storage, support of structural stability, motion, transport of lower-level elements within the system, defense, reproduction, metabolic processing, etc. [22, 30]. Functional diversity of elements expanded dramatically with evolution of living systems but even for systems like atoms and molecules we see differentiation in function between elements that are positioned at the boundary of these systems vs. internally positions, the elements with high structural flexibility vs. low, etc.

Space does not permit to provide more complete analysis of functional differentiation, however the three operators with the FD-3 functions of selection, integration and maintenance of performance can be traced most universally within natural systems of various levels of complexity [22, 30]. An example of the FF related to these three functions is provided in the last part of this Section.

4.3 Saving outliers, or how to feed the wolves and to save the sheep

If there is a tendency towards automatizing prfs, then eventually prfs should become similar to one another, risking death should a change in the environment challenge their capacities. Here is a tough choice for nature: (1) to excel in the consistency of structures, multiplying extensively in a stable environment but with high odds of dying in an unstable environment, or (2) to sacrifice consistency by allowing deviations from the means, settling for less than perfect, "good enough" prfs but improving survival in uncertain environments. In simulations the tradeoff between sociability and diversity can be seen as a cycle between a totalitarian mode based on high interconnectivity and compatibility between agents and a structural decomposition mode based on the expansion of horizontal diversity [18, 31].

Diagonal subordination and contingency cycles are two solutions to the problem of "saving outliers vs. maintaining the mean" that appeared in evolution. Nature partly solved this problem through sex differentiation, which uses gene recombination. Such recombination is one of the examples of diagonal subordination as the parts of DNA molecules (configurations of "cruise controls" at *i-1* level) are exposed to selection by different environmental conditions (*i+1* level) in production of the phenotypes (*i-* level). What is rarely discussed in regards to sex differentiation is that it provides two partitions of a biological species: one that stores beneficial features of the species, i.e. cruise controls, and another that allows the species to experiment with outliers [32, 33].

The Evolutionary Theory of Sex (ETS) that analysed sex differences in the shape of phenotypic distributions described sexual dimorphism as a functional specialization of a species into two partitions, (1) conservational (female sex) and (2) variational (male sex). Trofimova [34] analysed sex differences in communicative and exploratory abilities and mental disabilities from the perspective of the ETS and suggested that male superiority in risk- and sensation seeking, physical abilities, higher rates of psychopathy, dyslexia, autism, higher birth and accidental death rates reflected of the systemic variational function of the male sex. Female superiority in verbal abilities, lawfulness, socialisation, empathy and agreeableness reflected the systemic conservational function of the female sex. Trofimova also suggested that the male sex (variable partition) plays an evolutionary role of pruning of the redundant excesses in a species' bank of beneficial characteristics in spite of resistance from the conservational partition.

Constructive Neutral Evolution (CNE) theory also describes a "two step" origin-fixation process [13, 14]. Similar to the ETS, it suggests that the first stage of evolution is the origin of mutations (the ETS describes these emerging in males more often than in females) during which selection processes reinforce and fixate beneficial characteristics (by passing these to female genes). The difference between the CNE and Dievolution theory is that the latter treats the origin of performances and their fixation into cruise controls not just as stages but as *two simultaneous directions of evolution working on a diagonal across a vertical hierarchy and horizontal diversity of living systems.* It is suggested here that prf-s at several lower (*i–n*) levels, and not just one level, determine the shape of the Zone of Proximate Development (ZPD) (i.e. the properties of the emerging upper levels of organization). Making cruises" down the diagonal and shaping the ZPD up to the diagonal of levels of organization are two closely integrated processes, even though they happen in different time frames.

5 AN EXAMPLE OF NEUROTRANSMITTER SYSTEMS REGULATING HUMAN TEMPERAMENT

Neurotransmitter systems (NT) regulating human temperament provide an especially good example of diagonal evolution as clearly the morphology of their projections in the human brain is a product of coevolution of these systems with the specifics of human activities regulated by several upper levels of complexity, as well as changing, unpredictable and complex environments [26–29]. The function of the brain NT (*i-2* level) is commonly attributed to a transmission of signals, i.e. informational (Λ_{i+1}-type) functions tuning the integration of a performance (*i*-level) by an individual (*i-1* level) to a best fit between his/her abilities ($\{O_{i-n}O_{i-1}\}_j$) and environmental demands (Λ_{i+1}). The same Λ_{i+1}-type chemical composites or their close relatives implicated in somatic functions or peripheral (i.e. not brain) nervous system serve a function of behavioural integration of a physical motion ($[]_i$-type) "(as in acetylcholine, ACh and prolactin PRL) or the maintenance of physical activities (O_{i-1}-type) (like in neuropeptides). Such associations with different functionalities in the same body illustrate the concept of diagonal subordination and overlapping associations.

Let us focus now on the composition of the (Λ_{i+1}-type) function of NT and look at the functional differentiation between the main NT groups. Many ancient NT regulate the behaviour of modern humans: amino acids (with initial role as body nutrition, O_{i-n}), neuropeptides (integrating capacities and the needs of the body, $[]_{i-n}$) and serotonin (5-HT), including its predecessors (with an initial role of a homeostatic regulator which integrates mind and body with the outside world, including initial light sensitivity, Λ_{i+n} [35]. Azmitia [35] estimates that the 5-HT precursor tryptophan and its metabolites were present in unicellular organisms nearly 3 millions years ago. To make 5-HT from tryptophan, oxygen is needed, and in the earliest geological times the Earth's atmosphere had little oxygen. Azmitia points out that 5-HT emerged specifically in unicellular systems capable of photosynthesis and the cellular production of oxygen.

Recent research shows that the differences in functional roles within each of these three families of brain neurotransmitters can be also classified using the FD-3 operators. Thus, among amino acids such FD-3 functions can be seen in three main groups: Glu (O_{i-n}), ACh ($[]_{i-n}$) and GABA (Λ_{i-n}); among hypothalamic neuropeptides: GH (O_{i-n}), PRL ($[]_{i-n}$) and SOM (Λ_{i-n}), orexigenic neuropeptides: orexins (O_{i-n}), SubP ($[]_{i-n}$) and NPY (Λ_{i-n}) and neuropeptidic opioid-receptor systems: MOPr (O_{i-n}), KOPr ($[]_{i-n}$) and DOPr (Λ_{i-n}), as well as among monoamine neurotransmitters: 5-HT (O_{i-n}), DA ($[]_{i-n}$) and NE (Λ_{i-n}) [36, 37].

Reciprocal relationships between pairs within the same families of neurotransmitters were noted above in a context of contingency cycles. Interestingly, in spite of the mutual inhibition of each others' release, in many cases one member of such pairs is a "chemical parent" to another member: Glu to GABA, DA to NE, NE to E, PRL to GH, etc. Within each of these pairs the derivatives appear later in evolution from a "parent" and suppress this parent-neurotransmitter's release. This "production of their own rivals" is clearly beneficial for self-regulation of the amount of NT produced within one system and such ability to self-regulate illustrates that evolution likely continues at several levels of organizational complexity in an iterative and simultaneous manner.

Finally, most consistent biologically-based traits (temperament) regulating human behaviour also exhibit FD-3 specialization (shown as 3 columns in Table 1). Four horizontal groups of traits are given in an evolutional progression of regulatory systems. The 1st row relates to the ancient opioid receptor systems that emerged entangled with the immune system [38] and later developed into the regulation of the most recent monoamine neurotransmitters [38, 39]. Recently, it was found that the body is capable of producing endogenous opioids and of changing the density of endogenous ligands, inducing dispositional emotional states. As the activation of these receptors leads to the release of monoamines, these opioid receptor systems likely play an even more important role in emotionality than do monoamines as they amplify a sense of security (MOPr), alertness and chronic anxiety (KOPr) or an initiation of action (DOPr) [39].

The second group of traits relates to the aspects of actions requiring lesser cortical involvement: either pre-made habits and/or orientation to explicitly present reinforcers. Behavioural regulation in known and/or predictable situations uses habits (learned or explicit elements, either physical or cog-

Table 1: The contribution of neurotransmitter systems in most stable (temperament) traits according to the neurochemical model Functional Ensemble of Temperament (FET) [36, 37], with notations of the FD-3 operators. *Note*: 5-HT: serotonin; DA: dopamine; NE: noradrenalin; ACh: acetylcholine; Glu: glutamate; GH: Growth Hormone; SOM: Somatostatin; PRL: prolactin; OXY: oxytocin; SubP: Substance P; NPY: Neuropeptide Y; AdrR, KOPr, MOPr, DOPr: adrenergic, kappa-, mu-, delta-opioid receptors.

Functional aspects:	*Behavioral orientation to types of rein-forcers:* *NE+...*	*Dynamical aspects* Preferred speed of integration of actions: *DA+...*	*Energetic aspects:* mainte-nance of pro-longed and/or intense activities: *ACh, 5-HT+...*
Regulation by the opioid receptors systems, amplification/appraisal of: orientational, dynamical and energetic aspects			
Emotional amplifier	$\Lambda_i ([]_i)$ **Neuroticism** KOPr→NE-HPA KOPr > MOPr	$[]_i = \{\{\Lambda_{i+n}^*, O_{i,i-n}\}\}$ **Impulsivity** DOPr→(DA, MOPr)	$O_i(\Lambda_{i-1})$ **Self-confidence** MOPr→(5-HT,DA) KOPr < MOPr, SOM
Regulation by neuropeptide and basal ganglia monoamine systems as a tuning between bodies' capacities and explicitly present, more determined degrees of freedom in behavior			
Physical-motor aspects	$\Lambda_i = \{O_{i-1}{*}O_{1-n})$ **Sensation-seeking** NE+NPY/SubP, AdrR	$[[]]_i = []_{i+1}\{\{O_{i-n}\}\}$ **Motor Tempo** DA+PRL+GABA/Glu	$O_{i+1}[[[]]]$ **Endurance** 5-HT+ACh, GH Orexins
Regulation by cortical monoamine systems as a tuning between developed blocks of explicit ac-tions and implicit, more probabilistic aspects of situations			
Mental aspects	$\Lambda\, []_i = (\Lambda_{i,i+n}([]_i(O_{i,i-1}), \Lambda_{i+2,}\, O_i)$ **Sensitivity to probabilities** NE+DA+ACh	$[]O_i = []_i(\Lambda_{i,i=l,i+n}\,(O_{i,i-1}))$ **Plasticity vs. rigidity** DA+5-HT	$O\Lambda_i = O_i(\Lambda_i, []_i, (\Lambda_{i=l}{*}O_{i-1}))$ **Mental Endurance (attention)** NE, ACh
Regulation by monoamine and hormone systems tuned to socialization			
Social-verbal	$\Lambda_{ij} = \{[]_{ij}\{O_i{*}O_k\}\}$ **Empathy-autism** NE+OXY, VSP	$[]_{ij} = \{\{O_{ik}, {*}O_{jk}\}\}$ **Social-verbal Tempo:** DA+ PRL, OXY	$O_{ij} = [[[]]]_{ij}\{O_{ij}{*}O_{kj}\}_{i,i-l}$ **Social-verbal Endurance** 5-HT+NP, OXY

nitive) that serve as "cruise controls" of behaviour, representing recursive prf components "down the diagonal. The traits of the 3rd row use more cortical involvement for processing socio-cultural context and complex probabilistic environment in general. They developed in evolution as a capacity for "internalization of external reinforcers", helping to map probabilities of events and to create anticipatory behavioural dispositions prior to events. Therefore, these traits are involved in the ZPD of the upper part of the diagonal emergence in building performance of single individual. In this sense, the

3rd row illustrates the ZPD for human analytic capacities, also distinguished for energetic maintenance of sustained attention (mental endurance, O_i), integration and shifts in actions under highly changeable and unpredictable conditions (plasticity, $[]_i$) and abilities to process information about not immediately present, implicit, abstract and probabilistic features or causes of events (Sensitivity to probabilities, Λ_i) [36, 37].

In summary, this article briefly described principles of the theory of diagonal evolution (Dievolution) that uses at least three operators that could be linked to the functionality of elements composing complex systems. These three operators act at minimum three different levels of complexity and might lead to Fractal Functionality within the functional differentiation of elements. The Dievolution theory suggests that the evolution and emergence of living phenomena are a result of the processes happening *on a diagonal of a vertical hierarchy and horizontal diversity of living systems, and these processes happen in two directions.* Processes from the up-down direction of the diagonal emerge as "cruise controls", or elements with the highest stability at the lower level of organization, and processes from the bottom-up direction of the diagonal emerge in the shaping of the Zone of Proximate Development of upper-level systems.

REFERENCES

[1] Whitehead, A.N., *Process and Reality. An Essay in Cosmology. Gifford Lectures Delivered in the University of Edinburgh During the Session 1927–1928*, Cambridge University Press: Macmillan, New York, 1929.

[2] Rosen, R., Effective processes and natural law. In *The Universal Turing Machine. A Half-Century Survey*, ed. R.Herken, Oxford University Press, pp. 523–537, 1988.

[3] Goldstein, J., Emergence as a construct: history and issues. *Emergence: Complexity & Organization,* **1**(1), pp. 49–72, 1999.
http://dx.doi.org/10.1207/s15327000em0101_4

[4] Trofimova, I., Universals and specifics in psychology. In *Nonlinear Dynamics in Life and Social Sciences,* eds. W. Sulis & I. Trofimoval, IOS Press: Amsterdam, pp. 286–307, 2001.

[5] Trofimova, I.N., Who is in charge of science: men view "Time" as more fixed, "Reality" as less real, and "Order" as less ordered. *Cognitive Systems Research*, **15–16**, pp. 50–56, 2012.
http://dx.doi.org/10.1016/j.cogsys.2011.07.001

[6] Trofimova, I., Observer bias: how temperament matters in semantic perception of lexical material. *PLoS One*, **91**, p. e85677, doi:10.1371/journal.pone.0085677, 2014.
http://dx.doi.org/10.1371/journal.pone.0085677

[7] Cairns-Smith, A., *Genetic Takeover*, Cambridge University Press, 1982

[8] Gould, S.J., *The Structure of Evolutionary Theory*, The Belknap Press of Harvard University Press: Cambridge, MA. pp. 365–382, 2002.

[9] Vrba, E. & Gould, S., The hierarchical expansion of sorting and selection: sorting and selection cannot be equated. *Paleobiology*, **122**, pp. 217–228, 1986.

[10] Trofimova, I., How people of different age, sex and temperament estimate the world. *Psychological Report*, **85**, pp. 533–552, 1999.
http://dx.doi.org/10.2466/PR0.85.6.533-552

[11] Trofimova, I., Understanding misunderstanding: a study of sex differences in meaning attribution. *Psychological Research*, **776**, pp. 748–760, 2012.
http://dx.doi.org/10.1007/s00426-012-0462-8

[12] Dawkins, R., *The Blind Watchmaker*, Norton & Company: New York, 1986.

[13] Stoltzfus, A., On the possibility of constructive neutral evolution. *Journal of Molecular Evolution*, **49**, pp. 169–181, 1999.
http://dx.doi.org/10.1007/PL00006540
[14] Stoltzfus, A., Constructive neutral evolution: exploring evolutionary theory's curious disconnect. *Biology Direct*, **7**, p. 35, 2012.
http://dx.doi.org/10.1186/1745-6150-7-35
[15] Gould, S.J. & Lewontin, R.C., The spandrels of San Marco and the Panglossian paradigm: a critique of the adaptationist program. *Proceeding of the Royal Socirty London B*, **205**, pp. 581–598, 1979.
http://dx.doi.org/10.1098/rspb.1979.0086
[16] Dawkins, R., *Climbing Mount Improbable*, Norton & Company: New York, 1996.
[17] Barkow, J.H., *Darwin, Sex, and Status: Biological Approaches to Mind and Culture*, University of Toronto Press: Toronto, ON, 1989.
[18] Trofimova, I., Principles, concepts and phenomena of Ensembles with variable structure. In *Nonlinear Dynamics in Life and Social Sciences*, eds. W. Sulis & I. Trofimoval, IOS Press: Amsterdam. pp. 217–231, 2001.
[19] Sulis, W., Stochastic phase decoupling in dynamical networks. *Nonlinear Dynamics, Psychology, and Life Science*, **124**, pp. 327–358, 2008.
[20] Wilson, D.S. & Wilson, E.O., Rethinking the theoretical foundation of sociobiology. *Quarterly Review of Biology*, **824**, pp. 327–348, 2007.
http://dx.doi.org/10.1086/522809
[21] Sulis, W., Collective intelligence: observations and models. In *Chaos and Complexity in Psychology*, eds. S.J. Guastello, M. Koopmans & D. Pincus, Cambridge University Press, pp. 41–72, 2009.
[22] Trofimova, I., Functional differentiation in developmental systems. In *Unifying Themes in Complex Systems*, ed. Y. Bar-Yan, Perseus, pp. 557–567, 2000.
[23] Trofimova, I., Potapov, A. & Sulis, W., Collective effects on individual behavior: in search of universality. *International Journal of Chaos Theory and Applications*, **3**, pp. 53–63, 1998.
[24] Thompson, J.N., *The Coevolutionary Process*, University Chicago Press, 1994.
http://dx.doi.org/10.7208/chicago/9780226797670.001.0001
[25] Heckman, T.M. & Kauffmann, G., The coevolution of galaxies and supermassive black holes: a local perspective. *Science*, **333**(6039), pp. 182–185, 2011.
http://dx.doi.org/10.1126/science.1200504
[26] Sober, E. & Wilson, D.S., *Unto Others: The Evolution and Psychology of Unselfish Behavior*, Harvard University Press: Cambridge, MA, 1998.
[27] Leontiev, A.N., *Problems of the Development of the Mind*, Trans. M. Kopylova. Moscow: Progress Publishers, 1981.
[28] Altman, J., Three levels of mentation and the hierarchic organization of the human brain. In *Psychology and Biology of Language and Thought*, eds. G.A. Miller & E. Lennerberg, Academic Press: New York, pp. 90–107, 1978.
[29] MacLean, P., *The Triune Brain in Evolution*, Plenum: New York, 1989.
[30] Trofimova, I.N., Universal principles of evolution as a foundation for analysis of human nature. Philosophy Research, 3, pp. 5–23, 1995 in Russian.
[31] Trofimova, I., Sociability, diversity and compatibility in developing systems: EVS approach. In *Formal Descriptions of Developing Systems*, eds. J. Nation, I. Trofimova, J.D. Rand & W. Sulis, Kluwer Press. pp. 231–248, 2003.
http://dx.doi.org/10.1007/978-94-010-0064-2_13

[32] Geodakyan, V.A., Sexual dimorphism. In *Evolution and Morphogenesis*, eds. J. Mlikovsky & V.J. Novak, Academia, Praha, pp. 467–477, 1985.
[33] Geodakyan, S.V., *Two Sexes. Why? The Evolutionary Theory of Sex*, Wilmington, USA, 2012.
[34] Trofimova, I., Do psychological sex differences reflect evolutionary bi-sexual partitioning? *American Journal of Psychology*, **1284**, pp. 485–514, 2015. http://dx.doi.org/10.5406/amerjpsyc.128.4.0485
[35] Azmitia, E., Evolution of serotonin: sunlight to suicide. In *Handbook of Behavioral Neurobiology of Serotonin*, eds. C. Muller & B. Jacobs, Elsevier: New York, 2010. http://dx.doi.org/10.1016/S1569-7339(10)70069-2
[36] Trofimova, I. & Robbins, T.W., Temperament and arousal systems: a new synthesis of differential psychology and functional neurochemistry. *Neurosciences & Biobehavioural Reviews*, doi: 10.1016/j.neubiorev.2016.03.008, 2016.
[37] Trofimova, I., The interlocking between functional aspects of activities and a neurochemical model of adult temperament. In *Temperaments: Individual Differences, Social and Environmental Influences and Impact on Quality of Life*, ed. M.C Arnold, Nova Science Publishers, Inc.: New York, pp. 77–147, 2016.
[38] Stefano, G.B. & Kream, R., Endogenous opiates, opioids, and immune function: evolutionary brokerage of defensive behaviors. *Seminars in Cancer Biology*, **18**, pp. 190–198, 2008. http://dx.doi.org/10.1016/j.semcancer.2007.12.001
[39] Bodnar, R.J., Endogenous opiates and behavior. *Peptides*, **32**, pp. 2522–2552, 2011. http://dx.doi.org/10.1016/j.peptides.2011.09.020
[40] Barnsley, M., *Superfractals*, Cambridge University Press: Cambridge, 2006. http://dx.doi.org/10.1017/CBO9781107590168

TOWARDS THE AUTONOMY OF URBAN MANAGEMENT: SIMPLE SYSTEM AS A SOLUTION FOR COMPLEX URBAN ENVIRONMENT

R. BARELKOWSKI[1], K. BARELKOWSKA[2], L. CHLASTA[2], J. JANUSZ[2] & L. WARDESKI[2]
[1]West-Pomeranian University of Technology in Szczecin.
[2]University of Technology and Life Sciences in Bydgoszcz.

ABSTRACT

There was an unprecedented development of cities, both in 19th and 20th century. Especially, late 20th century faced the moment in which urban population exceeded rural population only to accelerate this trend at the turn of the 21st century. Cities have become extremely fragile, technologically densified, extremely saturated with men and their artifacts, and too much extent very random, chaotic in its emergent patterns, either expanding the city or transforming its existent parts.

Simultaneously, these complex urban patterns encouraged the elaboration of sophisticated systems of urban management. But these systems work only in some areas of the world, most often unable to follow the quick rate of expansion. Therefore, substandard urban areas, undeveloped or overdeveloped, poor areas, segregation, exclusion zones, economically inefficient areas which cannot gain resources to get renovated, refurbished, and renewed.

The work discusses small-scale intervention in chaotic urban areas – Ariadne's Thread system intended to provide signage, telecommunication support, emergency reactive platform with surveillance options, and lighting integrated and dedicated to areas currently having none of these systems appropriately developed. The cluster of pylons connects to permanent local node, compulsory hard infrastructural element. The role of the latter, tasks to perform, the equipment used to analyze signals from cluster and respond in an appropriate manner is a significant supplementing and coordinating hub of the system. This paper will present the structure of the system, its working patterns, and methodology applied.

Keywords: design for communities, development and management, infrastructural systems, planning, sustainability, urban strategies.

1 INTRODUCTION

Rapid growth of cities worldwide exceeds the ability of many communities to react accordingly and be prepared for such unprecedented scale of simultaneous processes, resulting in expansion and densification of urban tissue, fluctuations in attractiveness, land value, land availability and accessibility, among a myriad of factors. Undoubtedly city can be seen as the pinnacle construct of civilization, the ultimate concentration of highly specialized people, the most sophisticated functions serving the community, infrastructure and transportation systems subservient yet indispensable to the city as an entity. However, its ideal image is being distorted by the forces of society and economy, its growth isn't harmonious and fluent. Sophisticated urban entity, despite all complex systems and tools implemented by various societies in attempts to control and moderate its performance, too often transforms into hardly manageable, unpredictable field of tensions, co-created by contradictory tendencies. While the discourse on sustainable development leads to more advanced theoretical findings and some successful, but local and fractional implementations, the overall result of the processes of urbanization becomes more incompatible with sustainability standards [1].

DOI: 10.2495/DNE-V11-N4-522-531

Poor, underdeveloped areas or areas, where rapid development is caused by conditions encouraging chaotic growth, whether it will be typical sprawl of North American or European cities, favelas in South America, or shanty towns in Africa or Asia, provide multiple examples of how cities generate processes which being the result of their temporary success counterbalance many if not all positive achievements. The city is the epitome of tendencies to seize control, manage, regulate, and organize have inherent mechanism, which prevents reaching that status, and which rather cannot be easily overcome by structures acting in top-down modes of operation. As Jean-Claude Bolay describes one of these phenomena – slums, these poor or extremely poor districts of a city are expressions of its nature, and therefore not only every attempt to eradicate them from urban structures failed, but increased social tensions, problems, as well as led to less sustainable parameters of urban environment [2]. When a city cannot withstand a further increase of the inhabitants, quite often they migrate not so far away to connect sub-agglomeration villages and towns closing the gap between agglomeration center and its satellite settlements, eradicating green belts, severing ecological corridors, increasing polluted areas and spreading negative impacts [3]. The reason for this situation, however, is much more complex as Ooi and Phua are attempting to diagnose for Asia – it's not the inability of local authorities to provide affordable housing, healthcare and sanitary conditions [4]. These are nothing, but symptoms of social inability to share environmental assets and inefficiency to provide access to them in an equal way.

2 AREAS OF INTERVENTION

Historically, urban management has reached its maturity quite late, in the 19th century, when the foundations for the contemporary theory of Urbanism have been laid. While throughout the centuries, from the ancient times of Varanasi (Benares), Kirkuk (ancient Arapha), Faiyum, or Jericho, cities were devised, designed, founded purposefully, arranged, organized, it was in the 19th century, when cities increased exponentially level of complexity of infrastructure introducing the issues of mass transport, communication, and multiple infrastructural subsoil or suspended networks. Multiplicity of systems affecting the quality of life in urban environment immediately exposed the issue of coherence between the systems.

Figure 1: La Quiebra squatter settlement in Medellin, Colombia, photo: RB.

Contemporary cities have been developing informal settlements for many years, and this process is ongoing and independent of the policies applied by vast majority (if not all) municipalities. Development, as a process directed among administrative bodies and private or corporate owners, offers high rate of unpredictability, resulting in some forms of spontaneous or counter-intentional transformations (from authorities' perspective). Random or spontaneously generated settlements not necessarily form extremely poor environments – their typologies differ depending on culture, geographical and environmental properties, and social behaviors [5]. There are substandard houses or shelters barely giving any protection, but there are also illegal settlements or overdensified and overcrowded areas, or simply improperly maintained urban tissue, now dilapidated and insufficiently fitted out. There are also legal, but unplanned housing areas, emerged from administrative inability to control legal gaps – these settlements, usually chaotic, are perfect examples of chaos inflicted not by the poverty, but by ill-placed thriftiness, seeming "optimization" leading to the devaluation and extensive exploitation of the environment [6]. Some of these settlements or specific districts appear in the central parts of the cities, but the respective majority is observed in peripheral urban areas, called peri-urban zones [7].

The tissues of contemporary cities are dependent on the veins of infrastructure. The capacity and efficiency mark not only the ultimate level of potential densification, but the maturity of the development of urban organism as well. Any attempt to diagnose conditions and quality of life in the city is heavily affected by the quality of infrastructure, which quite often is a primary diagnosed factor of sustainability [8, 9], and the rate of infrastructural coherence and integration, which, unexpectedly, is often overlooked.

Informal settlements, shanty towns, favelas, slums are very special areas, in which many parameters discourage planners, designers, and managers from employing standard concepts and typical organization of infrastructure. Low quality of houses or apartments, spontaneously created semi-public spaces unfitted to play its role as integrating spaces, particularly poor materials, which are far

Figure 2: Informal "street" of Santa Marta favela in Rio de Janeiro, photo: RB.

from being durable, high density and low area per inhabitant – all these factors constitute dense urban habitat with inappropriate, underdeveloped services. There is little room to plan usual spacing for infrastructure, or to provide full scale public squares or sometimes even basic facilities required for local communities [10]. Multiple programs to improve the quality of life in these substandard districts are devised and attempts to implement these programs are underway in countries where above specified problems have become a major issue. The understanding of the necessity to blend these settlements instead of struggling with their emergence and existence has been both the result of inefficiency of previous policies [11] and of the observation that in many cases informal settlements are inherent components of the city not contrary, but because of its nature [12, 13].

Contemporary strategies to improve living conditions in shanty towns focus on minimum intervention – maximum effect concept, and hardcore, noncompromising interventions are considered as justified only as extraordinary means to achieve extraordinary goals. As Banerjee et al. see the problem, every settlement has the opportunity to be upgraded, if not transformed, as a result of gradual improvement focused on crucial spaces, which in this case means street network [14]. Seeing truly multifaceted role of the streets one must notice that space here has many "dimensions" in comparison to traditional street – its rich geometry, its ambivalence becomes an immediate vessel to convey meanings, information, interactions, namely more than just communication / transportation. These informal streets or inserted additional streets, build a minimal skeleton – both functional and social – of these parts of the city, what sometimes requires pulling down some structures, an unpopular move, but necessary and acceptable if its scale is moderate to preferred small [15].

3 ADDING WITHOUT INTERFERENCE

Street spaces, including those informal ones, provide – obviously – the access to settlements, but these multifunctional corridors quite often have significant impact on privacy, safety, accessibility, availability, and mobility of local community members, to provide but the few possible interpretations of "street interface". Considering any kind of upgrading intervention, one must take into account the impact of any such introduction of any kind of infrastructure within the environment – the lesser, the more compact the infrastructure can be. Pearlman, who says, that the investment made in people rather than engineering and architectural infrastructure is always better [16], brings forward important, social argument, but one must not forget, that to aid people it is sometimes necessary to invest in people by investing in the infrastructure, people will be able to benefit from and investing in (architectural) space local community will be able to reclaim, appropriate and use as organizing space for all its members – how otherwise wishes of favelas' inhabitants of having access to health care, schools, or some socializing places can be fulfilled [17].

That is true, that concluding substandard quality of life as being the result of lack of basic services mostly would be naïve, but at the same time what is crucial it is the assessment of negative impacts infrastructural or architectural interventions may have. Assuming they are harmless for the quality of local community ties, its autonomy, its ability to maintain local identity, such a project positively fulfills criteria of "Basic Services for All", the United Nations Human Settlements Programme (UN-HSP). Basic services – BS (or unsatisfied basic needs – UBN) which giving the example of India, according to Dhote, Singh and Silakari, include social security among others [18], are not only those, community is aware of – these are also security issues related to regionally considered risks, like landslides in Medellin [19], flooding in Dhaka [20], local administration is responsible.

Many municipalities employ sophisticated management tools to operate the city in an attempt to control its behaviors, to monitor its phenomena. Often GIS systems supported by the commercialization and availability of telecommunication technologies allow the administration to diagnose and react, to collect data, analyze, and plan, at the same time giving the inhabitants access to multiple services, mainly making information available for mobile users. IT applications supported by the satellite control facilitate city management and either process information or become an interface to efficiently direct and coordinate any required reactions and municipal service performances. However, this performance may be limited by particular features of geographic environment and significantly hampered by the structure and quality of urban (transformed) environment. Squatted settlements are exceptionally prone to the risk of uncontrollable events and phenomena which may be harmful, if not lethal to local communities, in other cases, these areas are simply hard to be accessed easily in emergency cases. And there is one more significant issue, absent in the majority of urban management systems – its unidirectional structure, limiting the ability to receive social feedback.

But these systems require infrastructure themselves, they are managed independently in an uncoordinated manner (commercial vs municipal), they are fragile and usually disconnected from local lives of communities. Ariadne's Thread is a proposal of an infrastructural system to integrate several spatial and social functions, being used as available-to-all urban system. Its spread and non-serial, autonomous components are highly redundant to maintain connection and services in case of local malfunctioning or damage to the system [21].

4 SIMPLICITY VERSUS COMPLEXITY

In case of any urban system, it is unavoidable to implement techniques and subsystems, which are highly dependent on technology, but this technology may successfully tackle two typical problems: high costs of implementation for single-mode system, multilevel hierarchical structure making the system prone to errors, crashes, and similar, and difficulties in use application within diverse social groups, including elderly people, handicapped, youth, or simply foreign to the place and culture, and thus excluded by the lack of appropriate communication skills.

The project is treated as a pilot project, with its first stage focused on the attempt to integrate four main urban systems: signage system, urban lighting, local telecommunication facilitating devices, and security system components. Lighting and security systems (cameras) provide required and often expressed feeling of safety, lighting as well as signage are also crucial for accessibility, telecommunications provide the ability to exchange information, socialize, but also to get informed of significant events, risks, disasters if necessary. This idea of integration of various infrastructural systems is being researched in multiple ways, including that of French [22], whose seemingly correct proposals should be however adjusted to the specificity of every local infrastructural system involved in the process of integration. What's more, despite the fact one may completely agree with the French's idea of joint development of infrastructural systems, at least some of these systems are incompatible: power supply or water distribution. The most important division here is that of inherent permanence of the infrastructure. Technological constraints discard any attempts to abandon the

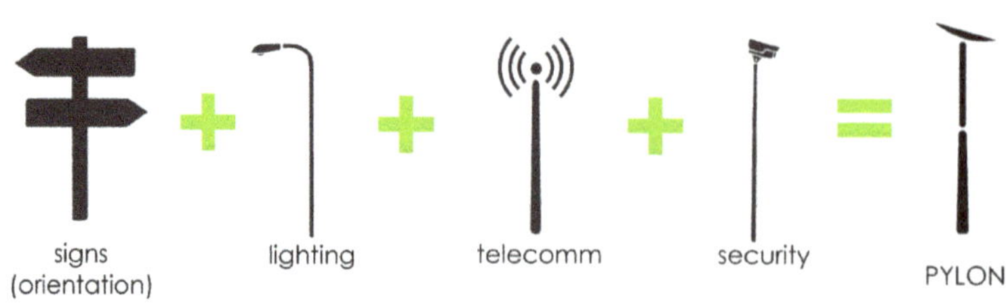

Figure 3: Conceptual scheme of infrastructural integration in Ariadne's Thread.

typical power supply grid, while other systems, like lighting, particularly in well-isolated regions of the world, may become completely independent of any connecting elements. The concept of reduction the basic component to locally installed single, detached object makes such an object potentially movable, dynamic in use in a sense of retaining the ability to be replaced, transferred to another, site better for optimal performance, to react to changes in the environment (with the help of urban servicemen, of course).

Usually considered permanent, the infrastructure often inadequately adjusts to the new conditions, thus becoming extensively distributed. This increases losses of any system, and complicates fixed, immovable structure. As Gandy notices, a more flexible form of infrastructure is required nowadays, when society is more mobile, and environments change so rapidly [23]. Ariadne's thread relies on the idea of light infrastructure, with local, autonomous devices, mutually amplifying the performance of the system.

The system is supposed to be built with the use of locally centralized facilities and cluster of pylons. The composition of Ariadne's thread physical infrastructure is as follows:

- pylon – basic device, a structure containing basic interactive components of the system,
- cluster – group of pylons operating within a particular area,
- permanent local node (PLN) / center – a privileged location, central for an area; optimally, the best site for location of PLN would allow to host the majority or all emergency / municipal units assuming 5 to 10 minutes reaction time, depending on specific requirements and health and safety analyses,
- base of operations (BoO) – local or a central base of operations in which all events and occurrences may be monitored and reactions directed,
- Ariadne's Thread Hub (ATH) – integrated local infrastructural center composed of PLN, BoO, and additional seats of local service and emergency units; it shall be noticed, that in case of any PLN or BoO malfunction another adjacent ATH resumes the operation instantaneously reconfiguring newly adopted cluster to indicate itself as new center and new reference point.

Ariadne's Thread Hub composed of permanent local node, base of operations, and seats of various municipal units or services is a reconfigurable site. Usually it should provide health care, police station, fire station, environmental and/or infrastructural reaction units. Social program includes

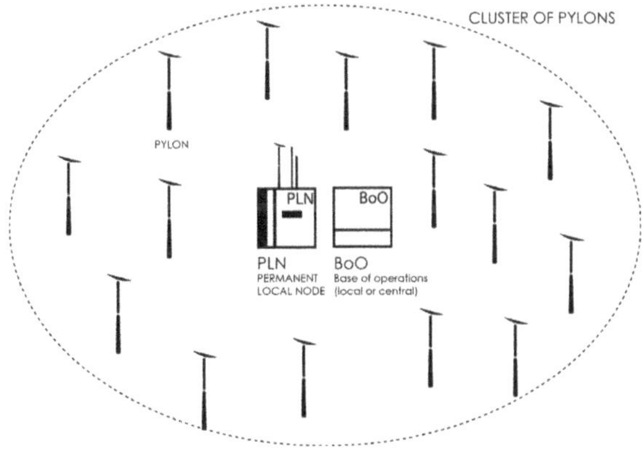

Figure 4: Simple structure of Ariadne's thread system.

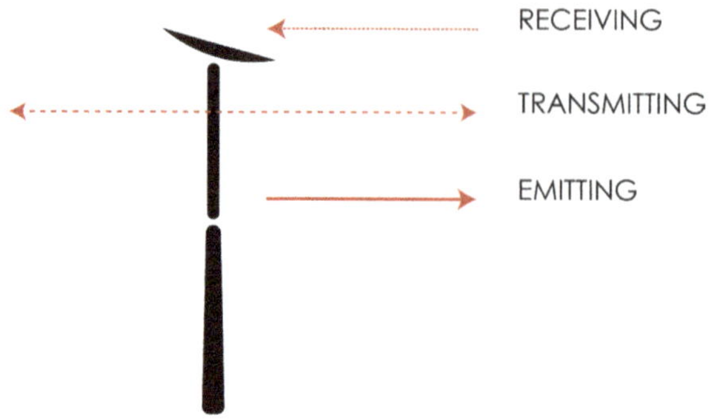

Figure 5: Pylon – Ariadne's Thread basic device modes of operation.

kindergarten, school, library or cultural center, naturally depending on the capacity of the available site, and needs as well as the preferences of the local community. Physical infrastructure, apart from Ariadne's Thread Hub and a cluster of pylons requires selected orientation point – intentionally it is PLN, but it may be substituted by an existing object to be easily visually recognized.

Experiencing the most densest informal settlement brought forward the issue of space availability for pylon location and mounting. The initial concept of only one typology of a pylon seems to be inappropriate and thus, while self-standing pylon remains the default solution. Pylons may take several forms:

- pylon type T – default, standing, foot-based device on reinforced concrete foundation, with steel column supporting dish with POV, cameras, and additional inner equipment,
- pylon type L – devised for narrow spaces and streets, cantilevered and mounted (wall or roof anchored),
- "octopus" (box with receptive and signaling LED-based tentacles) – destined to be mounted in extremely narrow areas or maze-like dense environments.

Pylons are multimodal devices acting as receivers, transmitters, and emitters at the same time. Each one is an independent object, acting as a part of local cluster, a bidirectional vessel for communication between the one emitting signal or initiating signal, and the one that receives the signal. This multimodal performance makes great difference, because when traditional systems of urban management do not require a physical infrastructure and operate based on supervision, on analyzing simulated representations of the city, Ariadne's Thread can not only react, but also communicate. It decentralizes basic municipal services flow of information, and despite its partial dependency on the permanent local node and base of operations.

It could seem that the proposal of AT is more complicated, but the principles of the system to make it less vulnerable to any errors or breakdowns, partially thanks to this particular proposal being a dispersed one, constituted by autonomous, individual devices. When urban space requires a basic infrastructure, it is often built as underground or suspended power cabling, ICT cabling, and piping hidden below roads and streets in these technical corridors of the city. In case of AT system, its simplicity and ability to deliver services through a network of multiple and interchangeable structures and devices. AT has similar ICT elements, but the principal difference lies in ability to move freely parts of the system.

Ariadne's Thread (AT) is based on the concept of saturating informal or spontaneously created settlements (including those legally designed and built, but being the result of unwanted urban sprawl) with theoretically movable objects with the only part of the system – PLN – as immovable. Naturally, other urbanized areas may be also subject to AT performance. A good example of the application may be seen in environments in which natural changes blur or camouflage spatial distinctions. Similarly, it can be used in overdeveloped high-density residential estates where the introduction of infrastructure has been delayed. Economic conditions for the use of AT system is to have affordable, if not cheap, the components being able to evolve according to the needs of the community, supporting emerging districts risen as a result of urban sprawl, assuming the authorities were unprepared for this rate of development. Informal settlements increase costs of a standard infrastructure and thus make it extremely difficult to supply the inhabitants contrary to autonomous components of AT. Flexibility of Ariadne's Thread is directly converted into a potentially autonomous status of community, whose members live in the area, contributing to self-awareness, social activity, and cooperation. The system broadcasts signals within the urban environment and increases social ties – emergency calls encourage people to help others, make them more sensible and aware of the fact that something significant happens in the area. Signals are much easier delivered to multiple users.

Information is transferred by different ways. First of all, there is a mobile application working with smartphones. Easy to use application provides the opportunity for registered users to inform others on any risk or danger, to alert municipal services in emergency case, by simple selection of the option. However, this mode of operation is currently achieved without necessity to use AT system. In fact, AT is more oriented toward working within urban space, to avoid excluding e.g. technically inapt persons, elderly people who don't use modern communication tools. Thus, the use of credit card (as identification, not as payment device) also allows to trigger signals on the pole, directly, and in the case of type L pylon, by signal-box located within reach of any person on the street or an urban square.

Exemplary cases to use the system may be described as fire case scenario, in which two various modes of pylon reaction may be triggered, or police call, or environmental disaster evacuation route indication. The signal is processed by the system in a following way:

stimuli response > recognition > decision making / activation > processing

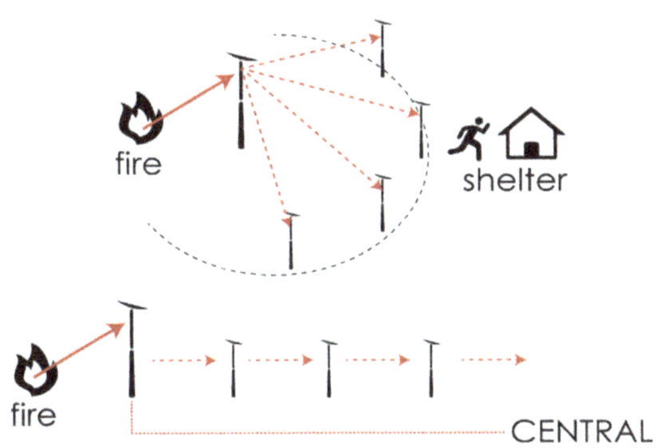

Figure 6: Fire case scenario – cluster of pylons *modus operandi* scheme.

and the cluster may work in a centripetal way, sequential way leading emergency unit to the required place, sequential in an opposite way, when the user is to be expected and helped in base of operation. It may also work in zonal mode, particularly in case of environmental hazards.

5 CONCLUSIONS

Ariadne's Thread is currently more of a proposal of a system, than something that could be field-tested. Its performance, however, may be easily simulated and consulted with local communities, because it does require as little physical connections as possible. Weinstock and Gharleghi argued that the contemporary city must be (entirely) reactive and responsive, adjusting its structure to urban flux, to changes in time-space, and the AT system is a step toward these principles [24]. Despite the technical inclination of equipment used in a pylon, system is relatively simple, yet it may be easily plugged-in into urban management systems, acting semi-independently. It is also a great lesson in the design process, as well as methodological aspects of joint reflection on particular environmental qualities of informal settlements. Information technology becoming more private issue results too often in alienation, social insensitiveness – therefore AT focuses on visibility, simple signs that are recognizable by the majority, if not all members of the local community, and let people act and also being judged for their actions or inactions. Marginalized slums, favelas, and squatter settlements are good reminders of how people can overcome disparities, how they are determined despite discouraging conditions – and for them simple solutions should be designed and made available.

REFERENCES

[1] Bithas, K.P. & Christofakis, M., Environmentally sustainable cities: critical review and operational conditions. *Sustainable Development*, **14**, pp. 185–187, 2006. http://dx.doi.org/10.1002/sd.262

[2] Bolay, J.C., Slums and urban development: questions on society and globalization. *The European Journal of Development Research*, **18**(2), pp. 287–289, 2006. http://dx.doi.org/10.1080/09578810600709492

[3] Ooi, G.L. & Phua, K.H., Urbanization and slum formation. *Journal of Urban Health: Bulletin of the New York Academy of Medicine*, **84**(1), pp. i28–i29, 2007.

[4] Ibid. (30).

[5] Tsenkova, S., Self-made cities. In *Search of Sustainable Solutions for Informal Settlements in the United Nations Economic Commission for Europe Region*, United Nations Publications: New York and Geneva, 2009.

[6] Ibid. (16-17).

[7] Nilsson, K., Nielsen, T.S., Aalbers, C., Bell, S., Boitier, B., Chery, J.P., Fertner, C., Groschowski, M., Haase, D., Loibl, W., Pauleit, S., Pintar, M., Piorr, A., Ravetz, J., Ristimäki, M., Rounsevell, M., Tosics, I., Westerink, J. & Zasada, I., Strategies for sustainable urban development and urban-rural linkages, research briefings. *European Journal of Spatial Development*, 2014.

[8] Ajayi, O.O., Oviasogie, F.O., Azuh, D.E. & Duruji, M.M., Urban design and sustainable development: a case of Makoko area of Lagos State, Nigeria. *European Scientific Journal*, **2**, pp. 90–91, 2014.

[9] Ali, M.H. & Sulaiman, M.S., The Causes and Consequences of the Informal Settlements in Zanzibar, *Shaping the Change*, XXIII FIG Congress Munich, Germany, pp. 1–5, 2006.

[10] Marx, B., Stoker, T. & Suri, T., The Economics of Slums in the Developing World. *Journal of Economic Perspectives*, **27**(4), pp. 192–193, 2013. http://dx.doi.org/10.1257/jep.27.4.187

[11] Pearlman, J., *Favela. Four Decades of Living on the Edge in Rio de Janeiro*. Oxford University Press; New York, 74–77, 268-271.
[12] Xavier, H.N. & Magalhaes, F., The case of Rio de Janeiro. *Understanding Slums: Case Studies for the Global Report on Human Settlements*, 21–24, 2003.
[13] Godwin Premsing, J. & Philip, S., Improving living conditions in Slums Dwellers. *IOSR Journal of Humanities and Social Science*, **19**(6), pp. 63–64, 2014.
[14] Banerjee, B., Acioly, C., Gebre-Egziabher, A., Clos, J. & Dietrich, K., Streets as tools for urban transformation in slums: a street-led approach to citywide slum upgrading. UN-Habitat Working Paper, United Nations Human Settlements Programme , UNON, 13–15, 18–22, 2012.
[15] Ibid. (22).
[16] Op. cit., Pearlman (2010: 22).
[17] Ibid. (84).
[18] Dhote, K.K., Singh, P.O. and Silakari, R., Social implication assessment framework for basic services to an urban poor program in India. In *The Sustainable City VIII. Urban Regeneration and Sustainability*, eds. S.S. Zubir & C.A. Brebbia, WIT Transactions on Ecology and the Environment, Vol. 179, Wessex Institute of Technology, WIT Press, Southampton & Boston, 1321–1330, 2013.
[19] Restrepo, A.E. & Orsini, F.M., Informality and Social Urbanism in Medellin, *in* M. Hermelin, A. Echeverri and J. Giraldo (eds.), *Medellin: Environment, Urbanism, and Society*, Universidad EAFIT, Fondo Editorial, 148–149, 2010.
[20] Jabeen, H., Johnson, C. & Allen, A., Built-in resilience: learning from grassroots coping strategies for climate variability, *Environment & Urbanization*, **22**(2), pp. 419–420, 2010.
[21] Barelkowski, R., Designing Ariadne's thread for the urban maze, 10th International Conference on Design Principles and Practices, Common Ground Publishing + Pontificia Universidade Catolica de Rio de Janeiro, Rio de Janeiro, pp. 12, 2016.
[22] French, S.P., Designing more sustainable cities by integrating infrastructure systems. In N. Marchettini, C. A. Brebbia, R. Pulselli and S. Bastianoni (eds.) *The Sustainable City IX. Urban Regeneration and Sustainability*, Vol. 1, WIT Transactions on Ecology and the Environment, Vol. 191, Wessex Institute of Technology, WIT Press, Southampton & Boston, 507, 2014.
[23] Gandy, M., Urban flux. *Architectural Design*, Vol. 79, No. 5, Profile No 201, Sep/Oct 2009, 14–15, 2009.
[24] Weinstock, M. & Gharieghi, M., Intelligent cities and the taxonomy of cognitive scales. *Architectural Design*, No. 4, Profile No 224, Jul/Aug 2013, 60–62, 2013.

ON A MATURITY MODEL FOR COMPLEXITY, COMPLEX SYSTEMS, AND COMPLEX SYSTEMS ENGINEERING

B.E. WHITE
Complexity Are Us←Systems Engineering Strategies, USA.

ABSTRACT

Complexity, complex systems, complex systems engineering, and other related terms are defined. As yet not enough about these topics is broadly known, understood, and applied to claim the existence of a maturity model. Nonetheless, one can contemplate what might constitute such a model and to begin its development. Ways of stimulating a more effective maturation process are suggested. Criteria are outlined for recognizing when maturity is approached. An initial draft maturity model is offered that includes examples of complexity, a collection of complex system characteristics, behaviors, and problems; a set of complex systems engineering principles for attacking and making progress in alleviating the world's more difficult problems, and an assessment of systems engineering practices. The main objective is to stimulate further thinking on these topics with the goal of constructing an improved maturity model through the collective interactions of interested parties.

Keywords: *complex systems, complex systems engineering, complexity, maturity models, system of systems, systems engineering.*

1 INTRODUCTION

"The term *systems engineering* (SE) can be traced back to at least the 1940s" [1, 2]. The practice of SE is now roughly 65 years old. Currently there are three recognized forms of SE, viz., (1) classical, conventional, or traditional SE; (2) System of Systems (SoS) Engineering (SoSE); and (3) Complex Systems (CSs) Engineering (CSE) [3] representing regimens for addressing problems of increasing engineering difficulty.

1.1 Definitions

To appreciate the SE context, it is useful to consider some definitions.

1.1.1 System and system of systems

A system is an interacting mix of elements forming an intended whole greater than the sum of its parts. These elements may include people, cultures, organizations, policies, services, techniques, technologies, information/data, facilities, products, procedures, processes, and other human-made (or natural) entities. The whole is sufficiently cohesive to have an identity distinct from its environment.

A SoS is a collection of systems that functions to achieve a purpose not generally achievable by any subset of the individual systems. Each system can operate independently (in the same environment as the SoS) and is generally managed primarily to accomplish its own separate purpose. There are four different forms of SoSs, directed, acknowledged, collaborative, and virtual [4].

DOI: 10.2495/DNE-V11-N4-532-542

1.1.2 Complexity and complex systems

Complex means more than complicated, the simplest form of complexity. Complicated problems are typically attacked by first trying to subdivide or partition the entity into separate pieces in a hierarchical structure until the pieces are sufficiently small that they can all be understood in a rather thorough fashion.

If a portion of a CS is separated from the CS, that component is often not able to operate on its own within the environment of the CS.

A complex entity may be characterized by a number of attributes such as

- Uneven, varied, diverse, and changing compositions of multi-sized elements
- Both tightly-coupled and loosely-coupled regions of network connectivity
- Elements and connections among them that continually evolve through self-interactions and exposure to the entity's environment
- Rich textures and multiple colors, sounds, and patterns
- Perceived as having different approximations of its underlying reality.

The more such attributes apply, the more complex the entity.

A CS is an adapting and evolving open (unconstrained by artificial means) system with continually cooperating and competing elements and unlimited interactions among its components (often including autonomous or semi-autonomous agents) and its environment. Relationships among CS elements and the effect of their interactions on the whole CS or its external environment are difficult to describe, manage, design, and/or change, let alone understand or predict.

1.1.3 Engineering, systems engineering, and complex systems engineering

Engineering consists of methodically conceiving and implementing viable solutions to existing problems, usually of the human-made variety.

SE is an iterative and interdisciplinary management, development, and implementation process that defines the needs and/or desires of interested stakeholders and transforms them into an operational system. Activities include conceiving, researching, architecting, utilizing, designing, developing, fabricating, producing, integrating, testing, deploying, operating, sustaining, and retiring system elements. There is emphasis on selecting, analyzing, synthesizing, and applying the appropriate scientific, mathematical, and other technological knowledge.

CSE is the deliberate leadership and accelerated management of natural and human processes that shape useful CS developments. CSE involves more environmental, economic, political, philosophical, social, organizational, ethical, moral, and other non-technological aspects. Accordingly, it is recommended that all key stakeholders be included in the CS intentionally from the start of any CS project. Because people are willful, unpredictable, and difficult to lead and manage, one is not in control of a CS and cannot prespecify or predict CS outcomes with any high degree of confidence.

1.2 Examples

Here are some examples of SoSs, complicated entities, and CSs.

1.2.1 Systems of Systems

- The Space Shuttle (not counting the crew) (a directed SoS).
- A satellite communications system (an acknowledged SoS).

- The internet (a collaborative SoS).
- A terrorist organization (a virtual SoS).

1.2.2 Complicated entities

- Wiring diagram for a large electrical installation.
- Architectural plans for a building complex.
- Technical specifications of a military electronic system.
- Instructions for conducting elementary school classes.
- Rules for driving roadway vehicles.
- Engineering standards handbook.

1.2.3 Complex systems

- An anthill, beehive, bird flock, or fish school.
- A seashore.
- The weather.
- The human body.
- The auto industry.
- Any group of people called to work together for a common purpose.
- A concert orchestra.
- A professional society.

2 BACKGROUND

In this section, we claim only an immaturity model for complexity, CSs and CSE. Then we explain by reviewing the current state-of-the-art. Next we suggest some challenges and what might be done to develop such a maturity model. Finally, we characterize how we might recognize when maturity has been attained.

2.1 Immaturity Model

CSs are well understood by those that study complexity science.

However, most engineers often say complex when they mean complicated. Surprisingly, relatively few systems engineers grasp the richer meanings complexity; most apply SE as if they live in a merely complicated world.

If we hope to deal effectively with our most difficult problems, CSE must deal with the multifarious spectrum of complexity. We have formidable challenges and have a lot more work to do before we can honestly state that we have a mature model for complexity, CSs, and CSE.

What we have presently can only be labeled an immaturity model!

2.2 Where we are now

Significant progress is being made in the past decade in learning about complexity, offering definitions of CSs, characterizing CS behaviors, and offering principles for CSE. More professionals are being exposed to these ideas and some are becoming sufficiently interested and motivated to join early adopters of this burgeoning field.

CSE practitioners and CS theorizers enjoy interacting with each other and preaching to the choir, as it were. Unfortunately, much of this effort enters the ether and has little influence in furthering progress in modernizing the practice of SE. As an example, many senior members of the International

Council On Systems Engineering (INCOSE), the world's largest SE society (currently about 9312 members as of December 2015, http://www.incose.org/about, are still wedded to classical, traditional, or conventional SE, which also tends to hamper the education and development of more junior systems engineers concerning complementary techniques that might have greater impact in solving our most difficult problems.

INCOSE has published a wonderful forward-looking vision of SE [5] but there is precious little offered on just how to get there. People are not considered part of the system, and processes and technology are relied upon to solve the problems. However, the importance and relevance of the social sciences is recognized.

INCOSE's latest SE handbook version [6] is a voluminous and detailed exposition of what is meant by SE. Below we highlight a few areas that still need fixing despite many recent handbook revisions.

The handbook advocates a hierarchical approach of partitioning subsystems indefinitely until they are small and simple enough to be understood. Best is used indicating a mindsight of optimization, rather than balance and "satisficing" more appropriate in CSE. Although recognizing that reductionism may not work well for a CS, no more cogent CSE approach is offered.

There is an overemphasis on the life cycle and sequential methods, although incremental and iterative techniques are also discussed. Too much is devoted to defining requirements and mitigating risks which are, respectively, quite hard to know and treated thoroughly at the expense of pursuing opportunities.

Also, there is little advocacy for the early construction of true guiding architectures. In practice, reference architectures tend to change with system development, and this area typically degenerates into presenting so-called architectural views before an underlying architecture is firmly established.

But many good things are included, e.g., systems thinking, Checkland's soft systems methodology [7] (which emphasizes understanding the problem as opposed to assuming requirements are knowable), Forrester's system dynamics [8, 9], Boardman's and Sauser's systemigrams [10], SE leadership vs. management, the importance of case studies, and several case study examples are rightly highlighted.

2.3 The challenge

Many purveyors of SE methodologies believe they are in control, know how to solve complex problems, and advocate a straightforward step-by-step process they think is sufficient to make practitioners successful. Unfortunately, this is inadequate for CSs, as many failed SE attempts documented in case study examples amply illustrate.

So how can these systems engineers be dissuaded from such fruitless pursuits and convinced to learn about or explore and adopt more promising methods? Even their experience with failed approaches does not help when they continue to repeat their standard methods expecting different results!

What about the rest of the world? CSE is something that can be applied in many domains, not just engineering fields. Those wrestling with formidable problems in many diverse non-engineering fields likely already understand complexity and bring a healthy dose of humility when trying to make progress. Why can't engineers embrace the same mindsight when encountering and dealing with their challenges?

2.4 What we might do

Reaching out to and establishing collaborative interactions with practitioners in other complexity domains about what works and what does not should help improve CSE.

So how might one ignite this endeavor? Systems engineers should be encouraged to participate in other forums to learn similar concerns, make personal contacts, and begin dialogs for mutual benefits. Protagonists from other fields should be invited to participate in SE forums for the same reasons. These interactions can stimulate greater understanding and fresh ideas for accelerating progress in SE through diversity, taking advantage of the Medici effect [11], where innovative ideas derive from seemingly unlikely sources. Another suggestion is to get systems engineers involved with the Club of Rome and their offshoots, to motivate and orient them and their colleagues to better address big world problems within the SE profession as part of their job.

2.5 How to recognize maturity

It would be nice to know when we might be attaining enough maturity to finalize and accept a viable maturity model. Here are a number of areas (with brief explanations) to watch in ascertaining the extent to which certain CS behaviors and CSE principles are being acknowledged and applied in a sufficiently widespread manner; and, clearly, there may be several other meaningful indicators.

- Intentionally including stakeholders in the system to be improved: this makes the system unpredictable because people are willful and difficult to control.
- Investing significant time and effort up-front to understand the problem: charging ahead with perceived solutions impatiently is often counterproductive.
- Embracing transdisciplinary fields to garner additional valuable insights: take advantage of the Medici effect which often suggests fruitful innovations
- Sharing information freely to illuminate the underlying reality collectively: this fundamental means of establishing trust and understanding should be rewarded.
- Modifying the system boundary in establishing a proper context: this offers the potential of establishing a more attractive solution space leading to a simpler objective while recognizing the original complex problem may remain unsolved.
- Focusing on system capabilities instead of requirements: initial requirements are largely unknown by the stakeholders or users and are continually changing; as desirable and useful capabilities are developed they should be introduced incrementally while watching how well they work over time, and then adapting.
- Developing and adopting a guiding (and layered) architecture early on; a true underlying architecture that will not change much compared to subsequent system configurations must be established before promulgating popular descriptive architectural views that otherwise may have little behind them.
- Recognizing and pursuing opportunities as well as identifying and mitigating risks: opportunities typically abound in CSs, and the biggest risk may be in not pursuing an opportunity, thereby missing a great solution, while trying to get back on a rigid, preconceived development track or path.
- Planning for emergent properties comprised of the unexpected overall system outcomes that result from the confluence of interactions within the system or with its environment; surprises that cannot be explained can be illuminating.
- Concentrating on a holistic approach instead of resorting to reductionism and collectivism: subdividing the problem and optimizing subsystem solutions does not work well in CSs; by the time one performs this cycle the CS has moved on.
- Striving for balance among subsystems rather than optimization: treat competing objectives together to fairly represent the interests of each effect.

Table 1: Complex system characteristics.

1	Unique: irreproducible; no other exact copies exist
2	Ill-Defined: ambiguous shifting boundary and content
3	Unpredictable: cannot foresee or prespecify outcomes
4	Emergent: unexpected results; multiple scales involved
5	Evolutionary: concurrent/continual development and operation
6	Un-Stable: transitory; largely non-hierarchical structure
7	Self-Organized: highly interactive and re-integratable
8	Collaborative: thrives on cooperation and competition
9	Experimental: explores and tests new possibilities
10	Adaptable: learns from and responds to current conditions
11	Open: influenced by and influences external environment
12	Inter-Relational: long-range effects but short-ranges dominate
13	Robust/Inefficient: relies on diversity and effective holistically
14	Anti-Fragile: strengthened by small random perturbations

Until most of the above aspects are explicitly, broadly, and routinely addressed, how can our understanding of complexity and approach to CSE be considered mature?!

3 COMPLEX SYSTEMS

In this section, we elaborate on CS characteristics, behaviors, and problems.

3.1 Characteristics

Table 1 lists 14 CS characteristics along with a brief explanation of each.

Highly CSs are unique in that no exact replicas exist or can be reproduced. CSs are difficult to define and have no fixed boundary; one needs to discuss and agree on where the "fuzzy" boundary may lie. They are unpredictable and have (unexpected) emergent properties. CSs evolve at several different scales which need to be viewed concurrently to be understood. Although a CS may become relatively stable at times, it will soon move on, fluidly reaching and passing through other states. CSs are self-organized, collaborative, adaptive, and open. They exchange energy and information with their environments, and are heavily interdependent on the multifarious interactions of their component portions. CSs tend to be inefficient in the usual sense because they possess extra capabilities which ensure overall effectiveness. Healthy CSs are robust and can become even stronger from small random perturbations that induce some stress but create a greater ability to face future threats.

3.2 Behaviors

Table 2 lists 14 CS behaviors along with a brief implication of each.

A common behavior is the stimulation of different observer perspectives of the multiple scales of a CS's evolution. A CS thrives on diversity, self-organization, and freewheeling interactions. All this helps to integrate the many factors at play to ensure adaptable and survivable, i.e., robust operation; if possible stimulate anti-fragility. The CS is ever changing, evolving on its own as a whole, and

Table 2: Complex system behaviors and recommended responses.

1	Stimulates Different Perspectives: seek understanding
2	Thrives on Diversity: study inter- and external relationships
3	Self-Organized: note collaboration, competition, and cooperation
4	Internal and External Relationships are Key: learn from experiments
5	Acts Robustly: recognize adaptability, survivability, and anti-fragility
6	Many Factors are at Play: seek to build upon areas of common ground
7	Evolves on its Own as a Whole: appreciate holistic un-predictability
8	Ever Changing: endeavor to pre-prepare in managing uncertainties
9	Unexpected Properties Emerge: welcome them and wrestle with surprises
10	Sensitive to Small Effects: carefully record and mind initial conditions
11	Informs the Observer: rein in emotions and practice objectivity
12	Exhibits Tight and Loose Couplings: map and categorize interactions
13	Performs Openly: attempt to interpret causal relationships with humility
14	Operates at Edge of Chaos: marvel at reflected health where best happens

exhibiting emergent properties that are largely unexpected and even surprising. A slight change of initial conditions can yield quite different results (the "butterfly effect"). CSs generally perform openly but it is challenging to observe objectively, especially when one is inside the CS. A healthy (chaordic) CS operates at the edge of chaos.

3.3 Examples of complex systems problems

In Table 3, we have offered specific CS examples, suggested to what degree they are complex, and have indicated what some solution approaches might entail.

Since people are complex, systems that include them are also complex. Challenging systems, even those that "boggle the mind," but that do not include people as part of the system, are only complicated, or at least not as complex, compared to those that do. Fundamentally people are semi-autonomous, semi-independent CSs [12]; their interactions are what make situations especially complex.

4 COMPLEX SYSTEMS ENGINEERING

In this section, we (1) propose CSE principles that may be more effective than those of classical, conventional, or traditional SE; and (2) discuss SE practice vs. theory.

4.1 Complex systems engineering principles

Table 4 lists this author's latest CSE principles and short explanations. The first 13 principles have existed for years [13, 14] but there is a new 14th principle.

First, we must suppress hubris in trying to quickly solve the problem; instead devote considerable effort to understand the problem by gathering facts and speaking with experienced practitioners within the domain of interest.

Table 3: Examples of complex system problems and potential approaches.

Complexity	CS Challenges	Remedy?
Very High	Combat International Terrorism	Address Why They Hate Us
Very High	Marginalize Iran and N. Korea	Create Economic Incentives
High	Accept Global Climate Change	Reduce Harmful Emissions
High	Bolster Renewable Energy	Transfer Fossil Fuel Efforts
Medium	Update Government Acquisition	Move to Paying for Results
Medium	Develop New Military Systems	Focus Only on What is Needed
Medium	Improve Air Traffic Control	Modernize Existing Technology
Medium	Rebuild National Infrastructure	Stimulate United States to Invest
Low	Fix Privacy and Cyber Security	Give Devices Unique Passwords
Low	Innovate in Social Networks	Concentrate on Educating Public
Complicated	Complete Income Tax Forms	Maintain Accurate Records
Complicated	Keep Expense Budget	Create Thorough Spreadsheet
Complicated	Use Technical Paper Format	Pay Attention to Detail

Table 4: Complex systems engineering principles.

1	Bring Humility: acknowledge shortcomings and seek understanding
2	Follow Holism: focus on the whole system rather than just the pieces
3	Achieve Balance: shun optimality but improve overall capabilities
4	Utilize Trans-Disciplines: apply good techniques of related domains
5	Embrace POET*: define the CS boundary and keep the big picture in mind
6	Nurture Discussions: listen to others and reward information sharing
7	Pursue Opportunities: take informed risks by pursuing alternative solutions
8	Formulate Heuristics: formulate rules of thumb and improve decision taking
9	Foster Trust: brave personal vulnerability in gaining help from others
10	Create Interactive Environment: open and facilitate meaningful interchanges
11	Stimulate Self-Organization: nurture, reward, and celebrate collaborations
12	Seek Simple Elements: deal with the most pervasive common concerns
13	Enforce Layered Architecture: establish early a powerful guide to progress
14	Elevate Future Goals: endeavor with others to help solve world problems

*politics, operations, economics, and technology

Concentrate on the whole CS and the interactions among its components and with the environment. Unless the problem is just complicated, avoid the usual SE approach, reductionism and constructionism; these techniques are fine but only when they apply!

Rather than trying to optimize the parts or even the whole, instead be mindful of achieving balance among competing system aspects to improve overall capabilities.

Utilize trans-disciplines to learn from other, perhaps widely separated, domains.

Nurture discussions and share relevant information (leaders should reward but not punish this!) to garner different points of view so collectively there is a better understanding of the problem's underlying reality and the CS boundary definition.

CSs opportunities often abound and they should be pursued while being informed of potential associated risks. If one only identifies and tries to mitigate risks, attractive solutions that might be found by different paths could be missed.

One way of helping decision takers make better decisions is to recall, develop, or propose good heuristics. Sometimes postponing decisions works well because more information is available on what is happening with previous interventions.

One engenders trust by increasingly sharing information that is reliable and safe in both directions. This takes some courage to show vulnerability in sharing but the prospect of gaining greater understanding from others' knowledge should be worth it.

An interactive environment is critical in facilitating meaningful interchanges. Those in charge should stimulate self-organization by creating attractive conditions for collaboration, competition, and cooperation.

Concentrate on those things that everyone cares about rather than relatively minor "stovepipe" concerns. This can create compelling simplicity within complexity which can achieve results that have widespread benefits such as horizontal interoperability.

The early building of a guiding architecture that does not change much compared to the system being developed is recommended. To the extent possible the architecture should be layered to be more assured of simple and stable interfaces between layers.

Elevate future goals, hopefully with appropriate incentives, so as to reach for more ambitious achievements and even greater success in solving world problems that could benefit all humankind. What higher goals might one's organization aspire to!

4.2 Systems engineering practice vs. theory

In science, which is about understanding the universe, often theory drives practice. For example, based upon their objective observations physicists first develop a theory which gains some degree of acceptance, at least among fellow scientists. This holds until the theory is proven wrong by the practice of conducting experiments.

In engineering, which is about solving human made problems, the reverse tends to be true, i.e., practice drives theory.

Suppose 1950 is when classical, traditional, or conventional SE began. In this era M.I.T.'s Jay Forrester invented system dynamics [9], for example. Assume that SE was practiced from the late 50s through about 1990 (and beyond) when the National Council on Systems Engineering (NCOSE) (now INCOSE) was formed in the US. So 1990 might be the demarcation where SE theory began.

The practice of SoSE began circa 1998 with Mark Maier's definition of SoSs [15]. Efforts toward a theory of SoSE started about 2008 with a seminal paper by Judith Dahmann, et al., on the four different forms of SoSs, mentioned earlier [4].

The scientific field of complexity also extends back into at least the 1940s. The practice of CSE began circa 1993 with Yaneer Bar-Yam's efforts [16] in making the study of complexity more practical and actionable. Theoretical methodologies for CSE began in earnest prior to 2005 with, for example, regimen proposals by Doug Norman, Mike Kuras, and Brian White [17, 18].

Table 5: Essential elements of a complexity, CSs, and CSE maturity model.

Benefits	Sample indicators
Development/Provision	Informed strategies to better influence difficult outcomes
Consistency/Relatedness	Synergistic characteristics, behaviors, and principles
Effectiveness/Usefulness	Greater CSE impacts than traditional methodology
Alignment/Progress	Education, understanding, and acceptance of approach
Action/Awareness	Widely proselytized, promulgated, and publicized ideas
Impact/Value	Measurements of successful applications and practices

Just recently, two interesting works included in the conference, Complex Systems 2016, have appeared: [19] is a literature review assembling thoughts on launching a science of (CS) emergence; and in [20], although the author seems to limit complexity to natural, not human or human-made CSs, he advocates doing more to help solve critical problems the world faces.

5 CONCLUSION

We summarize this paper by suggesting what we might expect from working toward a maturity model. Six essential milestones, each of which provides some perceived benefits, are envisioned and described briefly in Table 5.

The first step is to develop an understandable collection of ideas that will provide the impetus for realizing a signature maturity model. The focus should be on strategies for facilitating desired outcomes in helping to solve our most difficult problems.

The descriptions of the characteristics and behaviors of CSs, and the espoused principles of CSE must be synergistic, internally consistent, and interrelated to propel a cohesive way forward.

Of course, whatever is advocated should be significantly more useful and effective than the traditional SE approach.

As people are exposed and educated to these ideas their understanding of complexity CSs and CSE will likely improve, and they should become more aligned and receptive to such a modernization of SE.

As the elements of the model are explained and distributed in ever widening arenas, in time more practitioners will become acutely aware of how they can apply CSE in their normal work, and motivated toward acting in a fashion consistent with trying to solve more challenging problems.

Naturally, the ultimate assessment of the viability of CSE will depend on its positive impact and value based on measurements of its application in practice, supported in parallel by case studies.

REFERENCES

[1] Systems Engineering, Wikipedia, the free encyclopedia, https://en.wikipedia.org/wiki/Systems_engineering (accessed 14 January 2016).

[2] Schlager, J., Systems engineering: key to modern development. *IRE Transactions*, **3**(3), pp. 64–66, 1956.
http://dx.doi.org/10.1109/IRET-EM.1956.5007383

[3] White, B.E., A complex adaptive systems engineering (case) methodology — the ten-year update. *IEEE Systems Conference*, Orlando, FL, 2016.

[4] Dahmann, J.S., Rebovich, G. Jr. & Lane, J.A., Systems engineering for capabilities. *CrossTalk, The Journal of Defense Software Engineering*, pp. 4–9, 2008, http://www.stsc.hill.af.mil/crosstalk/2008/11/index.html

[5] A world in motion — systems engineering vision 2015. International Council On Systems Engineering (INCOSE), 2015.

[6] Systems engineering handbook — a guide for system life cycle processes and activities. International Council On Systems Engineering (INCOSE), 4th edn, TP – 2003-002-004, Wiley: Hoboken, NJ, 2015.

[7] Checkland, P., *Systems Thinking, Systems Practice — Soft-Systems Methodology : a 30-year Retrospective*, Wiley: New York, 1999.

[8] Forrester, J.W., The beginning of system dynamics, banquet talk. *International Meeting of the System Dynamics Society*, Stuttgart, Germany, 1989, http://web.mit.edu/sysdyn/sd-intro/D-4165-1.pdf

[9] Forrester, J.W., System Dynamics, circa 1958, http://en.wikipedia.org/wiki/Jay_Wright_Forrester

[10] Boardman, J. & Sauser, B., *Systems Thinking – Coping With 21st Century Problems*, CRC Press: Boca Raton, FL, 2008.

[11] The Medici Effect: Breakthrough Insights at the Intersection of Ideas, Concepts, and Cultures, Book by Frans Johansson.

[12] Gorod, A., White, B.E., Ireland, V., Gandhi, S.J. & Sauser, B.J., *Case Studies in System of Systems, Enterprise Systems, and Complex Systems Engineering*, CRC Press, Taylor & Francis Group: Boca Raton, FL, 2015.

[13] White, B.E., A personal history in system of systems, special session on system of systems (SoS). *International Congress on Ultra Modern Telecommunications and Control Systems (ICUMT-2010)*, Moscow, Russia, 2010.
http://dx.doi.org/10.1109/ICUMT.2010.5676561

[14] White, B.E., Managing uncertainty in dating and other complex systems. *Conference on Systems Engineering Research* (CSER) 2011, Redondo Beach, CA, 2011.

[15] Maier, M.W., Architecting principles for systems-of-systems. *Systems Engineering*, **1**(4), pp. 267–284, 1998.
http://dx.doi.org/10.1002/(SICI)1520-6858(1998)1:4<267::AID-SYS3>3.0.CO;2-D

[16] Bar-Yam, Y., *Making Things Work – Solving Complex Problems in a Complex World*, 1st edn, Knowledge Press: New England Complex Systems Institute (NECSI), 2005.

[17] Norman, D.O. & Kuras, M.L., Engineering complex systems. Chapter 10. *Complex Engineered Systems—Science Meets Technology*, eds. D. Braha, A. Minai & Y. Bar-Yam, New England Complex Systems Institute (NECSI), Springer: Cambridge, MA, 2006.
http://dx.doi.org/10.1007/3-540-32834-3_10

[18] White, B.E., Complex adaptive systems engineering. *8th Understanding Complex Systems Symposium*, University of Illinois at Urbana-Champaign, IL, 2008, available at http://www.howhy.com/ucs2008/schedule.html

[19] Lichtenstein, B., Complexity at a crossroads—toward a science of emergence. *Oral Presentation, Complex Systems 2016*, New Forrest, UK, Subsequent version is being submitted to a journal for publication, 2016.

[20] Motloch, J.L., Unlocking complexity: big science project and research agenda. Complex systems 2016, new forrest, uk, 1-3 june 2016, to be published by the Wessex Institute of Technology (WIT) press in a special issue of the. *International Journal of Design & Nature and Ecodynamics,* ISSN:1755-7437; Digital ISSN:1755-7445.

EMERGING PATTERNS IN CROWD STREAMS AND THE AID OF ABM FOR EGRESS MANAGEMENT

F.A. PONZIANI[1,2], A. TINABURRI[1,2] & M. ANGELINO[3]
[1]Central Direction for prevention and technical safety – Department of firefighters, of the public rescue and civil defence – Ministry of the Interior - Italy.
[2]University of Rome Tor Vergata - Italy.
[3]University of Loughborough - United Kingdom.

ABSTRACT

The aim of this study is to illustrate the advantage of using agent based models (ABM) in order to deepen the understanding of crowd phenomena and help manage design strategies and egress operations on the field.

The crowd stream, recognised as a proper nonlinear system, is in fact able to produce characteristic patterns that interact with the environment in recurring shapes. In particular, the free stream and the multi-leader stream are chosen as two mostly representative patterns that develop under typical conditions in a real life egress environment, so that the ABM engineering may bring an added value also in the development of the environment layout.

The first part of the study describes the features of these emerging patterns in one geometrical enclosure typical of a Mall. Such issues as queuing, clogging generation, paths of crowd spreading are, for instance, the primary goal upon which to compare the modelling. The overall outcomes, in terms of travel times and people fluxes, complete the scene.

The second part of the study describes the connected evolution of these emerging patterns in an environment typical of the landside space in an Airport terminal.

Such issues as people advancement, emotions transport, and change of direction are, for instance, the primary goal for the modelling. The overall outcomes of the travel times and people fluxes are the remaining issues to be assessed.

The comparison between the results coming from the various ABM packages (Evac, Pathfinder) completes the study, showing how a proper choice in modelling can be based depending on the foreseen goal of the design stage.

The procedure developed in this study, then, may be used and form the basis for an advanced design method that can be of practical use both in Fire Safety Engineering and in Egress Management.

Keywords: agent based model, egress, mall, terminal.

1 EMERGING PATTERNS IN COMPLEX SYSTEMS: CROWD STREAMS

One of the most astonishing phenomena occurring in complex systems is the appearance of patterns that emerge from the interactions arising in the system, due to internal and external constraints. Depending on the system, these patterns may come from some sort of interaction, like chemical or emotional reactions.

The crowd streams represent one such system, where the interactions among the people have the character of both physical and cognitive nature: think of the constraints due to the free space available for moving or due to the emotional states appearing.

Several approaches have been used to understand and model the crowd streams, from Hydraulics to Particle Dynamics to Cybernetics, and the Agent Based Model (ABM) this research is about may

DOI: 10.2495/DNE-V11-N4-543-552

use a blend of them. Other models may be considered, from Genetic Algorithms to Cellular Automata, but these are not implemented in the ABM packages used in this study.

While the Hydraulics gives an idea of the global flow, the Particle Dynamics gives an idea of the single entities, the Cybernetics gives an idea of the interactions of various nature, the ABM gives an idea of the behaviours of the agents.

Besides strengths and weaknesses of each approach, the effect of nonlinearity is one of the most important to be captured, as can be underlined considering how some nonlinearity from first principles have been represented.

We can mention three models that have been employed in the realm of crowd dynamics: reaction–diffusion, pseudo-chemical kinetics, wave propagation.

In the reaction–diffusion model [1] the parameter that may exhibit changes in time and space is the concentration of some species that, for the crowd streams, may be represented by the density. The temporal variation is determined by reaction terms that may be derived from a potential plus diffusion terms that let the concentration spatially vary. By a proper choice of the potential and of the diffusivity, the states of the system with nonlinear terms can be investigated, exploring the possibilities for steady state, equilibrium, instability and transition.

In the pseudo-chemical kinetics model [2] the reagents that react following rates in a set of reactions are the cognitive states that, for the crowd streams, may be represented by cognitive behaviours, such as emotions, beliefs and actions. By combining proper reactions sets of cognitive states, the dynamic of the system including nonlinear terms can be modelled and various transitions can be analysed in terms of activation, inhibition, and recovery.

In the wave propagation model [3] the signal that is transferred with a velocity of propagation is any feature of the disturbance that, for the crowd streams, may be represented by the density. The variation of the density in time and space includes the nonlinearity of the propagation velocity that is function of the density. By introducing the flow velocity as the ratio of the density to the flux, and considering a functional relation between the flux and the density, it is possible to express the propagation velocity as the variation of the functional relation of the flux with the density. This leads to the fundamental diagram relating density and flow. Higher order effects may also be included, such as diffusion due to non-local interactions expressed via a density gradient, and response time due to the lag expressed via a desired velocity, which the mover accelerates toward in a relaxation time. Further, shock structures can be introduced to fit the jump discontinuities that may appear in some density distribution.

The emerging patterns that can be observed in the crowd streams may appear as clusters of people and waves of density that form and dissolve in time and space, are more or less persistent in the initial shape appeared, define a gradient, mark some direction.

Depending also on the layout of the environment – especially in terms of space available to be filled and narrow sections to be crossed – and the condition of the situation – especially in terms of real or perceived time of safe movement available – these patterns show like queuing, clogging generation, paths of crowd spreading, people advancement, transport of emotions, abrupt change of direction.

2 ENVIRONMENT LAYOUT: MALL AND AIRPORT SPACES

The environment layouts considered in this research are typical of multi-service premises and of transport facilities.

Two layouts have been properly designed here to fit the purpose: one for a mall, one for an airport terminal. These layouts have been chosen since various emerging patterns in the crowd streams may show, due to the high number of people who assemble and move inside.

The mall is composed of one base floor level and a double height ceiling, representing the link of different spaces, with an overall square-like shape in plan. On the two sides at the floor level, four retail shop spaces align in the left while one assembly space, the restrooms and one bar align on the right, with the mall as the connecting space in between. The side spaces are single height, compared with the double height of the mall. One balcony fills the space at single height, protruding into the mall concourse from the line of the shops. Each space on the sides is connected to the mall through one door, the mall is connected to the outside at the front and the back of the floor with three doors on both, the assembly space has two doors and the bar has one door to the outside on the right side of the floor. The total length of the mall space is 32 m, the total width is 32 m, 8 m on the left, 16 m in the middle, 8 m on the right, the single height is 4 m and the double height is 8 m.

The following Fig. 1 show a rendering of the mall from one visualization tool used for the numerical simulation [4], with some surfaces made fictitiously transparent to let see through.

The airport is composed of one base floor level and a single height ceiling, representing the land-side space of the terminal, with an overall strip-like shape in plan. On the left side at the floor level, two partially open square spaces connect to a central concourse, while on the slightly narrower right side a couple of linear walls with turning end corners mark the line. There are four doors to the outside, two in the front part of the concourse and one for each end side, while there are four passageways for each square space that partially open to the concourse.

The total length of the airport space is 60 m, 32 m on the left and 28 m on the right, the total width is 12 m on the left and 8 m on the right, the total height is 4 m. The following Fig. 2 show a rendering of the airport.

The dimension of the doors, passageways and exits – essentially the width – can be varied based on the code followed for the design, both for the mall and the airport.

Several dimensions have been considered in this study, but for the aim of reducing an over-control of the flux that could hinder the formation of patterns well away from the exits, the remainder considers sufficiently wide exits to let patterns develop more freely, at the same time letting possible clogging and queuing with high density of people inside the premises.

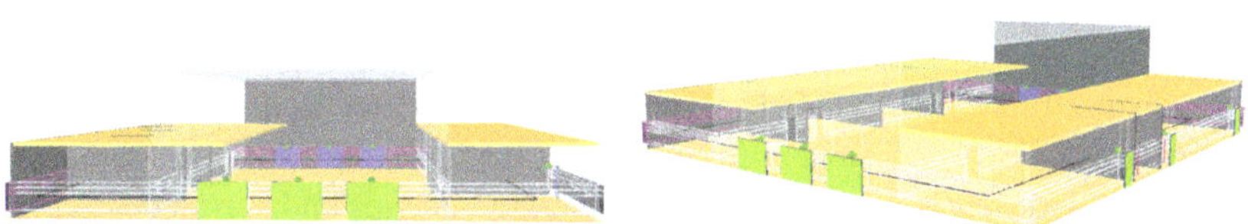

Figure 1: The mall space.

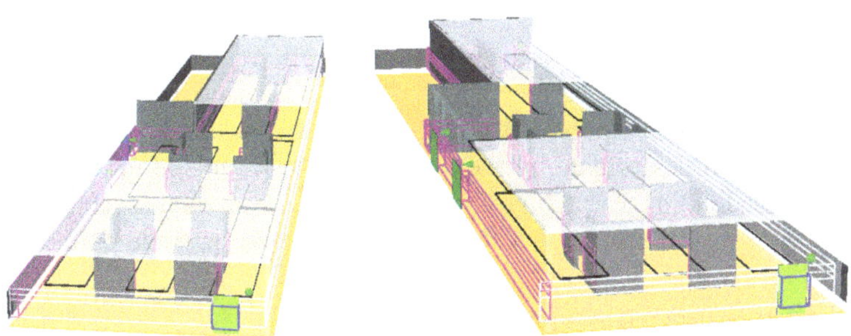

Figure 2: The airport space.

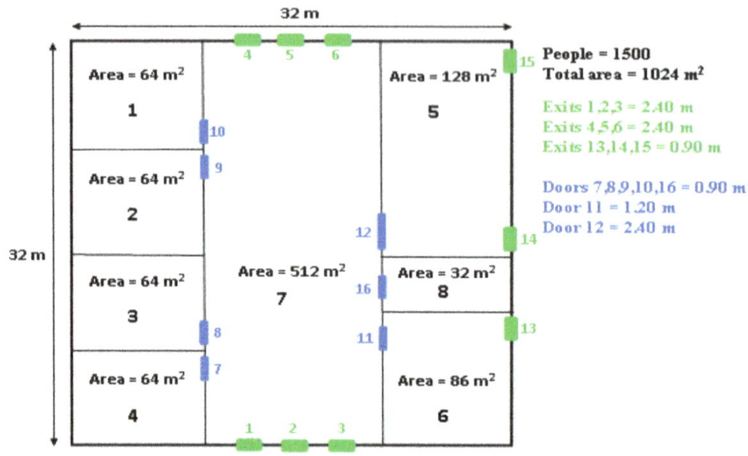

Figure 3: Plan view of the mall.

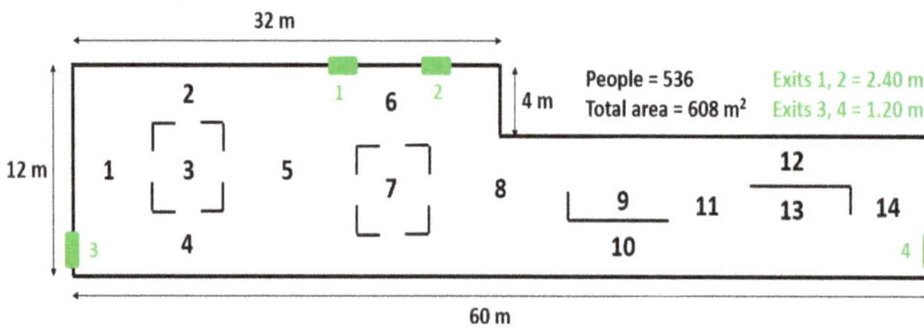

Figure 4: Plan view of the airport.

In order for the emerging patterns of the crowd streams to be coupled to the designed layouts, the virtual environment of the ABM is used here to study what kind of patterns may be represented and how this information may assist in developing proper design strategies and egress operations.

Figures 3 and 4 show the plan of the spaces with some information about spaces and people inside.

3 AGENT BASED MODELLING: EMERGING PATTERNS AND OVERALL FLUXES

The principal characteristic of the ABM is to define autonomous entities, called agents, who try to resemble the behaviour of real people in the same situation. Depending on the amount of knowledge of the parameters defining the system, and on the level of the implemented behavioural choices, the agents may wish to simulate some expected or anticipated behaviour in real life. Even if there is a certain variety in the possibilities offered by the ABM packages available to date, the following related to the packages used in this research give an idea of some important aspects: the agents follow some sort of artificial mental mapping, so to say, usually based on some optimization sets of rules for decision making.

The decisions are about the path to be followed from the current position to a target, like doors or exit passageways, considering the priority of reducing the total time of travel, evaluating the formation of queues, assessing the presence of dangerous effects, reacting to stimuli from the environment or from other agents, and so on. Two ABM packages have been used for this study: the EVAC (plus FDS) [5] and the Pathfinder [6].

The former is essentially based on the interactions deriving from a Social force model [7] and the latter is essentially based on the interactions from a Steering model [8].

The following Table 1 summarizes the behavioural characteristic assigned to the agents for the numerical simulation of the research.

The assigned names are primarily meant to be internal references to the research for identification, more than fixed behaviours in the ABM packages, and a combination of these names is in fact used sometimes with all the packages depending on the parameters assigned for the simulations.

The numerical simulations refer to the Mall and the Airport layouts designed to highlight the emerging patterns of the crowd stream and have an idea of the overall fluxes of the movement.

In order to capture the essentials for the pattern formation, the runs only consider the movement of the crowd streams in the environment layout with no external source of danger – meaning external to the crowd stream itself – and with practically quick responding time to start.

For future developments of the research, instead, external sources of danger such as fire and unforeseen blockage of exits will be added and the pattern formation will be contrasted with the ones

Table 1: Behavioural characteristics.

Autonomous Agent		
Assignment Identification	***ABM Package***	***Behaviour Highlight***
Conservative	Evac	Known door first then the visible ones with variable known door probabilities
Conservative 1	Evac	Known door first then the visible ones with fixed known door probabilities set to one *for all exits*
Active plus Herding	Evac	Visible doors and known doors equal with fixed known door probabilities set to one *for all exits* plus choose the doors that the others are using if no known doors given
Active staff plus Herding	Evac	Visible doors and known doors equal with fixed known door probabilities set to one *for some exits* plus choose the doors that the others are using if no known doors given
Active staff advanced plus Herding	Evac	Visible doors and known doors equal with fixed known door probabilities set to one *for some selected exits* plus choose the doors that the others are using if no known doors given
Speedy	Evac	As Conservative with increased velocities assigned to a group
Steering	Pathfinder	Steering and collision handling
Steering plus SFPE	Pathfinder	Steering and collision handling plus limited door flow rate

obtained in the same layout with no external source of danger. Also several distributions of reaction times will be considered.

Moreover, sufficiently total exits width available is considered as a first assessment for the layouts that will be restricted in the future development of the research, in order to contrast with different total times of egress and various design code provisions.

The adoption of a normalized time $\tau = \tau_{current} / \tau_{end\ of\ run}$ referred to the end time of the run is used for comparing the different simulations and the resulting patterns.

3.1 Numerical simulations for the Mall layout

Several runs have been launched varying the characteristics assigned to the agents, ranging from a general knowledge of the various exits present to a preferred knowledge of some particular exit to a slightly increased velocity of movement. For example, the Fig. 5 below shows a sketch of the preference for some exits for the people inside the space where the arrows start from, mimicking a guiding strength of leadership.

The emerging patterns appeared at one third of the runs are showed, for instance, in the following Fig. 6 that compares the patterns emerged by using the Evac ABM (pictures in yellow background), with all the six different behavioural characteristic assigned and the patterns emerged by using also the Pathfinder ABM (pictures in grey background), with two different behavioural characteristic assigned.

The comparison in terms of total time for exiting and usage of some doors is reported in the Figs 7 and 8.

3.2 Numerical simulations for the Airport layout

Several runs have been launched varying the characteristics assigned to the agents, in the same way done before. Figure 9 below shows a sketch of the preference for some exits mimicking a guiding leadership:

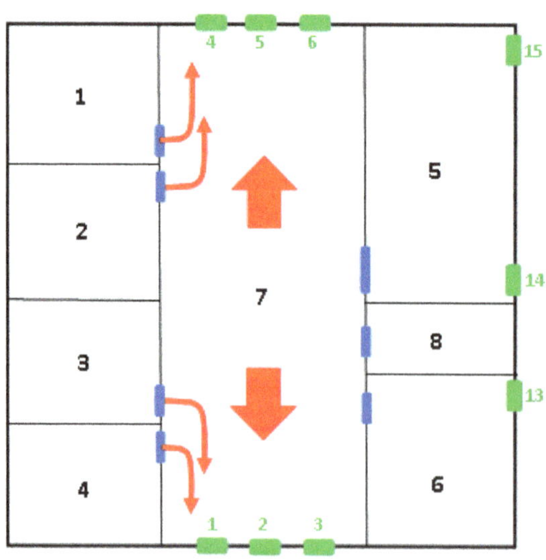

Figure 5: Preference for exiting.

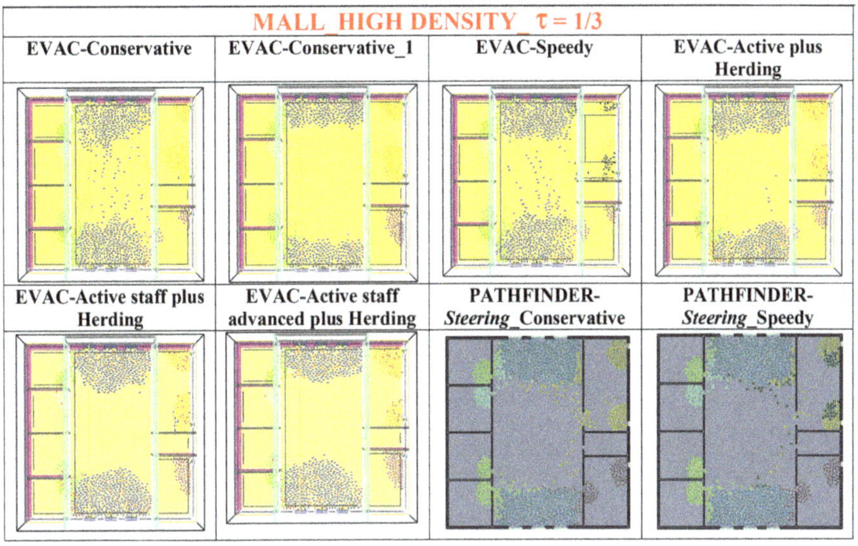

Figure 6: (Evac and Pathfinder): emerging patterns in the mall layout.

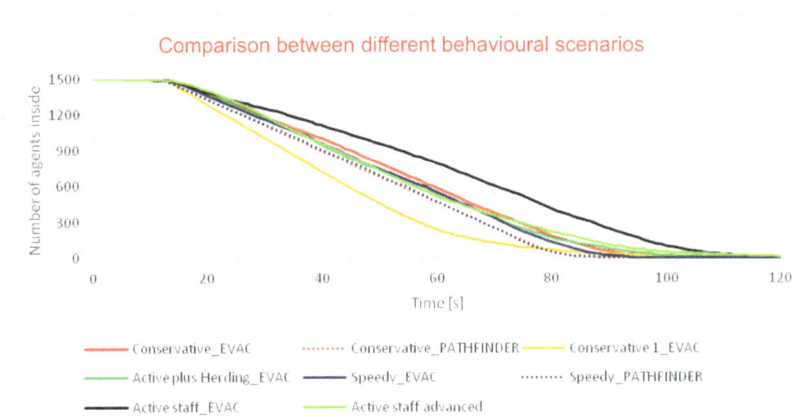

Figure 7: Total time for exiting in the mall layout.

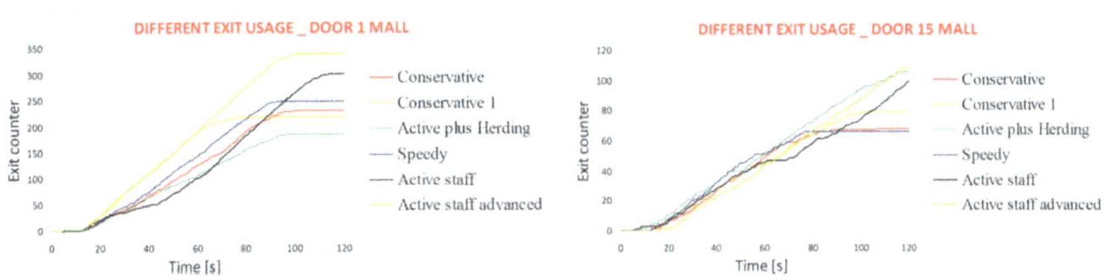

Figure 8: Exit usage trends of two different doors in the mall layout.

The emerging patterns appeared at one third of the runs are showed, for instance, in the following Fig. 10 that compares the patterns emerged by using the Evac ABM (pictures with purple walls), with four different behavioural characteristics assigned and the patterns emerged by using also the Pathfinder ABM (pictures with grey walls), with two different behavioural characteristic assigned.

The comparison in terms of total time for exiting and usage of some doors is reported in the Figs 11 and 12.

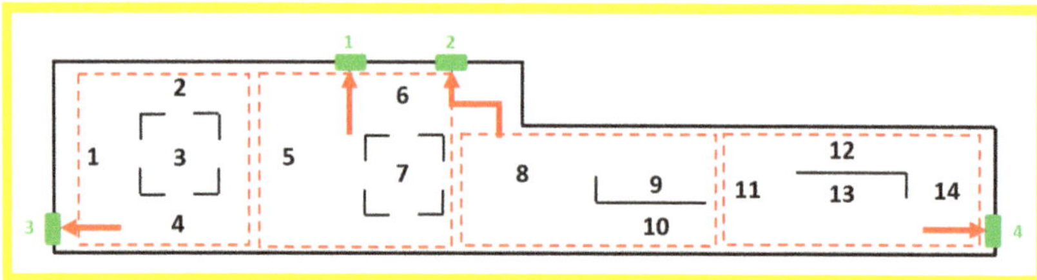

Figure 9: Preference for exiting.

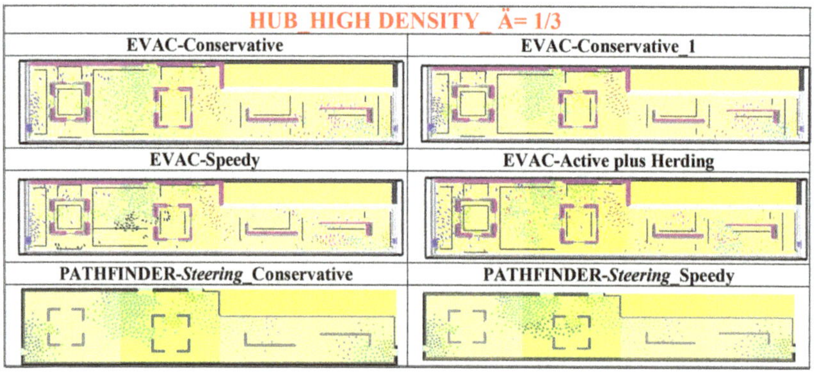

Figure 10: (Evac and Pathfinder) emerging patterns in the airport layout.

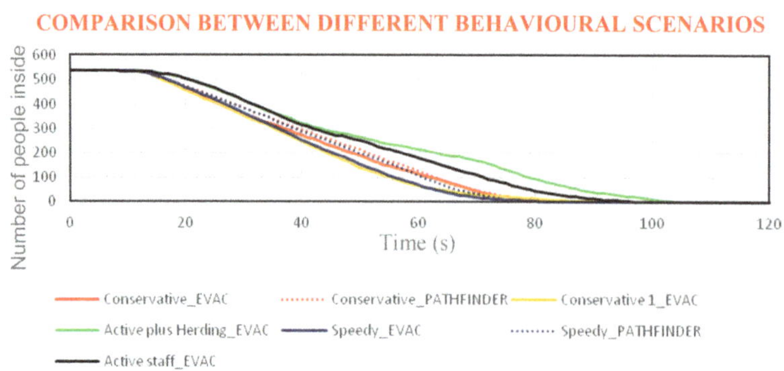

Figure 11: Total time for exiting in the airport layout.

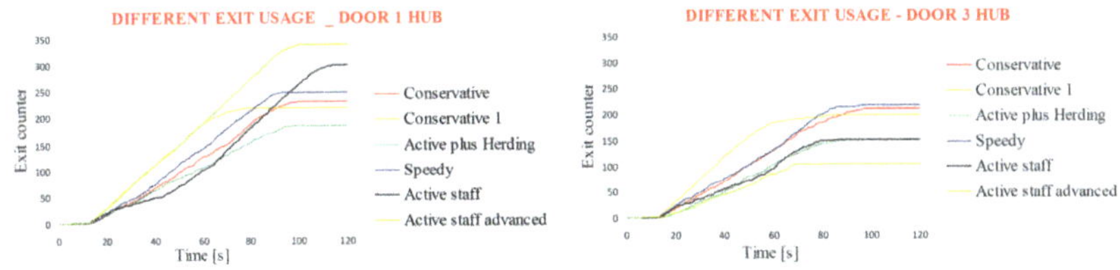

Figure 12: Exit usage trends of two different doors in the airport layout.

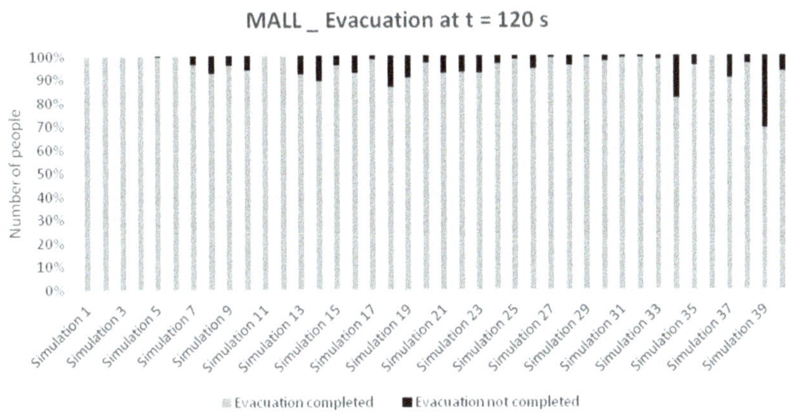

Figure 13: Evacuation state for the mall layout.

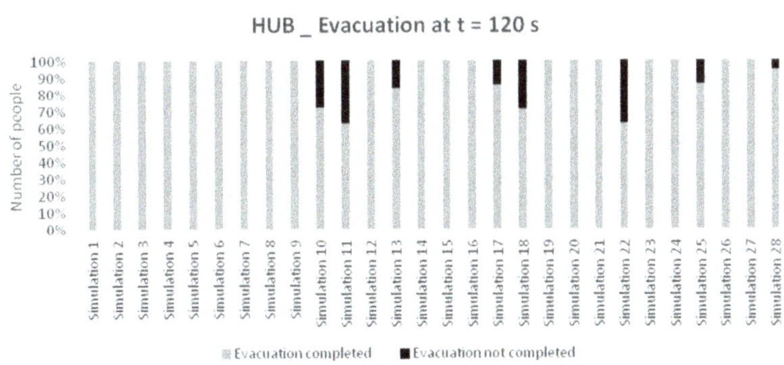

Figure 14: Evacuation state for the airport layout.

4 CONCLUSION: ADAPTIVE ENGINEERING

As a general assessment of the designed layout, it is useful to examine the way the simulations have dealt with the total evacuation capacity at the end of the run time scheduled. Figures 13 and 14 show, for the Mall and the Airport layouts, the percentage of people evacuated at the end time of 120 s scheduled for the runs, considering all the runs performed.

As can be seen, for the most part all the simulations give complete evacuation at the end time scheduled for the run, for any behaviour assigned to the agents, thus encouraging in the choice made for designing the layouts. Further runs could be performed to improve some characteristics, yet most important is to extend the research to include sources of danger external to the crowd streams.

The patterns generated resemble the emerging ones typical of crowd streams, where the main differences in the shapes are related to the layout of the space – different for the Mall and the Airport – with minor impact of the ABM package used.

A significant impact seems to have the behavioural characteristic assigned to the agents, especially in terms of usage of some exits and consequent slowing or speeding up of the overall crowd streams, with different times to exit.

Further developments of the research might explore the behavioural impact, considering also the internal rules of interactions among the agents and the presence of external sources of danger.

Apart from the practical need of varying the reaction times of the agents and the width of the passageways for applications using different design codes, it is worth-noting to remind the necessity of doing several runs for each type of characteristic implemented for the ABM in order to grasp the dispersions of the data obtained and to take a coefficient of safety in designing a solution.

Better still, experimental observations and databases are hoped in the near future to be available so to balance theory and experience, gaining added value for the design stage also.

In such a way, from Agent Based Modelling a leap could bring to Adaptive Engineering, anticipating the behaviours in a real-life situation through real-time pattern recognition, activating a space layout generation capable of changing in response to the event occurring, considering a way finding from active signals along the paths to guided paths unfolding.

The procedure developed in this study, then, may be used and form the basis for an advanced design method that can be of practical use both in Fire Safety Engineering and in Egress Management.

ACKNOWLEDGEMENTS

The Authors wish to acknowledge Giordana Gai for her contribution in setting up the simulations.

REFERENCES

[1] Haken, H., *Synergetics. Introduction and Advanced Topics*, Springer, 2004.

[2] Adamatzky, A., Dynamics of crowd-minds. *Patterns of Irrationality in Emotions, Beliefs and Actions*, World Scientific, 2005.

[3] Whitham, G.B., *Linear and Nonlinear Waves*, John Wiley & Sons, 1999. http://dx.doi.org/10.1002/9781118032954

[4] Smokeview, NIST Special Publication 1017-1, 2014, Online. http://www.nist.gov/el/fire_research/fds_smokeview.cfm

[5] Evac (plus FDS), VTT Technical Research Center of Finland, Online. http://www.vtt.fi/fdsevac/

[6] Pathfinder, Thunderhead Engineering, Online. http://thunderheadeng.com/pathfinder.

[7] Helbing, D., Farkas, I. & Vicsek, T., Simulating dynamical features of escape panic. *Nature*, **407**, pp. 487–490, 2000.

[8] Reynolds, C., Steering behaviours for autonomous characters. *Proceeding Game Developers Conference,* pp. 763–782, 1999.

THE IMPACT AND MITIGATION OF COMPLEXITY DURING PRODUCT DESIGN

X. ZHANG & V. THOMSON
Department of Mechanical Engineering, McGill University, Canada

ABSTRACT

Product complexity is driven by the interdependence of product functions, which in turn determines the interdependence of design tasks, and this is reflected in the complexity of the design process. Ever increasing product complexity has become an obstacle to effective product design. This paper introduces an agent-based model that was used to study the impact and mitigation of product complexity, where complexity was characterised by metrics defined from a knowledge perspective. In the model, a product was represented as a set of functions that required designer knowledge, component design and component integration. Designers were modelled as agents who learned knowledge through consultation and who applied knowledge to function design tasks. Variables that characterised different coordination mechanisms influenced the efficiency and quality of communication between designers and impacted the global behaviour of product design. The results from simulation experiments suggested that a growth in complexity increased effort and span time exponentially and that coordination mechanisms which quickly increased designer learning or which improved collaboration reduced overall effort. The implication for managers is that, for the design of complex products, attention should be paid to the effectiveness of coordination mechanisms, and how they reduce the time for designer learning. The implications with regard to complexity during product design can be applied to other activities where learning is a key performance factor.

Keywords: agent-based model, complexity, product design, coordination, simulation.

1 INTRODUCTION

Product design is the process by which a company enhances characteristics of existing products or introduces new products according to customer demand. Through this process, companies grow and adapt to evolving markets; thus, product design is essential for their viability and success [1]. Nowadays, fast-paced technology and globalisation create a very competitive market. As companies respond to challenges by increasing product diversity, introducing new technology and accelerating innovation, they are generally faced with continuously increasing complexity during product design [2, 3].

Product design is a complex activity that has intrinsic properties of multiplicity, dependency and uncertainty for the aspects of product and process. A product consists of numerous interconnected functions. Adding or modifying a function often involves more effort to adjust other existing functions and even requires a complete reconsideration of the whole set of product functions. Design work is usually broken down into manageable tasks where there is exchange of information in a complex network of design tasks. Since products are generally multi-disciplinary, tasks are disparate regarding knowledge base, required design tools, procedures, etc. Yet, designers are educated and trained discipline-wise [4]; so, effective inter- and intra-coordination among product functions, tasks and designers is essential for the satisfactory design and performance of products. Furthermore,

DOI: 10.2495/DNE-V11-N4-553-562

early phases of product design often confront considerable uncertainty due to incomplete product information, limited understanding of the required functions, and a lack of both technical and design knowledge. [5]. Greater uncertainty leads to a longer design or learning cycle, making it more difficult to manage the design process. Consequently, interdependencies of product functions, coordination of these interdependencies during design activities, required interactions among designers, and system uncertainty contribute towards increasing design complexity, which becomes more challenging when shorter development times are required.

Complexity, if not well managed, is a drag on project performance and can lead to project failure. Increasing costs, longer development cycles and higher risks of failure are observed with rising complexity [6]. Case studies have shown that a company's ability to handle complexity was positively associated with the coordination of designers [7] and designers' knowledge [8]. However, these studies did not give any suggestions on how to improve team performance in projects with high levels of complexity.

The research in this paper developed an agent-based model, which enabled us to simulate the impact of product complexity on the design process and to give suggestions on which coordination mechanisms to adopt for different levels of complexity. The paper is organised as follows: section 2 defines complexity from the knowledge perspective, which serves as the basis for the model; section 3 describes the logic of the model; section 4 presents the data that served as input to the model; section 5 presents the simulation scenarios and discusses the results; and section 6 summarises the paper and gives conclusions.

2 COMPLEXITY DEFINED IN THE KNOWLEDGE PERSPECTIVE

Product design has been recognised as a knowledge-intensive process [9–13]. The multi-disciplinary nature of product design was found to be a key factor driving complexity [4, 14], which indicated the important contribution of designers' knowledge towards handling complexity [15]. Knowledge is the link between product and designers, as knowledge is embedded into product functions, which require designers with corresponding knowledge to realise them. Thus, we characterised product and designers from the knowledge perspective. The set of knowledge used in design projects was classified, where the levels of knowledge required for designing a function were defined with a scale: minimum (numerical value = 1), general (numerical value = 2) and intense (numerical value = 3). Correspondingly, designers' level of ability to apply knowledge was defined with the scale: unskilled (numerical value = 1), proficient (numerical value = 2) and expert (numerical value = 3). Knowledge items were treated as being independent; thus, K knowledge items formed a K-dimensional space. The knowledge of a product function or a designer was represented as a vector in the knowledge space: $\vec{V} = (k_1, k_2, ..., k_n, ..., k_N)$ where k_n is the value of knowledge item n for a function or a designer.

During product design, designers do technical work to develop product functions and communicate to resolve function interfaces. The technical effort is greatly affected by the technical difficulty of functions, and the communication effort is mainly driven by interdependencies among functions. To model technical difficulty, we defined function complexity, which is the complexity of developing a certain function in the product. It is assumed that functions requiring more knowledge are more complex to design; so, the complexity of an individual function is calculated as the root mean square of the elements in its knowledge vector. To model functional interdependency, we introduced integration complexity, which is the complexity of integrating functions into a whole product. It is assumed that the integration of two functions becomes more difficult with more interfaces and with the involvement of different knowledge. Thus, the integration complexity between two functions is calculated as the product of their knowledge difference and the number of interfaces. The knowledge difference between two functions was defined as the intersection angle θ of their knowledge vectors. To keep the value consistent with function complexity in terms of the order of magnitude, we used $3^{\sin\theta}$ to measure the knowledge difference, since 3 was the upper limit of the knowledge scale. Defining product functions, designers and complexity from the knowledge perspective provided a basis for the development of the agent-based model of product design.

3 AGENT-BASED MODEL OF PRODUCT DESIGN

A research model of product design was built using the agent-based paradigm as it allowed variations in designer characteristics and in the interactions between designers. It also permitted the study of the effect of different coordination mechanisms on individuals. This model was able to track instantaneous complexity and dynamic effort during the design process.

A product was modelled as a group of functions that were knowledge intensive and interdependent. In the model, functions were agents requiring knowledge and effort to create or integrate them; designers were modelled as agents with different knowledge and experience aimed at finishing tasks such as function design, rework and functional integration. The designer agents had the autonomy to decide when to start, change or end activities such as the technical work of designing a function, consultation or coordination with another designer. Factors such as coordination efficiency, designers' experience and knowledge influenced the effectiveness and quality of communication between designers, which further determined a designer's decisions. Simultaneous design tasks as well as interactions among designer agents and function agents generated the complex behaviour of the design process, where change in effort and project duration versus different product complexity, designers' knowledge level and experience were analysed. The detailed model logic is explained in the following sub-sections.

3.1 Technical work

Function agents require a certain amount of technical effort determined by their function complexity. Designer agents devote effort to their assigned design tasks with different efficiency. The amount of work to do function i, which was denoted as r_i, was calculated with eqn (1), where M was a coefficient associated with the effort estimation for the product, W_i was the function complexity, $\mu_{p,i}$ was the technical efficiency of designer p developing function i, and $effort_{p,i}$ was the effort given by designer p. When r_i was reduced to zero, the function was considered completed and ready for integration.

$$r_i = M \cdot W_i - \sum_p \mu_{p,i} \cdot effort_{p,i} \tag{1}$$

Technical efficiency $\mu_{p,i}$ is determined by eqn (2), where $KA_{p,k}$ is the ability level of designer p applying knowledge k, $KR_{i,k}$ is the requirement level of function i for knowledge k, and N is the total number of knowledge items. It means that designers' technical efficiency increases with their level of knowledge and decreases as a function's requirement for knowledge increases.

$$\mu_{p,i} = \frac{1}{N}\sum_k \frac{KA_{p,k}}{KR_{i,k}} \tag{2}$$

When functions were completed, designers went to the integration phase. When designers integrated two functions, they could find incompatibility between two functions, which generated rework. It was assumed that incompatible work was more likely to happen when designers did not have enough experience on the project. A parameter, $c_{p,ij}$, defined in the range [0%, 100%] modelled a designer p's experience with the interfaces between function i and function j. The percentage of technical work that needed to be reworked was determined by eqn (3), where D_{ij} was the set of designers that were involved in the development of function *i* and function *j*, and d_{ij} was the number of designers in the set D_{ij}.

$$rework_{ij} = 1 - \frac{1}{d_{ij}} \cdot \sum_{p \in D_{ij}} \varepsilon_{p,ij} \tag{3}$$

3.2 Consultation on technical knowledge

Consultation in the model was defined as communication between two designers concerning certain knowledge involved in technical work. It was modelled such that, when the knowledge requirement of a task was beyond a designer's ability, the designer consulted with a designer with a higher ability of the required knowledge. The decision to go and consult was triggered when eqn (4) was satisfied, where $KR_{i,k}$ was the required level of function *i* for knowledge *k*, $KA_{p,k}$ was the ability level of designer *p* applying knowledge *k*, and *TK* indicated the tolerance of the knowledge gap.

$$\frac{KR_{i,k}}{KA_{p,k}} > TK \tag{4}$$

After a designer decided to consult on knowledge *k*, the designer searched for consultant designers whose knowledge *k* was higher than or equal to the requirement of the function ($KA_{cnslt,k} \geq KR_{i,k}$). If there were many consultants, the designer randomly picked one of them. If no one satisfied the condition, the designer stopped looking for a consultant, and did the work at the designer's own pace.

As a result of consultation, designers' technical knowledge improved. Generally speaking, the effect of consultation was better when more consultation effort was devoted and communication was more efficient. However, consultation efficiency could be impaired by the complexity of the function. Although a designer's ability to apply knowledge improved, the improvement was constrained by the consultant's knowledge level, i.e., a designer's improved knowledge could not exceed the consultant's level of knowledge. Thus, a designer's improved ability can be quantified by eqn (5), where $KA_{p,k}^{New}$ is designer *p*'s improved ability of knowledge *k* after consultation, $KA_{p,k}^{Old}$ is the ability before consultation, $KA_{cnslt,k}$ is the consultant's ability of knowledge *k*, and *c* is the communication efficiency that can be interpreted as complexity solved per hour, W_i is the complexity of function *i*, and $effort_{p,k}$ is the consultation effort in hours. The equation means that a designer's knowledge improves faster with a more experienced consultant, and a designer cannot surpass the consultant in terms of the consulted knowledge. Also, if a designer already has a high level of knowledge before consultation, it takes less effort to reach a certain improved level. The direct effect of consultation is the improvement of a designer's ability using knowledge k. Since technical efficiency increases with designers' knowledge level according to eqn (2), the technical efficiency of developing a function where there has been consulting is improved by the effort spent on consultation.

$$KA_{p,k}^{New} = KA_{cnslt,k} - \left(KA_{cnslt,k} - KA_{p,k}^{Old}\right) \cdot e^{-\left(\frac{c \cdot effort_{p,k}}{W_i}\right)} \tag{5}$$

3.3 Coordination on interface

Coordination in the model was defined as the communication between two designers who were developing different functions that were to be integrated. It was assumed that when designers were aware of a function's dependencies, they coordinated with other designers who were developing relevant functions. The decision to go and coordinate was triggered when eqn (6) was satisfied, where designer *p*'s experience, $\varepsilon_{p,ij}$, indicated how well a designer was aware of I_{ji}, the integration complexity between function *i* and function *j*, *c* was the communication efficiency, $effort_{p,ij}$ was the effort that designer *p* had spent on coordinating function *i* and function *j*, and *TI* was the tolerance of unsolved integration complexity. After designer p decided to coordinate with function *j*, the designer that was assigned to function *j* was randomly picked.

$$\varepsilon_{p,ij} \cdot I_{ij} - c \cdot effort_{p,ij} > TI \tag{6}$$

It was assumed that a designer's experience of the interfaces between two functions, $\varepsilon_{p,ij}$, improved after coordinating the two functions. A designer's improved understanding of two functions, $\varepsilon_{p,ij}$, was determined by eqn (7), where $\varepsilon_{p,ij}^{Initial}$ was designer *p*'s initial understanding of function *i* and function *j*, *c* was communication efficiency, I_{ij} was the integration complexity of function *i* and function *j*, and $effort_{p,ij}$ was the effort that designer *p* had spent on coordinating function i and function j in hours. From the equation, designer's experience improves faster with greater communication efficiency. Also, if a designer already has high experience before consultation, it takes less effort to reach a certain improved level. A designer's experience improves faster with lower integration complexity. Since rework percentage is determined by designers' experience according to eqn (3), the rework for functions that have been coordinated is reduced by the effort spent on coordination.

$$\varepsilon_{p,ij} = 1 - \left(1 - \varepsilon_{p,ij}^{Initial}\right) \cdot e^{-\left(\frac{c \cdot effort_{p,ij}}{I_{ij}}\right)} \tag{7}$$

4 INPUT DATA

The model needed input data of product functions, knowledge involved in the product design, knowledge requirement of functions, interdependency among functions, and designers' ability of applying knowledge. The functional data about a hydroelectric generator was obtained from GE Hydro [16]. Through functional analysis, the product was decomposed into five major functions: control environment, provide housing, provide monitoring, provide safety and control power. The five major functions were further decomposed into sub-functions, which gave 57 functions in total.

The knowledge involved during product design was obtained by consulting the personnel who worked on the project. Ten types of knowledge were identified, including HVAC (heating, ventilating and air conditioning), air circulation, water circulation, heat transfer, electric-heat generation, control, mechanical engineering, sensor technology, physics and electrical engineering.

The personnel also gave information on the knowledge requirement for each function as well as the interdependency among functions. With the information, we were able to calculate the complexity of each function and the integration complexity between pairs of functions using the method introduced in section 2.

The data concerning the designers in the project was unknown; so, several different scenarios were created and simulated. Twenty designer agents were created in the model to develop the product. Each designer had different ability with different knowledge. As a baseline, the designers were initially set as having sufficient knowledge to complete the design tasks, i.e., their knowledge matched well with the knowledge requirement of the functions. Also, they were initially set as having medium experience (70%) about the project.

5 SIMULATION

The objective of the simulations was to investigate the performance of projects when different mechanisms were applied under different product complexity. By examining the chart of dynamic effort obtained from each simulation, we were able to tell how complexity unfolded with time. The chart also gave the total effort and project span time, which indicated project performance.

Factors influencing project performance in the model included: a) product complexity, b) knowledge level of designers, c) experience level of designers and d) communication efficiency. Three levels were set for each factor as illustrated below.

a) Product complexity. To obtain a series of different levels of complexity, we set the complexity level obtained from product data as the baseline and obtained a higher level of complexity by increasing the integration complexity by 50% and a lower level of complexity by decreasing the integration complexity by 50%.
b) Designers' knowledge level. The level obtained from designer data was set at medium. A high level of designers' knowledge was obtained by increasing designers' knowledge level by 0.5 (but not exceeding 3, the upper limit) and a low level of designers' knowledge was obtained by decreasing designers' knowledge level by 0.5 (but not lower than 1, the lower limit).
c) Experience in project. The baseline was 70%, which meant that designers had 70% of the integration knowledge between functions. High experience was set at 95%, which meant that designers were familiar with almost all the integration knowledge. Low experience was set at 45%, which meant that designers knew less than half of the integration knowledge.
d) Communication efficiency. The baseline was set at 0.5, which meant that 0.5 units of integration complexity could be resolved in one-hour of coordination. The high level was set at 0.8 and the low level at 0.2.

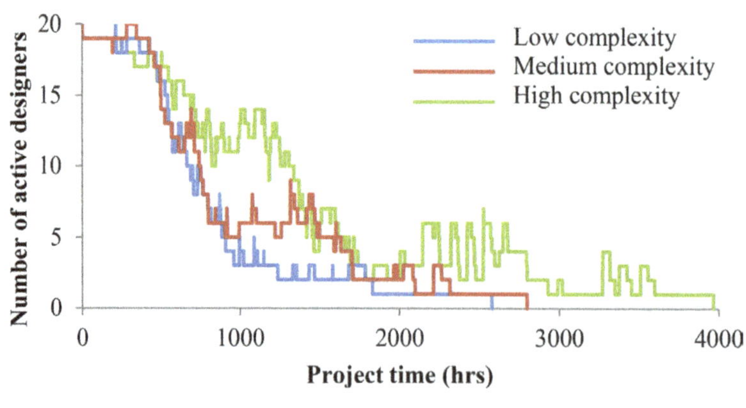

Figure 1: Effort versus project time for the design of a complex product, where complexity was low, medium or high. The data were created by modelling the activity during a multiyear project at GE Hydro.

To study the impact of product complexity, we varied factor a) and set factors b, c and d at the medium level. To study the effectiveness of mitigation mechanisms at different levels of product complexity, we varied factors b, c and d one at a time under different levels of factor a).

5.1 Impact of complexity

To study the impact of complexity, designer agents, who were communicating or working on functions, were considered to be giving effort, which the model captured, and then, generated the chart of instantaneous effort during the entire project.

Figure 1 shows the effort versus project time for the design of a product at GE Hydro, where complexity was low, medium or high. The horizontal axis is project time in working hours and the vertical axis is the number of designers contributing effort at any point of time. In the case of low complexity, dynamic effort reduces continuously, which indicates that complexity is continuously being reduced and that the design problem is converging quickly. Thus, the project finishes in a relatively short time. In the case of medium complexity, dynamic effort reduces at first, increases significantly in the middle of the project, and then, decreases towards the end. When complexity is high, more frequent and large increases in effort during product design are observed. This causes the scenario to have much greater span time than the low or medium complexity case. The increase in effort in the medium and high complexity cases is due to the greater effort required to resolve issues between interdependent functions.

5.2 Effect of mitigation mechanisms

We studied the effect of different mitigation mechanisms on effort and project duration by varying designers' knowledge level, experience and communication efficiency under different levels of product complexity. Figure 2 compares the effect of different factors on effort and span time under different levels of product complexity. In the baseline situation, designers develop a product with medium complexity; they have moderate experience, sufficient knowledge and communicate with medium efficiency. All the other scenarios are compared with the baseline situation regarding effort and span time. The first column shows the change of effort and the second column shows the change in span time. Different rows show the scenarios varying with the different factors.

It can be seen that a higher level of designers' knowledge helps to reduce total effort and to shorten project span time. A lower level of designers' knowledge increases effort, but does not necessarily prolong the project. This can be explained by the fact that other constraints such as coordination efficiency have a large effect on project span time.

Varying the factor of experience shows that a design team with less experience is more sensitive to increasing complexity; whereas, a highly experienced design team is more stable, i.e., when developing a more complex product, the designers can control effort and span time almost to the same level as for the design of a low complexity product.

As for communication efficiency, in the case of low and medium complexity, higher efficiency does not show obvious improvement in project performance. This implies that a medium level of communication is sufficient for the design of a product with low complexity. A product with high complexity is more sensitive to the effect of communication efficiency; both effort and span time vary greatly with this factor. When communication is very efficient, effort and span time for the high complexity case can be kept at the same level as developing a product with medium complexity.

To summarise, increasing complexity requires more effort and prolongs a project, especially for innovative projects about which designers have little experience. More knowledgeable designers

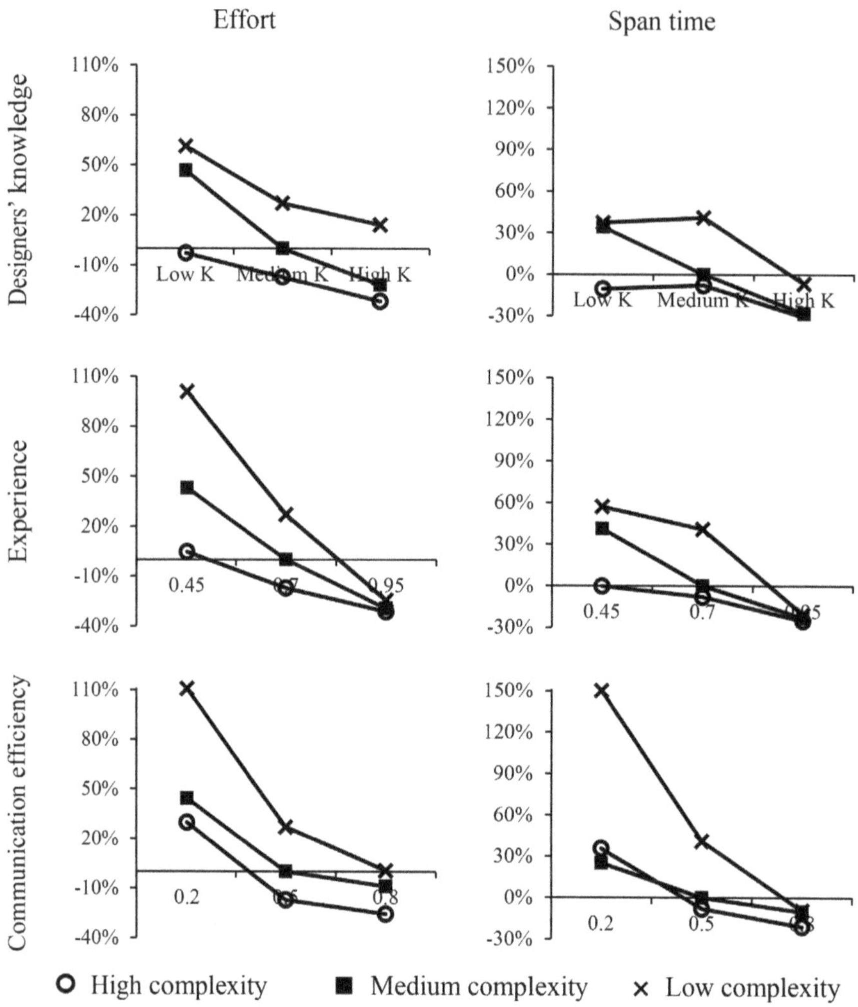

Figure 2: The effect of different factors on effort and span time under different levels of product complexity. Effort and span time are y-axis results, and knowledge, experience and communication efficiency are x-axis variables.

using efficient communication save effort and shorten project span time. This effect is more obvious when developing a highly complex product.

6 CONCLUSIONS

Ever increasing product complexity has become an obstacle to effective product design. In order to investigate how to manage projects better given greater product complexity, we defined complexity from the knowledge perspective and modelled the design process as a learning process, where designers communicate to improve knowledge about a product. Variables that characterised different mitigation mechanisms influenced the efficiency and quality of communication between designers and impacted the global behaviour of product design. The agent-based paradigm allowed us to model the micro-scale behaviour of designers and to observe the resultant macro-scale behaviour of the design process.

Using product data from GE Hydro, the paper simulated a series of scenarios to investigate the effect of designers' knowledge, experience and communication efficiency on project performance. Simulation results suggested that when product complexity increased, effort and span time increased. In addition, at high complexity, effort and span time increased exponentially, where the increase in effort was large and occurred towards the middle and end of a project. This increase in effort was caused by interdependency between functions. The increase in effort may not be obvious at the beginning of the project, and may be a surprise to design teams when it occurs. The simulation also showed that communication mechanisms which quickly increased designer learning or which improved collaboration reduced overall effort. The implication for managers is that, for the design of complex products, attention should be paid to the effectiveness of communication mechanisms and how they reduce the time for designer learning. Since required knowledge and learning are key factors in most projects, the implications with regard to complexity during product design can be applied to other activities where learning is a key performance factor.

REFERENCES

[1] Brown, S.L. & Eisenhardt, K.M., Product development: past research, present findings, and future directions. *The Academy of Management Review,* **20**, pp. 343–378, 1995.

[2] Tatikonda, M.V. & Rosenthal, S.R., Technology novelty, project complexity, and product development project execution success: a deeper look at task uncertainty in product innovation. *IEEE Transactions on Engineering Management*, **47**(1), pp. 74–87, 2000. http://dx.doi.org/10.1109/17.820727

[3] Lindemann, U., Maurer, M. & Braun, T., *Structural Complexity Management: An Approach for the Field of Product Design*, Springer, 2008.

[4] Tomiyama, T., D'amelio, V., Urbanic, J. & Elmaraghy, W., Complexity of multi-disciplinary design. *CIRP Annals - Manufacturing Technology,* **56**, pp. 185–188, 2007.

[5] Oberkampf, W.L., Helton, J.C., Joslyn, C.A., Wojtkiewicz, S.F. & Ferson, S., Challenge problems: uncertainty in system response given uncertain parameters. *Reliability Engineering & System Safety*, **85**, pp. 11–19, 2004. http://dx.doi.org/10.1016/j.ress.2004.03.002

[6] Braun, S.C. & Lindemann, U., The influence of structural complexity on product costs. *IEEE International Conference on Industrial Engineering and Engineering Management*, 2008. http://dx.doi.org/10.1109/ieem.2008.4737873

[7] Gokpinar, B., Hopp, W.J. & Iravani, S.M.R., The impact of misalignment of organizational structure and product architecture on quality in complex product development. *Management Science*, **56**, pp. 468–484, 2010. http://dx.doi.org/10.1287/mnsc.1090.1117

[8] Erhardt, N., Is it all about teamwork? Understanding processes in team-based knowledge work. *Management Learning*, **42**, pp. 87–112, 2011. http://dx.doi.org/10.1177/1350507610382490

[9] Nissen, M. & Levitt, R., Toward simulation models of knowledge-intensive work processes. *CIFE Working Paper*, **77**, 2002.

[10] Štorga, M. & Andreasen, M.M., Complexity of product development context. *DS 34: EDIProD 2004 – Fourth International Seminar and Workshop*, Zielona Gora, pp. 279–284, 2004.

[11] Madhavan, R. & Grover, R., From embedded knowledge to embodied knowledge: new product development as knowledge management. *Journal of Marketing,* **62,** pp. 1–12, 1998.

http://dx.doi.org/10.2307/1252283

[12] Davis, J., Subrahmanian, E. & Westerberg, A., *Knowledge Management: Organizational and Technological Dimensions*, Physica-Verlag HD, 2005.

[13] Mcgrath, J.E. & Argote, L., Group processes in organizational contexts. *Blackwell Handbook of Social Psychology: Group Processes*, Blackwell Publishers Ltd., pp. 603–627, 2008.

[14] Barclay, I. & Dann, Z., New-product-development performance evaluation: a product-complexity-based methodology. *Science Measurement and Technology, IEE Proceedings,* **147**, pp. 41–55, 2000.
http://dx.doi.org/10.1049/ip-smt:20000077

[15] Boisot, M., Knowledge management and complexity. *The SAGE Handbook of Complexity and Management*, eds. P. Allen, S. Maguire & B. Mckelvey, SAGE Publications, 2011.

[16] Bashir, H.A. & Thomson, V., Estimating design effort for GE hydro projects. *Computers & Industrial Engineering,* **46**, pp. 195–204, 2004.
http://dx.doi.org/10.1016/j.cie.2003.12.005

UNLOCKING COMPLEXITY: BIG SCIENCE PROJECT AND RESEARCH AGENDA

JOHN L. MOTLOCH
Sustainable Communities Institute LLC and Ball State University, USA.

ABSTRACT

This paper begins with the Big History story of complexity emerging through cycles of innovation and co-adaptation and humanity's role in this story. It addresses the current compromised status of complex adaptive systems, and potential for complexity science to re-empower these systems. It calls for a Big Science Project and Research Agenda that addresses symptomatic problems and the causal problem of loss of system co-adaptation needed to sustain complexity. Due to urgency, it calls for research and implementation that simultaneously promotes collaboration and co-adaptation at the scale and speed needed, while also addressing today's profound problems and need to re-provision the planet and humanity for a sustainable future. The paper has a meta-level focus on re-empowering complex adaptive systems by closing life-cycle loops, increasing process-level biomimicry, and other actions that increase whole-system health and resilience. It calls for systemic change that includes optimizing the energy-water-food nexus, bio-remediating hydrocarbon residuals, re-empowering resource cycles, replacing fossil-fuel and chemical-based agriculture with clean energy produced organic food and human behavior that interconnects more deeply with the complex adaptive system. The paper proposes a global network of centers and institutes where people learn to thrive into the future by collaborating with complex adaptive systems. It addresses these education-research-demonstration centers for learning and research focused on generating, applying, managing, and diffusing knowledge to transform dysfunctional ecological and social landscapes into fully-functional complex eco-social systems.

Keywords: big history, big science, centers and institutes, co-adaptation, complex adaptive systems, EWF Nexus, knowledge system, unlocking complexity.

1 COMPLEX SYSTEMS

Complexity occurs as agents at one scale produce behavior that lies at a higher level of complexity. As complexity grows through immense numbers of cycles of innovation and co-adaptation, complex adaptive systems emerge as networks of dynamic interaction [1]. Ecologic and social systems are complex and adaptive, as is humanity and its intelligent technology systems.

Complex adaptive systems have the co-adapted complexity needed to use photons of light to form chemical bonds and the carbohydrates that provide structure and feedstock to photosynthesis and living systems. Complex adaptive systems use genetic- and epigenetic-influenced co-adapted behaviors of immense numbers of participants at all stages of life cycles to sustain high levels of chemical, energy and material flows and to regenerate bio-capacity at increasingly higher levels. Whether people see complex adaptive systems as the millennia of evolutionary development, as bounty bestowed through creation, or both; more and more people are becoming extremely concerned with the escalating impacts of humanity's living in ways that compromise the ability of complex systems to fully regenerate their bio-capacity. This paper seeks to help humanity live within planetary boundaries [2] so as to re-empower systems to fully regenerate bio-capacity while allowing all of humanity as parts of a complex adaptive system to thrive.

DOI: 10.2495/DNE-V11-N4-563-572

1.1 Complex systems and the nature of humanity

Increasing inability of humanity to live within planetary boundaries and social foundations is driving change from fully-regenerating complex adaptive systems to systems that fail to sustain their bio-capacity and provide a safe and just place for humanity [3]. The profound challenge today is to adapt human behavior as agents in complex adaptive systems to interconnect physical, biological, and cognitive agents and re-empower full-regeneration of safe and just complex adaptive systems. Some people also see a need to reconnect science and spirituality and refocus the discourse from a reductive ecology/creation debate to a discussion of evolving ecosystems and unfolding complexity. Through the science lens, profound system change challenges us to appreciate humanity's role in fully-functional, regenerating ecosystems. Through the spirituality lens, it challenges us to live humanely in virtuous life cycles and unfolding complexity.

2 BIG HISTORY STORY

Big History is the story of complex adaptive systems from the Big Bang to modernity; and humanity's role in this story. It is the story of dynamic tension promoting complexification through massive numbers of cycles of innovation and co-adaptation that empower higher levels of system functionality, interconnectedness, and bio-capacity. Some see this complexification to include three transformations: physical transformation of the Universe, biological transformation of Planet Earth, and conscious transformation of the noosphere [4, 5]. Christian presents it as an eight-stage progression [6]. Both groups share appreciation of Big History as the story of the progression of complex adaptive systems.

2.1 Story of complexification

Swimme and Tucker see complexification as three transformations [4]. The first physically transforms the universe, with complexification powered by tension between expansion and collapse. The second transforms Planet Earth into living systems, with co-adapted behavior of massive numbers of participants at all stages of life cycles building and sustaining the planet's bio-capacity. The third on-going transformation of consciousness is seen by some to be driven by a universal need to know and understand [7] and by the dynamic tension, in the human mind, between individual and collective motivations [8, 9]

2.1.1 Story of humanity's role

In this current chapter of Big History, the mind of modern man uses symbols, language and recorded history as collective memory which, unlike genetics and epigenetics, includes innovations that successfully co-adapt and those that do not. This cognitive transformation is currently progressing into its third tradition of human engagement. The first tradition – *humanity living within local limits* – included most of human history, with humanity living within limited local resources and struggling to enhance the human condition. In the second tradition – *living unsustainably outside limits* – technology and science combined to enable humanity to mine resources at ever-increasing scales, externalize costs, and thereby cause contextual system degradation at increasing scales and rates. In this tradition, people isolated themselves from many of the impacts of their behavior, and externalized these impacts to other people, places and species. Empowered at ever-increasing levels, this tradition impacted complex adaptive systems at accelerating rates and scales. Resulting system change now exposes the profound threat science and technology pose to humanity [10] and the urgent need to learn to live within the complex adaptive system that sustains us. This realization is driving the emerging third tradition – *sustainable growth within global limits* – as an appreciative system that relates to the reality it helps create [11]. This tradition is challenging humanity to engage the systems upon which we depend – especially those that regenerate clean air, pure water, and productive soils – in ways that sustain system complexity, health, and regeneration. It challenges humanity to internalize all costs so engagement re-empowers system health and complexity and allows regenerative systems at all scales from local to global to sustain full-functionality. It challenges us to transform the landscapes of reduced functionality produced by centuries of false belief that humanity could live outside systems into healthy, fully-regenerating ecosystems.

2.1.2 Mapping the third transformation

Beck and Cowan [12], Ken Wilbur [13] and others have mapped levels of bio-psycho-social complexity as frameworks of consciousness. These levels are seen herein as stages in the third transformation wherein humanity learns to live appreciatively in the complexity it helps create. Since symbolic memory includes innovations that co-adapt and those that do not, human systems evolved by the second tradition of living unsustainably outside systems are not self-managing. In this third stage of transformation, we are beginning to discover and appreciate the immense interconnectivity needed to sustain the complexity upon which humanity depends, and the urgent need to shift to conscious actions that help regenerate complexity beyond our ability to fully understand, much less replicate.

2.1.3 Story of new potential

Humanity is beginning to imagine the immense opportunity inherent in re-empowering complexity and re-provisioning humanity in ways that can trigger immense bio-capacity. This potential lies in emergent behavior, where agents at one scale produce behavior at a higher level of complexity. Building from Roger's diffusion of innovations work [14], Motloch [15] sees four stages of engaging any growth potential. *Stage 1: Emergence* occurs when innovations adapted to system conditions positively address the problems created by previous decisions; *Stage 2: Growth* occurs and accelerates as innovation produces positive system feedbacks, thereby encouraging and rewarding further growth. *Stage 3: Maturity* is reached as maximum sustainable growth is achieved and system feedback begins to shift from positive to negative. *Stage 4: Decline* occurs and accelerates as mainly negative system feedback discourages further engagement.

The contention herein is that humanity has progressed into the fourth stage of its living unsustainably platform based on fossil fuel consumption. Unsustainable growth empowered by reductive science, powerful technologies, and fossil fuel consumption have converted fully-functioning regenerative ecosystems into systems with less than full-functionality. Fossil fuel residuals, heavy metals, and other constituents are serving as roadblocks to emergence of new system complexity. For example, fossil fuel residuals in soils prevent thriving of the complex living systems these soils could otherwise support.

Ramo refers to the present as *The Age of the Unthinkable* [16]; and discusses how the new world order continues to surprise us as actions taken for seemingly good reasons have unthinkable consequences. Motloch and Truex [17] see these outcomes as the profoundly negative feedback of engagement using what is now an outdated and inappropriate development tradition that mines resources and produces wastes beyond the capacity of the system to regenerate its complexity and sustainable yield. The contention herein is also that this unique time in history is an age of immense new potential; and that this potential can begin to be realized when humanity again commits to living appreciably in complex adaptive systems.

3 BIG SCIENCE STORY

Complexity science is the scientific study of complex adaptive systems whose parts interact to produce more-complex behavior [1]. Since Big History is the story of complexity through time, the scientific study of complexity is herein referred to as Big Science. Together Big History and Big Science, like Zukav's *Dancing Wi-Li Masters* [18] view of new physics, can help build the complexity science knowledge system for humanity to become an appreciate system that fully engages in the complex adaptive system that we help create.

3.1 Complexity science knowledge system

The current status of ecological and social systems as complex adaptive systems speaks to the urgent need to develop the complexity science knowledge system, including knowledge generation, application, management, and diffusion, needed to re-provision humanity to transform to sustainability. Building this knowledge system is a major shift from the science that empowered humanity with the ability to create today's profound problems that include but are not limited to pervasive poverty, global climate change, escalating crime, terrorism, nations on the verge of bankruptcy, and a myriad of other problems.

Building this knowledge system will require that we address complexity at three levels: object, system, and meta level. At the *object-level*, decisions will address immediate needs in ways that do not compromise complexity and interconnectedness. Interdisciplinary teams will investigate complex social and environmental problems and make appreciative decisions to address symptoms produced by previous unsustainable engagement, and will integrate their actions into complex systems while seeking to "do no harm". Object-level decisions will also be appreciative-ready, i.e., able to function in more appreciate future systems. At the *system level*, decisions will focus on whole-system complexity and functionality. Interdisciplinary teams will embrace diverse knowledge systems, broad awareness, and methodological flexibility to identify whole-system solutions and the complexity knowledge needed to function appreciatively. At the *meta level*, decisions will focus on relationships, conditions and processes that promote emergence of more appreciative system-level and object-level decisions. At this level interdisciplinary teams will focus on whole-systems, the synergy of diverse knowledge systems, process-level biomimicry, and embracing unknowns and uncertainty as agents of emergence rather than expressions of imperfection. Teams can use learning-forward strategies; and all fields of science can take ownership of integrating with complexity. Methods can avoid the specialized and self-referential nature of disciplines [19] and embrace interdisciplinary processes and team-generated solutions. Interdisciplinary teams can collaborate in centers or institutes as homes for emergence [20] using new administrative structures [21] and enhanced network information flow using transdisciplinary and team science approaches [22, 23] that re-provision universities [24] in ways that re-empower complex adaptive systems to fully-regenerate, co-adapt and complexify.

4 BIG SCIENCE PROJECT

To jump-start this new complexity science knowledge system, humanity needs a Big Science Project on the scale and urgency of the Genome Project. This Project will seek to build the knowledge system needed to empower humanity to live appreciatively within the complex adaptive system it helps create. The Project will be grounded in innovation and co-adaptation; and an understanding of the meta-problem of increasing inability of systems to co-adapt, fully function, and fully regenerate. As a complex adaptive system, this knowledge system can grow from small seeds in diverse domains, seeing what works and what doesn't, and co-adapting things that work within domains into a complex adaptive system.

The Big Science project will target the causal problem, seek to reverse the trend of reduced complexity and inability to co-adapt; and endeavor to build the understanding of complexity needed to empower humanity to co-adapt at the scale and speed needed. It will seek to optimize the ability of the current system to address complex problems, and facilitate emergence of the new system that produces less new problems. This complexity science knowledge system will focus on innovation [25] and co-adaptation. Due to the nature and urgency of the challenge, and need to re-provision concurrently with enhancing performance within existing provisioning, both thrusts will occur simultaneously.

4.1 Emergent status of the big science project

On-going initiatives around the world, if seen as a mosaic, look like the beginning of this Big Science Project, complete with a research agenda beginning to unfold as the nexus of several dimensions of complexity. In terms of *governance complexity*, the Widening Circle action campaign and related Global Citizens' Movement for a Great Transition [26] has identified principles that can inform proposals for world governance [27]. In terms of *social complexity*, the Network Nation collaborative project seeks to address concerns about social fragmentation and partisanship, and what is seen as chronic gridlock in governance. The Next System Project [28] is positioned within this nexus of complexity as part of Widening Circle 2.0 [29]; and as the Collaborative Backbone on the Next Nation [30]. In terms of *development complexity*, the Internet of Things and smart technologies offer immense ability to sense, store, interconnect and communicate the immense data needed to co-adapt, unlock complexity, empower the circular economy [31] and unlock full-spectrum capitals [32] to re-empower complexity.

As this Complexity Science Big Science Project continues to emerge, it can do so with governance appropriate for unlocking complexity. This governance can embrace principles similar to those of the Forum for a New World Governance that include power framed by the people being governed and sustained by a shared perception that decisions address individual and collective needs so as to promote an ethical, just society. They can include civility that builds a sense of shared destiny, excludes tyranny by individual or majority, and balances rights, powers, and responsibility. They can help provision society for competent and effective public and private institutions responsive to all needs of a diverse society; and embrace collaboration and co-adapted behavior among the full diversity of public and private entities, so governance helps sustain the commons. They may also embrace justice at all scales so societies from local to global can organize themselves and operate in ways compatible with social cohesion at all scales. [33]

4.2 Big science research agenda

Developing the complexity science knowledge system will require a massive action research agenda to generate, apply, manage and diffuse complex adaptive system knowledge. This agenda will embrace the co-adaptation needed to empower homeostasis, self–organization, and complexification; and can help humanity address today's myriad of symptomatic problems and, more importantly, the causal problem of growing inability of systems to co-adapt and fully-regenerate. This agenda can also focus on helping humanity collaborate with complexity; and co-adapt at the scale and speed needed to re-empower complexity; so humanity can address today's problems while re-provisioning economic and community development for a sustainable future.

This research agenda can integrate diverse knowledge systems – science, indigenous, vernacular, etc. – with a meta-level focus on re-empowering complex adaptive systems by closing life-cycle loops, increasing process-level biomimicry, enhancing fertility, accelerating ecological processes, empowering homeostatic, and increasing whole-system health and resilience. It can include system-level research focused on feeding good bacteria and starving bad, bio-remediating industrial hydrocarbon chains that have accumulated in systems, dissipating residuals, re-empowering resource cycles including nitrification, and replacing fossil-fuel powered chemical-based agriculture with clean energy-based, locally-grown organic, healthy food. Much research can occur within a global network of centers and institutions where people learn to thrive by collaborating with complex adaptive systems [34]. These centers can have an integrated training, research, and demonstration action agenda focused on how humanity can live within complexity. This action agenda can help build the complexity science knowledge system for generating, applying, managing, and diffusing knowledge that can transform dysfunctional ecological and social landscapes into fully-functional complex eco-social systems.

5 DESIGN SCIENCE ROLE IN COMPLEXITY SCIENCE

Design science will play a major role in the action agenda to address today's profound challenge for humanity to live within its fully-functional complex adaptive system. To serve this role, design science can sustain its focus on innovation and quickly embrace co-adaptation that completes each cycle needed for complexity. As such, design science can also play a major role in the nexus of design science, transdisciplinary team science, and complexity science [35].

Design science can inform decisions integrated with complexity to build ecological, physiological, and psychological health. Grounded in complexity science, design science can engage systems at the object, system, and meta levels. At the object level, projects can address local system needs. System level engagement can include management frameworks that set limits and provide guidance to object-level decisions. Meta-level decisions can pursue relationships, conditions and processes that promote transformation to sustainability. Design science can also embrace innovative models for inter-professional dynamics to make sustainable resource management decisions, while addressing the needs for specific projects. By so doing, design science can help integrate decisions made by many people with differing motives. It can help interdisciplinary teams integrate diverse systems – ecological, cultural, structural, infrastructural, visual, and economic – for maximum environmental and human wellness; and to help make appreciate decisions about the earth and its integrated systems [36].

In these and other ways, design science can play a major role in helping build the knowledge system – generation, application, management, and diffusion – needed to sustain interconnected energy, water, and food systems: the EWF Nexus. It can also play a major role in helping society replace the fossil fuel consumption that locks-up living systems with clean energy systems that use information encoded in the genetics and epigenetics of living systems. It can promote healthy water systems where water at all stages in life cycle addresses the needs of the diverse living and non-living components needed for full-functionality and regeneration of complexity. Finally, it can provide the vision needed to replace the current non-appreciative decisions that assault complex systems with dirty energy, contaminated water, unhealthy food, and disinformation with appreciative ones that ensure that all voices – human, other living, and non-living – needed for full-functionality and regeneration are heard.

5.1 Design science research

Rather than only targeting today's myriad of symptoms – climate change, species loss, organized crime, diverging incomes, failing states, etc. – design science research can also focus on the causal problem, i.e., loss of ability of systems to sustain complexity due to inadequately co-adapted behavior. It can seek to reverse the growing inability of complex adaptive systems to sustain co-adaptation of parts and behaviors, full-functionality, regeneration, and ability to coordinate behaviors in ways that sustain the capacity of the complex adaptive system. It can build understanding of the nexus of design science, team science and complexity science to empower whole-system functionality, co-adaptation and regeneration. It can help build understanding of how knowledge from diverse theories such as rapid co-adaptation theory [37] and relevant frameworks such as whole-system Indicator frameworks, can empower and accelerate co-adaptation of human behavior at the scale and speed needed to re-empower complex systems so humanity can thrive as part of these systems [33].

To help address the causal problem, application-focused design science action research can extend the time humanity has to become appreciative by increasing humanity's performance in addressing complex adaptive system needs, while accelerating the shift needed to re-empower complexity and avoid creating new complex problems. Furthermore, due to severity of the problem and urgency of the challenge, it can do both simultaneously and as quickly as possible. By so doing design science, as a subset of Big Science linked with Big History's unfolding complexity, can play a crucial role in the Big Science Project.

6 "UNLOCKING COMPLEXITY" AS GOAL OF BIG SCIENCE PROJECT

As stated earlier, fossil fuel residuals, heavy metals, and other impacts are often roadblocks to emergence of system complexity. For example, fossil fuel residuals lock-up agricultural soils and ecological systems, preventing them from supporting complex living systems. There is a profound need for this Big Science project and research agenda to focus on Unlocking Complexity; and for the complexity science knowledge system to re-provision global society in ways that promote co-adaptation, homeostasis, and self-organization. There is also the need to address symptoms and, most importantly, mitigate the causal problem of growing inability of systems to co-adapt and build complexity. This Unlocking Complexity Big Science Project and Research Agenda can build the knowledge system needed to unlock complex adaptive systems so they can build the complexity needed to move beyond their current condition that places humanity at profound risk. This Big Science Project will, of necessity, extend beyond unlocking ecologic complexity to unlocking cognitive complexity.

6.1 Unlocking ecological complexity

This Big Science Project and Research Agenda can include initiatives focused on unlocking ecologically complexity, closing life-cycle loops and increasing process-level biomimicry. It can include action items to empower para-magnetics and fertility, accelerate ecological processes, and re-empower homeostatic and self-regulation. It can include system-level initiatives to feed good bacteria and starve bad, bio-remediate industrial hydrocarbon chains that have accumulated in systems, dissipate residuals, re-empower resource cycles including nitrification, and replace fossil-fueled chemical-based agriculture with clean energy-produced organic food. By these and other actions, it can build whole-system health and resilience.

6.2 Unlocking cognitive complexity

This Big Science Project and Research Agenda can focus on unlocking the cognitive complexity arrested by the false belief that humanity could live outside systems, mine resources at ever-increasing scales, and externalize costs without profound consequences. Most recently cognition has been further arrested by culturally induced ignorance or doubt based on inaccurate or misleading scientific data. Unlocking this arrested consciousness can allow humanity to appreciate complexity beyond our current ability to understand; and to build the knowledge system needed to sustain healthy regeneration of all essential and interconnected systems. It can empower an appreciative humanity to use information to sustain clean and healthy energy, water, and food systems; and replace the fossil fuel consumption that locks-up living systems with *clean energy systems* that use information encoded in the genetics and epigenetics of living systems. It can empower *healthy water systems* where water conditions at all stages in life cycles address the needs of the diverse living and non-living components needed for full-functionality and regeneration. It can also empower implementation of *healthy local and slow food systems* that provide the food and fiber needed to support whole-system functionality and regeneration. In doing so, the Unlocking Complexity Big Science Project can replace the non-appreciative mindset and hierarchical communication strategies that assault complex adaptive systems with dirty energy, water, food, and disinformation with appreciative humanity that uses network conversation to ensure all voices – human, other living, and non-living – needed for full-functionality and regeneration are heard.

6.2.1 Global network of centers and institutes

There is a need for a global network or centers and institutes focused on unlocking complexity by building bridges between community development and the nexus of energy, water and food, as well as the nexus of design science, transdisciplinary team science, and complexity science. These centers and institutes can help humanity co-adapt at the scale and speed needed to re-empower the complex adaptive systems of which they are part. These centers can pursue the following principles of the Sustainable Communities Institute as triggers for building resilient, thriving communities [38]. The first is to live within the systems upon which people depend to sustain ecological, social, and economic prosperity. The second is to promote whole-system solutions that address the needs of people and the systems upon which they rely. The third is to build whole-system knowledge and the diversity of knowledge systems. The fourth is to engage all sectors to build shared-consciousness and the collective-will to be sustainable. The fifth is to collaborate among local and global experts to benefit from the full diversity of knowledge. The sixth is to build the full-spectrum of capitals needed to optimize whole-system health, prosperity, and growth.

In these centers and institutes, people can learn to thrive into the future by collaborating with complex adaptive systems. Each center can help people learn to live with system limits including optimizing the local food-water-energy nexus and re-provisioning communities to engage food, water and energy in ways that sustain people, place and prosperity. These can be education-research-demonstration centers for learning to live within complexity, build interconnectivity, and help transform dysfunctional ecological and human landscapes into fully-functional complex and co-adapted eco-social systems.

The heart of each of these centers can be integrated water-energy-food systems that produce more food for less money than traditional farming practices and that can grow food year round. They can produce fresh, healthy and great-tasting natural and organic food via an eco-friendly, green system that uses clean energy, conserves water and reduces waste [39]. In these centers, people can learn to use these systems and local food production to provide quality food, conserve water, and build economic opportunity, while removing agriculture loads from land areas to allow these areas to return to serving their vital role in the complexity of Planet Earth. Each can transform hydrocarbons into food for existing bacteria, allow these bacteria to better serve their role in complexity: and convert waste to food. They can transform benzene chains in soil into food for bacteria in the soil [40] to convert fossil fuel residuals that currently lock-up agricultural soils into food for bacteria in living systems so these bacteria can unlock the soil and provide healthy food for the complex web and food chains of complex systems.

The intent herein is that as education-research-demonstration centers for learning to live within complexity, each center is also a key participant in the Big Science Project and Research Agenda and action agenda to build the complex adaptive system knowledge system needed to unlock complexity.

REFERENCES

[1] Johnson, S., *Emergence: The Connected Lives of Ants, Brains, Cities, and Software*, Scribner, 2002.

[2] Rockström, J., *Planetary Boundaries: Exploring the Safe Operating Space for Humanity*, The Reliance Center, 2009.

[3] Raworth, K., *A Safe and Just Space for Humanity: Can we Live Within the Doughnut?* Oxfam International Discussion Paper, 2012.

[4] Swimme, B.T. & Tucker, M.E., *Journey of the Universe*, Yale University Press, 2011.

[5] Teilh.ard de Chardin, P., *The Phenomenon of Man*, Harper Torchbooks, 1961.

[6] The History of our World in 18 Minutes, David Christian, TED2011, available at www.ted.com/talks/david_christian_big_history?language=en

[7] Maslow, A.H. & Lowery, R. (ed0073), *Toward a Psychology of Being,* 3rd edn., New York: Wiley & Sons, 1998.

[8] Graves, C.W., 'Meet Mr. Graves' spiraldynamics.org/2012/06/meet-dr-graves/

[9] Wilbur, K., *A Theory of Everything: An Integral Vision for Business, Politics, Science and Spirituality*, Shambhala Pub. Inc: New York, 2000.

[10] Humanity is going to use science and technology to wipe itself out, S.W. Hawking, available at www.independent.co.uk/news

[11] Jantsch, E., *Design for Evolution,* Braziller Publisher: New York, 1975.

[12] Beck, D.E. & Cowan, C., *Spiral Dynamics: Mastering Values, Leadership, and Change*, Blackwell Publishing Ltd., 1996.

[13] Wilbur, K., *A Theory of Everything: An Integral Vision for Business, Politics, Science and Spirituality*, Shambhala Pub. Inc: New York, 2000.

[14] Rogers, E.M., *Diffusion of Innovations*, 5th edn., Simon & Schuster, 2003.

[15] Motloch, J.L., *Facilitating Social and Economic Change, Symposium to Increase Community Prosperity in the New Economy*, Ball State University, 2012.

[16] Ramo, J.C., *The Age of the Unthinkable: Why the New World Disorder Constantly Surprises Us and What We Can Do About It*, Little, Brown and Company Pub., 2009.

[17] Motloch, J.L. & Truex, S.I., Complexity Science and Architecture in the. *University of the Future 2nd International Conference on Social Housing and Sustainability*, PUCRS, Brazil, 2012.

[18] Zukav, G., *The Dancing Wu Li Masters: An Overview of the New Physics*, William Morrow and Company pub., 1979.

[19] Bursztyn, M. & Maury, M.B., The Brazilian experience with institutional arrangements for interdisciplinary graduate programs: I2S may provide a new way (Chapter 43). *Disciplining Interdisciplinarity: Integration and Implementation Sciences for Researching Complex Real-World Problems*, ed. G. Bammer, ANU E Press, 2013.

[20] Cram, L., Building I2S into an academic program (Chapter 41). *Disciplining Interdisciplinarity: Integration and Implementation Sciences for Researching Complex Real-World Problems*, ed. G. Bammer, ANU E Press, 2013.

[21] Nellis, M.D., Moving competitive integrated science forward: a US land grant university perspective (Chapter 46). *Disciplining Interdisciplinarity: Integration and Implementation Sciences for Researching Complex Real-World Problems*, ed. G. Bammer, ANU E Press, 2013.

[22] Lefroy, T., Interdisciplinary research is about people as well as concepts and methods (Chapter 47). *Disciplining Interdisciplinarity: Integration and Implementation Sciences for Researching Complex Real-World Problems*, ed. G. Bammer, ANU E Press, 2013.

[23] Falk-Krzesinski, I2S and research development professionals: time to develop a mutually advantageous relationship (Chapter 58). *Disciplining Interdisciplinarity: Integration and Implementation Sciences for Researching Complex Real-World Problems,* ed. G. Bammer, ANU E Press, 2013.

[24] Withers, G., Creating the new university, (Chapter 8). *Disciplining Interdisciplinarity: Integration and Implementation Sciences for Researching Complex Real-World Problems*, ed. G. Bammer, ANU E Press, 2013.

[25] Elsum, I., tackling integrated applied research: lessons from the management of innovation (Chapter 55). *Disciplining Interdisciplinarity: Integration and Implementation Sciences for Researching Complex Real-World Problems*, ANU E Press, ed. G. Bammer, 2013.

[26] Global Citizens Movement, available at www.kosmosjournal.org/programs/global-citizens-movement/

[27] Principles of Governance: A Guide to Developing Proposals for World Governance, available at www.world-governance.org/article152

[28] Democracy Collaborative; available at www.democracycollaborative.org/.

[29] Widening Circle 2.0, available at www.wideningcircle.org/ProgramPhase1/Index.html

[30] Network Nation, available at www. Networknation.net

[31] Intelligent Assets, *Unlocking the Circular Economy Potential*, Ellen Macarthur Foundation, 2016.

[32] Armistead, D., *Symposium on Community Social and Economic Change in the New Economy*, Ball State University, Muncie IN, 2011.

[33] Forum for a New World Governance, available at www.world-governance.org/article152

[34] Motloch, J.L., Living within humanity's life-support system. *International Conference on Sustainable Design Engineering and Construction*, 2015.

[35] Motloch, J.L., Review: G. Bammer (ed.): Disciplining Interdisciplinarity. *Journal of Environmental Studies and Science*, 2015.

[36] Motloch, J.L., *Introduction to Landscape Design*, 2nd edn., John Wiley & Sons, Inc., 2000.

[37] "Rapid Evolutionary Adaptation: Potential and Constraints". *ERACAPS Plant Science Database*, available at eracaps-psd.org/node/988

[38] Sustainable Community Institute, available at www.facebook.com/sci.thrive/

[39] ECSIA, Environmentally Controlled Integrated Sustainable Agriculture, available at www.ecsia.us/

[40] Fog Free Technologies, available at www.fogfreetechnologies.com

APPROACH TO EVALUATE COMPLEXITY IN NEW PRODUCT DEVELOPMENT PROJECTS

G. SCHUH, M. RIESENER & C. MATTERN
RWTH Aachen University, Laboratory for Machine Tools and Production Engineering (WZL), Germany.

ABSTRACT

Manufacturing companies respond to the customers' demand for more individualized products and enhanced functionality by increasing product variety and integrating multiple technologies in their products. Consequently, complexity in new product development (NPD) projects increases as more and more interdependencies and framework conditions have to be considered. A systematic evaluation of the parameters that drive the complexity of an NPD project prior to project kick-off would enable project management to identify critical issues up-front and adjust the project setup accordingly. There are several recent studies that analyze project complexity and identify complexity drivers. However, there is still a lack of a holistic approach that helps project managers to analyze the complexity of NPD projects and at the same time derive specific measures in order to adjust project complexity.

This paper presents an approach for complexity evaluation of NPD projects during project preparation. The approach is based on complexity drivers that have been determined through literature research, classified according to their impact on the complexity of NPD projects and influenceability by project management as well as on the analysis in terms of their interdependencies. Through individual assessment of the complexity drivers and through portfolio analysis, the focal points of complexity in an NPD project can be identified. Measures to level complexity according to the company's positioning can be derived from studying the portfolio of complexity drivers taking into account their impact on first order interactions between complexity drivers.

Keywords: complexity drivers, complexity evaluation, engineering, new product development, project management.

1 INTRODUCTION

Complexity in product development has been the subject of many research activities in the past decades. New product development (NPD) projects show various characteristics that are typical for complex systems. According to Ulrich and Probst, complex systems have multiple elements with a high number of interactions. [1] In an NPD project those elements are, for example, different parts of a component in the developed product, individuals or organizational units working together in the project as well as resources and technologies that are utilized to develop the product. All these elements interact either in a physical way like the components of the product or in a non-physical way like project staff sharing information with each other. Hence, a product development project can be considered as a complex system whose behavior is highly dynamic and whose future conditions are hardly predictable. [2]

Complexity in NPD projects has a significant impact on project performance. Recent studies show that around half the projects in product development are not completed on time and within the budget. [3] A prominent example of project delay and cost overrun as a result of high complexity in an NPD project is the development of the Boeing 787 Dreamliner. [4] Thus, complexity is an economic factor with considerable influence on a project's performance.

DOI: 10.2495/DNE-V11-N4-573-583

The analysis of an NPD project's complexity has to integrate various perspectives and dimensions. An isolated view on only specific elements of an NPD project is insufficient for the analysis of its complexity. An NPD project regarded as part of a system has various interactions with its environment. Therefore, elements from outside of the project's scope, which interact with its internal elements must be considered as well. [5] While elements located within the project's scope can mostly be influenced by the project manager, the elements located in the project environment are normally given and are beyond the sphere of the influence of project management. In order to achieve a balance between internal project complexity and external system complexity the interdependencies between external and internal factors have to be analyzed together. [6]

In this paper, a concept is presented, which allows to evaluate the complexity of an NPD project based on complexity drivers. Complexity drivers in general can be described as influential factors that cause complexity in the project. [7] Complexity drivers are valuable instruments to make complexity measurable and to ensure the practical applicability of the concept at the same time.

The concept aims at finding a balance between completeness of the evaluation on one side and accuracy of the measurement on the other side ensuring, at the same time, its practical applicability during project preparation. With a numerical assessment of the complexity drivers, the overall complexity of the project can be evaluated in different dimensions. This evaluation allows to identify critical elements within the project setup that determine the project's complexity. In addition, the interdependencies between the complexity drivers are analyzed. Complexity drivers with a high reinforcing impact on other drivers should be taken into consideration when defining measures to adjust project complexity.

2 RELATED WORK

Sources of complexity in projects were analyzed by a large number of authors, e.g. Baccarini [8] and Williams [9]. An extensive literature review regarding complexity in projects in general was conducted by Geraldi *et al.* [10]. NPD projects are a particular type of projects that are characterized by significant dynamics and complexity. [11] A framework for the analysis of complexity was contributed by Bosch-Rekveldt *et al.* based on semi-structured interviews. [12] However, the contributions of these authors aimed primarily at developing a holistic set of complexity drivers. They did not work on a concept for evaluation of those drivers, which could support project managers to identify critical elements of the project setup.

For the analysis of complexity in NPD projects, a holistic set of various complexity drivers is fundamental in order to cover all relevant aspects. Product-related complexity as one of the key underlying elements of project complexity is a fundamental part of this analysis. However, other aspects like organizational complexity have to be considered as well. Backlund presented a definition of organizational complexity regarding an organization's behavior, structure, and processes. [13] Daft presented a measuring method for both organizational complexity in the vertical and horizontal dimensions and spatial complexity that can be quantified by the number of hierarchical levels, the number of organizational units, and their geographical distribution. [14] Yet, these authors do not provide an integrated framework for complexity evaluation including aspects of product-related complexity as well as organizational complexity.

Some authors developed very accurate measuring methods with focus on particular aspects of NPD project complexity. Schlick *et al.* developed a design structure matrix-based approach using eigenvalues. [15] Nguyen *et al.* used the fuzzy analytic hierarchy process to quantify project complexity. [16] The approach of Wang *et al.* integrated entropy calculation and network theory in order to measure the complexity of organizations. [17] Qureshi and Kang analyzed the interrelations between complexity drivers using structural equation modeling. [18] These aforementioned approaches have a common factor in that they provide methods to measure the complexity of NPD projects on a high level of detail. However, they focus on specific aspects of complexity and do not provide a holistic approach to analyze the overall complexity of NPD projects.

Measurements of complexity in other areas were developed by Schuh *et al.* for global production networks [19] as well as a complexity driver-based approach for the evaluation of complexity in industrial services [20]. Complexity of products was the subject of the work of Sinha [21] Even though these approaches address other areas of application, parts of them can be applied to complexity evaluation of NPD.

This paper intends to present a concept that integrates the advantages and disadvantages of the aforementioned approaches with a holistic set of classified complexity drivers and an evaluation with medium accuracy to ensure practical feasibility of the approach in the early phase of project preparation.

3 APPROACH TO EVALUATE COMPLEXITY IN NEW PRODUCT DEVELOPMENT PROJECTS

The approach to evaluate complexity is presented in three sections. First, the overall framework to classify and evaluate complexity drivers is introduced. Then, the analysis of interdependencies between the complexity drivers based on an influence matrix is proposed. On this basis, complexity drivers with high strength of influence and influenceability are identified. The overall framework is an appropriate tool to visualize the effects of a shift in the rating of complexity drivers.

3.1 Framework for evaluation of complexity using complexity drivers

Complexity in NPD projects is a phenomenon that is difficult to determine. According to Bliss, there are exogenous and endogenous complexity drivers. Exogenous drivers characterize the complexity of the environment and endogenous complexity drivers describe the complexity of the interior of the project. [22] Referring to NPD projects, complexity drivers are classified into three groups. Complexity drivers describing complexity emerging from inside the project build the first group. In the second group are drivers located inside the company but outside the project. The third group contains drivers within the environment of the company, which can also have an impact on the complexity of the NPD project.

A thematic categorization to facilitate complexity analysis was developed in addition to the structuring by their localization in the project environment. For this purpose, the categories *Organization, Product, Resources, and Technologies* were defined to categorize the complexity drivers. For each of the categories, complexity drivers have been determined through literature review and complemented by own experience from accompanied development projects. Figure 1 shows the developed framework with these four categories.

The category *Organization* contains drivers that describe the interactions of an organization with its environment like public authorities or suppliers. Furthermore, the drivers characterizing the elements of the company's organization as well as the project organization are included in this category, e.g. organizational units. [23] The category *Product* contains complexity drivers that evolve from the product's perspective like market requirements or variety of functions. [24] Especially, a modern structure of the product program in a modular product platform causes additional complexity due to high interdependencies between modules and components. [25] The category *Resources* includes complexity drivers related to human, material, and financial resources. The availability of resources may have a significant impact on a project's performance. [3] Finally, the category *Technologies* covers drivers relevant to the technologies that are in use in the product being developed in the considered project. For example, using technologies with low experience may also result in additional difficulties in project execution. [26]

The framework allows to arrange the complexity drivers depending on their sphere of influence in a shell model. The drivers located in the inner circle are drivers from the inside of the project. Drivers regarding the project environment from within the company can be located in the middle shell, drivers from outside of the company are arranged in the outer shell. The identified complexity drivers were classified into the framework as shown in Fig. 1.

Complexity of a system in general can be measured on the basis of two dimensions multiplicity/diversity and variability/dynamics according to Ulrich and Probst [1] In order to describe the impact of a complexity driver on the overall complexity of the NPD project, each driver is evaluated through an individual rating, a company specific weighting, and factors describing the complexity driver's impact on the dimensions. [27] The results of the assessment are visualized using the four categories Organization, Product, Process, and Resources in a portfolio as shown in Fig. 2.

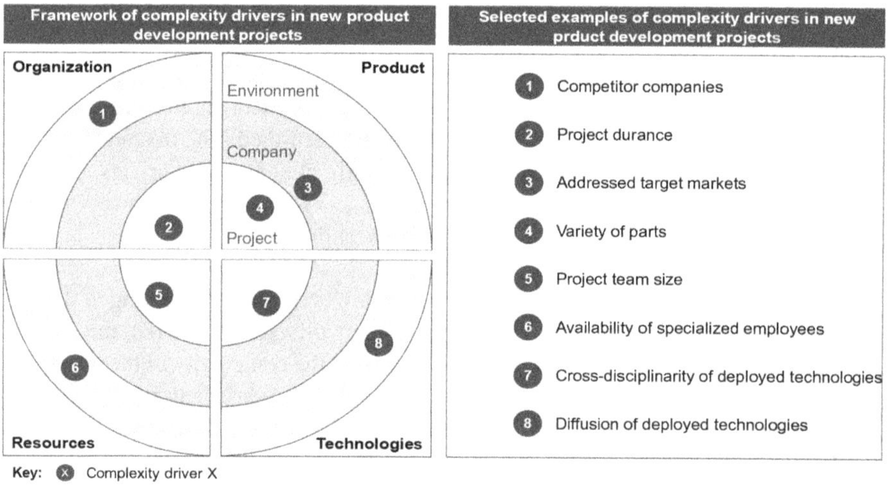

Figure 1: Categorization of complexity drivers.

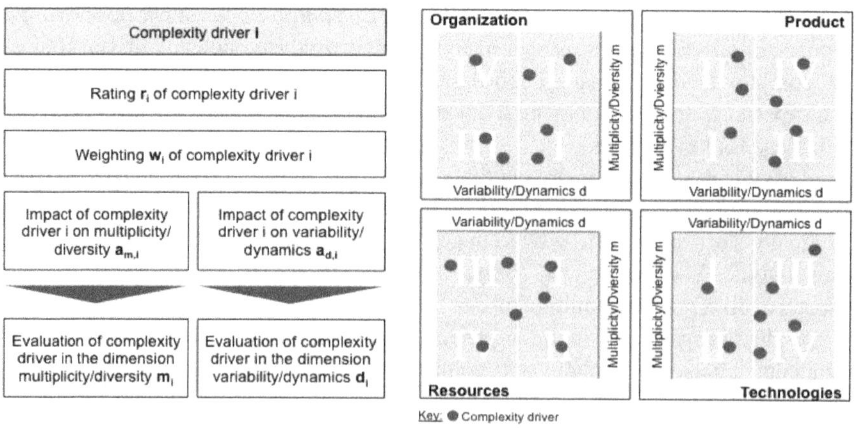

Figure 2: Visualization of evaluated complexity drivers.

The presented framework facilitates a large number of different analyses of project complexity. A comparison between the categories or between internally and externally caused complexity is possible. The evaluation of complexity drivers across the four categories allows to investigate cause–effect relationships e.g. between organizational complexity and product-related complexity. A recurrent use of the method allows to investigate how the complexity of a product develops over time. Measures to adjust product complexity should focus on elements with high evaluation. An example for such a measure to adjust project complexity is to reduce product complexity, e.g. number of different components in order to meet the capability of a project organization to handle complexity, e.g. number of project staff.

3.2 Analysis of interdependencies between complexity drivers

NPD projects are characterized by versatile interactions. [28] Thus, complexity drivers are also interdependent and their interdependencies have to be taken into account when adjusting the project setup as discussed in the chapter before. The starting point for the analysis of the interdependencies is to resolve the complexity drivers in specific indicators, that are quantitatively ratable. In order to analyze the interdependencies between the complexity drivers, a modeling of the complexity drivers and their indicators making their structure accessible for deeper analysis is required. Figure 3 shows an excerpt of the overall structure with a selection of complexity drivers in the category 'Product'. The influence factor *Product Requirements* is subdivided into the complexity drivers based on target markets, customer expectations, legal conditions, and others. A specific indicator is assigned to each of the complexity drivers. For example, the indicator *Number of relevant laws and norms* is assigned to the complexity driver *Legal requirements.*

This structure of complexity drivers and indicators serves as a basis for further investigation of their interdependencies. The interdependencies between the complexity drivers are determined by using the corresponding indicator. A shift of the rating of one complexity driver, e.g. *Target markets*, is expressed in a shift of the indicator, e.g. *Number of addressed target markets*. As the number of addressed target markets increases, the number of documents to be created for product documenta-

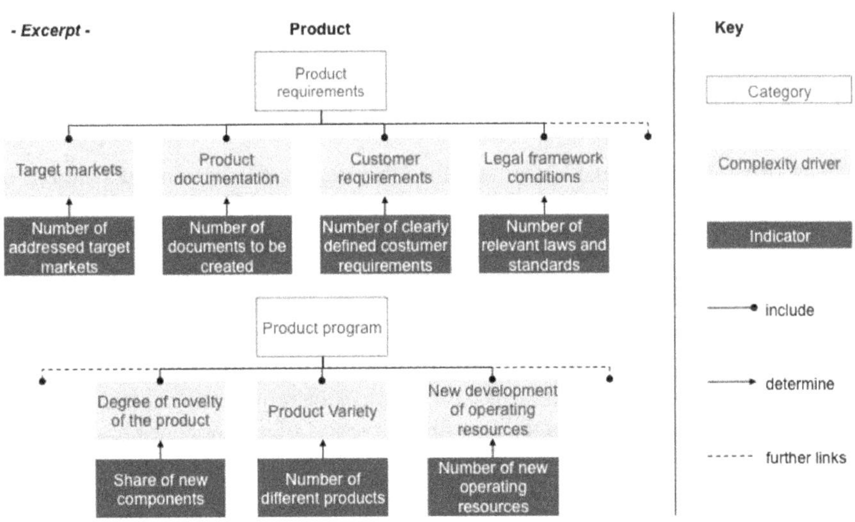

Figure 3: Extract of modeled structure of complexity drivers and indicators.

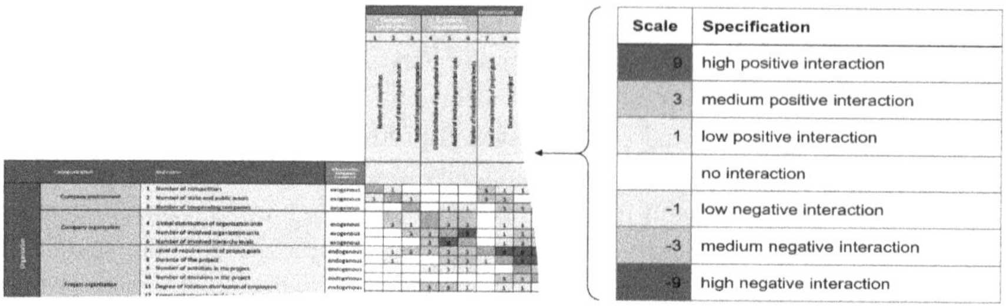

Figure 4: Analysis of interdependencies between complexity drivers.

tion may also increase due to diverse languages. Therefore, another indicator, e.g. *Number of documents to be created*, changes as well resulting in an shift of the rating of another complexity driver, e.g. *Product documentation*.

These interdependencies between the elements can be considered as elements of an interconnected system. [29] The interdependencies between the elements of such a system are analyzed by an influence matrix going back to Vester [30] Applied to complexity drivers, these interdependencies between them are analyzed in a matrix as shown in excerpts in Fig. 4.

The indicators of each complexity driver are listed on both axes of the matrix. Each of the cells of the matrix describes the effect of the respective indicator in the row on the respective indicator in the column. The degree and direction of the interaction is detected by a scale from –9 to +9.

3.3 Segmentation of complexity drivers in an influence portfolio

The complexity drivers are segmented depending on their strength of influence S_X and their influenceability I_X. According to Fink *et al.* [31] four segments are differentiated in an influence portfolio: system indicators, independent factors, system levers, and interactive nodes. The influence a complexity driver has on other drivers is indicated by the row total and the influenceability is indicated by the column total in the influence matrix. A high row total R_X means that the driver has a strong reinforcing effect on other drivers and, thus, a strong reinforcing effect on the complexity of the system in total. A high column total C_X means that the driver is subject to high reinforcing effects by the others. For the strength of influence and the influenceability, the absolute values of each cell $i_{a,b}$ are summed because positive and negative entrances would otherwise cancel each other out and thus falsify the result.

The complexity drivers are classified into the four aforementioned segments depending on the calculated strength of influence S_X and the influenceability I_X as shown in Fig. 5.

System levers are those complexity drivers that have a high strength of influence S_X and a low infuenceability I_X. As a change in the rating of system levers has a strong impact on multiple other complexity drivers, a system lever or an interactive node is an appropriate starting point for complexity adjustment of a project. On the other side, system indicators have low influence on other complexity drivers but they are strongly influenced by others. Independent factors, however, are characterized by only very few interdependencies. A change in the rating of such a complexity driver will have a low impact on the overall complexity of the project. Interactive nodes have both, a high strength of influence and influenceability. [31]

The effects of shifts in the rating of complexity drivers are again visualized in the introduced framework. The left side of Fig. 6 shows the initial evaluation of complexity drivers in the four dimensions.

In the presented example, three drivers are affected by a fictitious change in the project setup and thus their assessment in the dimensions Variability/Dynamics and Multiplicity/Diversity changes.

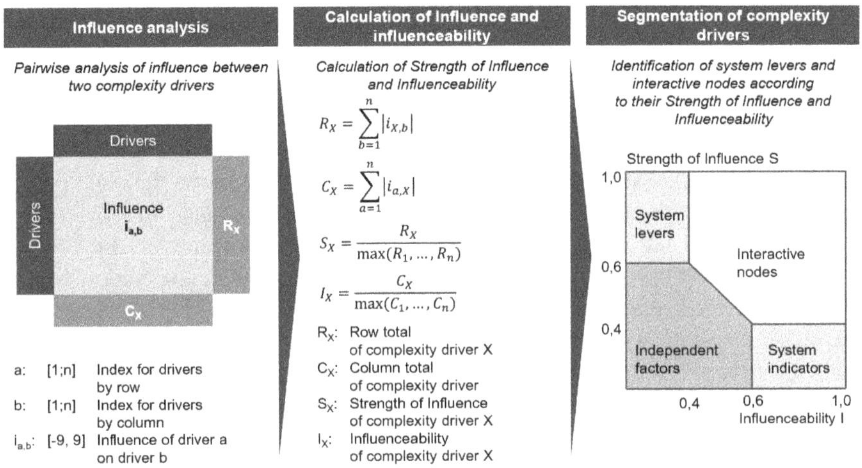

Figure 5: Segmentation of complexity drivers in an influence portfolio.

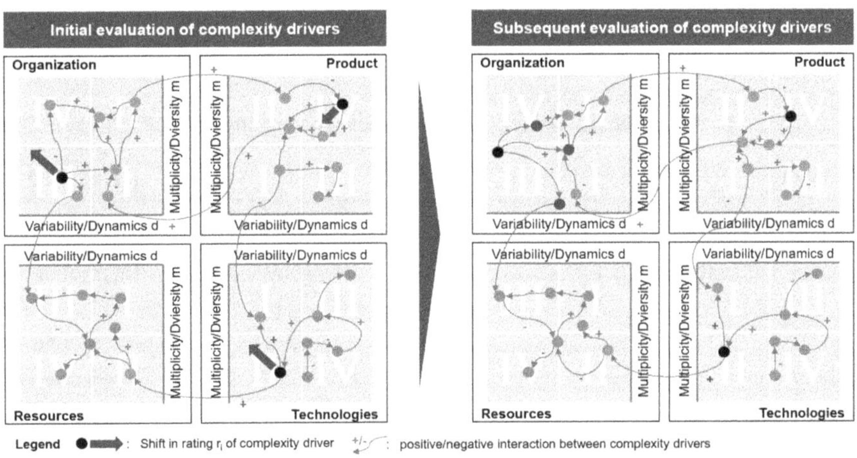

Figure 6: Effects of shift in the rating of complexity drivers.

Due to the interdependencies between the complexity drivers represented by arrows, the assessment of other complexity drivers will change as well. Figure 6 shows the subsequent evaluation taking into account first order interactions between the complexity drivers. The rating of complexity drivers with interactions will increase or decrease according to the sign of the interaction resulting in a corresponding shift within the framework.

3.4 Application example

The core elements of the developed concept were applied to a set of complexity drivers. First, approximately 50 complexity drivers were identified through an extensive literature review and classified into the defined categories. Then, a dedicated indicator was assigned to each of the complexity drivers. An excerpt of the resulting structure was introduced earlier in Fig. 3. The influence matrix was created with these indicators for an analysis of the interdependencies. The matrix was distributed to and filled out by several senior researchers with multi-year experience in accompanying product development projects in machine and plant manufacturing. For each cell of the matrix, the average was calculated and a merged matrix was created. The entries in the merged matrix were used to calculate the strength of influence S and the influenceability I. With the help of the influence portfolio, system levers and interactive nodes were identified.

The results of the analysis are shown in the following tables. Table 1 shows the eight complexity drivers with the highest strength of influence S. Especially, the *aspiration level of project goals* and the *interdependencies within the product architecture* have a high influence on other complexity drivers and determine an NPD project's complexity. Thus, they are 'interactive nodes' which have strong impact on other drivers, but are also highly dependent on others. The drivers *legal framework conditions, addressed target markets* and *variety of technologies used* also have a high influence on other drivers but are less affected by other drivers. Thus, they are system levers' This means that narrowing the number of addressed target markets with a project will have a mitigating impact on the overall project complexity.

A selection of complexity drivers with both low strength of influence S and low influenceability I is shown in Table 2. The results of the analysis show that the values for S and I are remarkable but relatively low compared to the other drivers. The listed complexity drivers like the *availability of specialists* or *competitor companies* are relatively less affected by other drivers and have only a low influence on other drivers at the same time.

Table 1: Selection of complexity drivers sorted by descending strength of influence S.

Complexity driver	**S**	**I**	**Driver segment**
Aspiration level of project goals	1,00	0,76	Interactive nodes
Interdependencies within the product architecture	0,97	0,69	Interactive nodes
Project durance	0,83	0,92	Interactive nodes
Functions of the product	0,81	0,75	Interactive nodes
Legal framework conditions	0,80	0,45	Interactive nodes
New development of processes and operation resources	0,75	0,72	Interactive nodes
Addressed target markets	0,74	0,39	System levers
Variety of technologies used	0,74	0,40	System levers

Table 2: Selection of complexity drivers identified as independent factors.

Complexity Driver	**S**	**I**	**Driver segment**
Availability of specialists	0,49	0,35	Independent factors
Buildings and rooms	0,25	0,52	Independent factors
Capacity utilization of IT systems	0,27	0,41	Independent factors
Competitor companies	0,49	0,39	Independent factors
Job tenure of project staff	0,43	0,50	Independent factors

4 CONCLUSIONS AND FUTURE WORK

Increasing variety and enhanced functionality of products lead to increasing complexity in product development. Complexity in NPD requires a deep understanding of multi-layered contexts in the project setup. Handling of complexity is a decisive factor in product development that has a significant impact on the success of a project. This paper presents an approach to evaluate complexity in NPD projects using complexity drivers. Based on an assessment of individual complexity drivers, critical factors that determine project complexity are identified and classified. Interdependencies between the complexity drivers are analyzed using the influence matrix and they are segmented in an influence portfolio. The complexity drivers and their interactions serve as a basis for the analysis of the impact of measures to adjust project complexity in the early phase of a project.

Future work should focus on the systematic identification of measures to adjust project complexity based on the results of the evaluation. The definition of generic measures and their impact on the complexity of a project could be an objective. A promising field for research is complexity in the dimension organization and the linkage between organizational and product-related complexity.

ACKNOWLEDGMENT

The presented results have been developed within the research project "Complexity evaluation of development projects" (SCHU 1495/87-1) funded by the Deutsche Forschungsgemeinschaft (DFG).

REFERENCES

[1] Ulrich, H. & Probst, G.J., *Anleitung zum ganzheitlichen Denken und Handeln*, Bern: Paul haupt, 1988.

[2] Schuh, G., *Produktkomplexität managen: Strategien-Methoden-Tools*, München, Wien: Carl Hanser Verlag, 2005.

[3] Lorenz, M., Jentzsch, A., Andersen, M., Noack, B., Waffenschmidt, L., Schuh, G. & Rudolf, S., *The Lean Advantage in Engineering. Developing Better Products Faster and More Efficiently*, The Boston Consulting Group & WZL of RWTH Aachen, 2015.

[4] Kotha, S. & Srikanth, K., Managing a global partnership model: lessons from the boeing 787 'dreamliner' program. *Global Strategy Journal*, **3**(1), pp. 41–66, 2013. http://dx.doi.org/10.1111/j.2042-5805.2012.01050.x

[5] Kirchhof, R., *Ganzheitliches Komplexitätsmanagement: Grundlagen und Methodik des Umgangs mit Komplexität im Unternehmen*, Wiesbaden: Deutscher Universitätsverlag, 2003.

[6] Ashby, W.R., An introduction to cybernetics, 1957.

[7] Berens, W. & Schmitting, W., Controllinginstrumente für das komplexitätsmanagement: potentiale des internen rechnungswesens. *Komplexitätsmanagement*, pp. 97–110, 1998.

[8] Baccarini, D., The concept of project complexity - a review. *International Journal of Project Management*, **14**(4), pp. 201–204, 1996. http://dx.doi.org/10.1016/0263-7863(95)00093-3

[9] Williams, T., The need for new paradigms for complex projects. *International Journal of Project Management*, **17**(5), pp. 269–273, 1999. http://dx.doi.org/10.1016/S0263-7863(98)00047-7

[10] Geraldi, J., Maylor, H. & Williams, T., Now, let's make it really complex (complicated). *International Journal of Operations & Production Management*, **31**(9), pp. 966–990, 2011. http://dx.doi.org/10.1108/01443571111165848

[11] Karniel, A. & Reich, Y., *Managing the Dynamics of New Product Development Processes: a New Product Lifecycle Management Paradigm*, Springer-Verlag: London, 2011. http://dx.doi.org/10.1007/978-0-85729-570-5

[12] Bosch-Rekveldt, M., Jongkind, Y., Mooi, H., Bakker, H. & Verbraeck, A., Grasping project complexity in large engineering projects: the TOE framework. *International Journal of Project Management*, **29**(6), pp. 728–739, 2011.
http://dx.doi.org/10.1016/j.ijproman.2010.07.008

[13] Backlund, A., The concept of complexity in organisations and information systems. *Kybernetes*, **31**(1), pp. 30–43, 2002.
http://dx.doi.org/10.1108/03684920210414907

[14] Daft, R.L., *Organization Theory and Design*, Mason, Ohio: South-Western Cengage Learning, 2010.

[15] Schlick, C.M., Duckwitz, S., Gärtner, T. & Schmidt, T., A complexity measure for concurrent engineering projects based on the DSM. *DSM 2008: Proceedings of the 10th International DSM Conference,* Stockholm, Sweden, 2008.

[16] Nguyen, A.T., Nguyen, L.D., Le-Hoai, L. & Dang, C.N., Quantifying the complexity of transportation projects using the fuzzy analytic hierarchy process. *International Journal of Project Management*, 2015.
http://dx.doi.org/10.1016/j.ijproman.2015.02.007

[17] Wang, Y., Chen, X.H. & He, X.W., The measurement of complexity of organization coordination network of MSTP. *Advances in Information Sciences and Service Sciences*, **4**(23), pp. 334–342, 2012.
http://dx.doi.org/10.4156/aiss.vol4.issue23.41

[18] Qureshi, S.M. & Kang, C., Analysing the organizational factors of project complexity using structural equation modelling. *International Journal of Project Management*, **33**(1), pp. 165–176, 2015.
http://dx.doi.org/10.1016/j.ijproman.2014.04.006

[19] Schuh, G., Potente, T., Varandani, R.M. & Schmitz, T., Methodology for the assessment of structural complexity in global production networks. *Procedia CIRP*, **7**, pp. 67–72, 2013.
http://dx.doi.org/10.1016/j.procir.2013.05.012

[20] Schuh, G., Rudolf, S., Breunig, S. & Lukas, M., Evaluation of complexity in industrial services. *Technovation*, 2014.

[21] Sinha, K., *Structural Complexity and its Implications for Design of Cyber-Physical Systems*, Massachusetts Institute of Technology, Cambridge, 2014.

[22] Bliss, C., *Integriertes Komplexitätsmanagement: Ansätze und Lösungsmöglichkeiten*: Wiss. Ges. für Marketing und Unternehmensführung, 1998.

[23] Ponn, J. & Lindemann, U., *Konzeptentwicklung und Gestaltung technischer Produkte: Systematisch von Anforderungen zu Konzepten und Gestaltlösungen,* 2nd edn., Springer: Berlin, Heidelberg, 2011.
http://dx.doi.org/10.1007/978-3-642-20580-4

[24] Schuh, G., Gerlach, M. & Rudolf, S., Methodology for optimizing product differentiation in product ranges. In *2015 Proceedings of PICMET '15: Management of the Technology Age*, ed. D.F. Kocaoglu, Portland, Oregon, pp. 1541–1548, 2015.
http://dx.doi.org/10.1109/picmet.2015.7273117

[25] Schuh, G., Arnoscht, J. & Rudolf, S., Integrated development of modular product platforms. *Technology Management for Global Economic Growth (PICMET)*, 2010.

[26] Lindemann, U., *Methodische Entwicklung technischer Produkte: Methoden flexibel und situationsgerecht anwenden,* 3rd edn., Berlin, Heidelberg: Springer, 2009.
http://dx.doi.org/10.1007/978-3-642-01423-9

[27] Schuh, G, Rudolf, S. & Mattern, C., Conceptual framework for evaluation of complexity in new product development projects. In *2016 IEEE International Conference on Industrial Technology (ICIT)*.

[28] Liso, N.D. & Leoncini, R., *Internationalization, Technological Change and the Theory of the Firm*, Routledge: New York, 2011.

[29] Patzak, G., *Systemtechnik - Planung Komplexer Innovativer Systeme*, Springer-Verlag: Berlin, 1982.

[30] Vester, F, *Die Kunst vernetzt zu denken: Ideen und Werkzeuge für einen neuen Umgang mit Komplexität ; ein Bericht an den Club of Rome,* 3rd edn., München: Deutscher Taschenbuchverlag, 2003.

[31] Fink, A., Schlake, O. & Siebe, A., *Erfolg durch Szenario-Management: Prinzip und Werkzeuge der strategischen Vorausschau*, Campus-Verl: Frankfurt, 2001.

TECHNOLOGY PLATFORMS AS A TOOL FOR SOLVING COMPLEX INNOVATION PROBLEMS

L.D. GITELMAN, M.V. KOZHEVNIKOV & D.G. SANDLER
Ural Federal University, Russia.

ABSTRACT
The article analyses the experience of using the concept of technology platforms for solving innovation problems of a complex multidimensional nature that stem from a lack of information and knowledge. Two cases are described when platform tools are applicable. The first one has to do with creating breakthrough technologies in promising industries, mostly high-tech ones, as well as with forecasting and developing competencies that are required for the new technology agenda. Second, they are applicable when major multinational corporations build their market infrastructure that requires stable system connections within value chains. The authors showcase their experience of engineering specialised university technology platforms that focus on introducing flexible and up-to-date educational products based on the integration of science, education, innovation and close cooperation with businesses.
Keywords: cooperation, education, high-tech industries, innovation, technology platform, technology transfer, uncertainty.

1 INTRODUCTION

In many countries, the transition to a new economy following the information technology revolution, which is conventionally referred to as 'digitalisation', or Industry 4.0, is accompanied by the development of the forms and mechanisms of innovation. The experience of other countries (for example, the USA, Great Britain, Germany, France) shows that it is impossible to create smart manufacturing systems and products without re-engineering intercorporate science and technology cooperation among the parties involved in the process of innovation: high-tech companies, infrastructure companies, universities, research centres, and service solutions suppliers [1].

Technology platforms (TP) often serve as a structural tool for such cooperation. They are usually built in new knowledge-based industries (biomedicine, smart energy, nuclear industry, urban infrastructure, robotics etc. See Fig. 1). The prospects of applying this tool to address poorly structured innovation problems are of significant scientific interest from the point of view of organisation and methodology in the first place. The analysis done by the authors enabled them to identify a number of conditions that ensure the viability of the platforms, and to come up with a project of a technology platform aimed at furthering anticipatory learning for the new industrial revolution.

2 METHODOLOGICAL ASPECTS OF DESIGNING TECHNOLOGY PLATFORMS

The consolidation of the best practices of operating technology platforms makes it possible to single out two main approaches to building them.

Under the first approach, a technology platform is a consortium (a global or a regional one) *that is created to solve super complex problems, typically in the field of science and technology, on the basis of a single ideology and research methodology.* The problems that are addressed on such

DOI: 10.2495/DNE-V11-N4-584-592

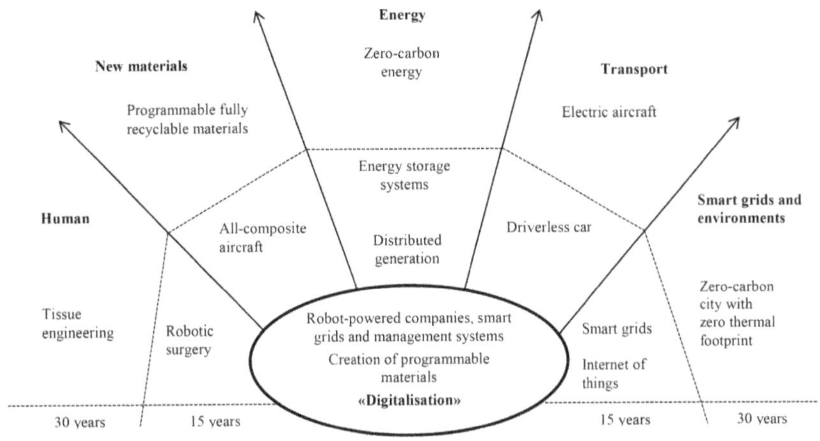

Figure 1: Examples of new industries and areas where the density of innovative businesses is growing (according to Technospark.ru).

platforms are always unique and one-of-a-kind. What makes them peculiar is their novelty, a non-linear structure, a lack of knowledge, first and foremost of interdisciplinary kind, and almost complete absence of required analytics.

An example of such a problem is the development of smart grids that are sought after by companies from various industries related to the energy sector [2]. The scale of the problem, its organisational, technical, financial, economic and legal features require that energy companies establish cooperation and explore and expand the potential for cooperation with partners from various fields of business (Fig. 2). Appropriate technology platform can be called 'Intelligent Energy'.

The approach was common between 1990s and 2000s in Europe due to the need to boost innovative development and create an advanced technology market in the European Union. The first technology platform was built in 2001 for the purpose of organising and coordinating cooperation between aircraft manufacturers and scholarly communities in planning the programmes of aeronautics research. Today, there are some 40 technology platforms in Europe that conduct research to further the development of high-tech industries – machine building, energy, the space industry, telecommunications, bioeconomy, transport [3, 4].

The key prerequisite for building the European technology platforms was the need to reorganise the market of research intensive technologies. It was, therefore, the need to reduce R&D costs in promising industries that spurred the creation of technology platforms in Europe. That was accomplished by drawing up a unified long-term innovation strategy of the territory that contained a package of appropriate legal and financial measures, by bundling companies that were working on identical products, by promoting cooperation among companies, universities and research centres.

The second approach is typical of the digital industry, although it does not contradict the first one, but complements and furthers it. **A technology platform is understood as an 'innovation ecosystem' that is built around a 'keystone company'** such as Google, Apple, or Facebook. A major firm acts as a provider of technologies of the future, with numerous smaller contractors, research centres, technology and logistics services, and, most importantly, customers gravitating towards it. The main 'product' of the platform is new markets founded on a new technology architecture [5–7].

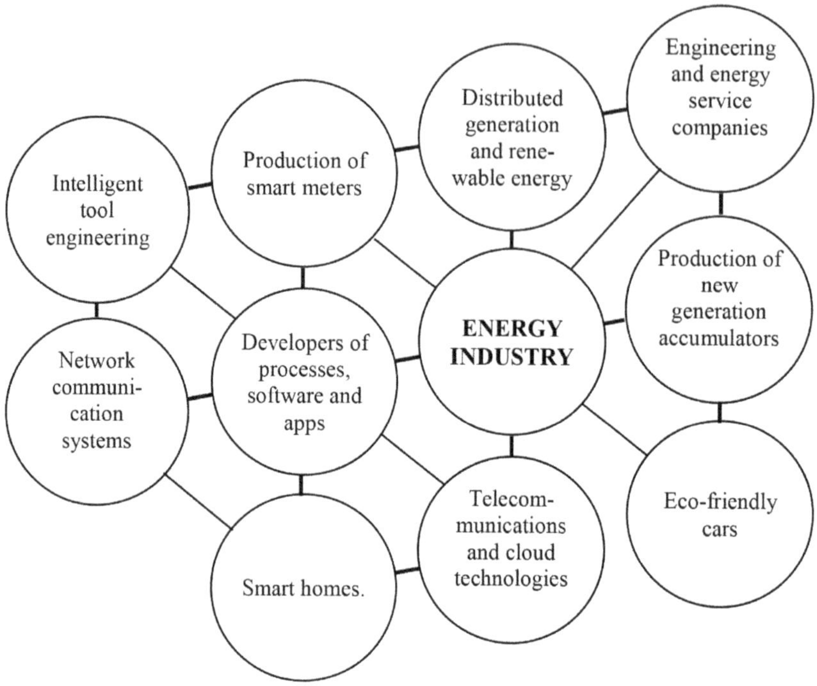

Figure 2: Platform and network structure of industry convergence in the course of building smart energy systems.

Technologies that are centrally developed by the keystone company are adapted to suit various groups of customers through the diffusion of innovations. As a result, the goal of the platform is to shape future preferences of a large group of potential customers. Such anticipatory impact ensures the keystone company's market stability for the next 5 to 10 years. The value of the technology platform grows as long as it attracts an ever increasing number of customers by generating a unique hybrid range of products to offer to them, and at the same time attracts new partners (entrepreneurs, developers, investors and other business entities) (Fig. 3).

One can, therefore, speak about the strategic importance of technology platforms for ensuring the competitive edge of individual research-intensive organisations as well as whole industries, regions and countries. It's no wonder that some scholars observe that competition among individual businesses will soon transform into competition among technology platforms [5, 8].

3 COOPERATION BETWEEN UNIVERSITIES AND BUSINESS AS KEY DRIVER OF TP DEVELOPMENT

The two approaches described above make it possible to define a technology platform as an alliance in which cooperation, especially on R&D, plays a key role in the effectiveness and viability of the platform. Today, high-tech corporations tend to outsource to universities, research centres, think tanks and consulting agencies. This trend is described in a study by The Organisation for Economic Cooperation and Development (OECD) [9].

The ongoing changes in corporate R&D models (from centralised R&D functions to predominantly non-captive R&D projects) have brought about a significant transformation of the nature of

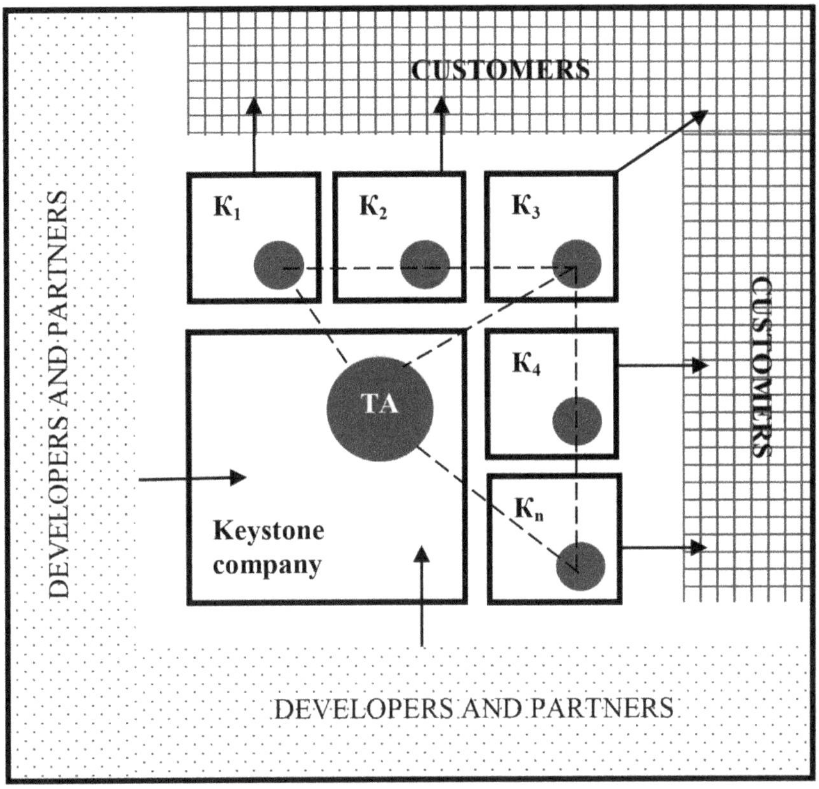

Figure 3: Visualisation of an 'innovation ecosystem'-type technology platform, $K_1 \ldots K_n$ are business structures of the keystone company that offer customers hybrid products built upon a single technology architecture (TA).

cooperation between university and business. Nowadays business seeks *limited, but highly targeted* partnerships with university structures. Such an attitude is due to two factors. First, the bulk of research being carried out in emerging industries is poorly structured and experiences an immense shortage of knowledge, which calls for cooperation with specialist research organisations, including universities. On the other hand, when a large number of external entities are involved in research, it gets harder to manage them; intellectual security risks intensify; problems arise with testing the market for innovations.

The empirical evidence from abroad [10–13] yields an observation that universities participating in technology platforms:

- act as developers and suppliers of new technologies for specific goals of business (so-called 'institutional agents');
- always take on the duty of designing an effective business model for commercialisation of the technologies;
- are actively engaged in handling the issues of intellectual property protection and patenting. For this purpose, they set up special university units ('transfer technology offices').

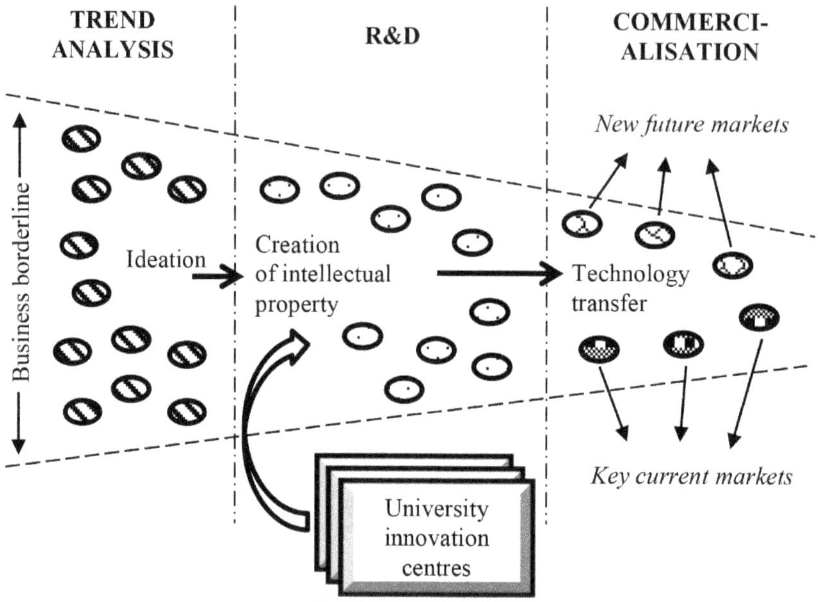

Figure 4: Place of university innovation centres in the chain of creating breakthrough innovation.

In most technology platforms, the participating university does not have much autonomy and rarely *initiates* the creation of a platform. One success story here is the case of the University Of Manchester (UK) that managed to bring together multinational enterprises from a range of industries (Astra-Zeneca, IBM, BP, Syngenta). The university proceeded from an idea that without cooperation with big businesses it could not continue to be a high-performing educational and research institution, to enhance the staff's competencies, to conduct a profound analysis of social, political, and economic trends, and finally, to improve its educational content [14].

The initiative, however, came from major agribusiness Syngenta AG that in 2007 made it a priority to expand cooperation with universities, and within a short period of time it established six so-called university innovation centres at research universities of Great Britain, China and Australia. The partnerships enabled the company to intensify efforts to create open innovations that are essentially aimed at shaping a long-term strategy of developing new technologies and markets for them (Fig. 4).

The involvement of universities in technology platforms opens opportunities for improving their core business – that of designing and implementing up-to-date interdisciplinary educational programmes using the latest achievements in science; creating new technologies for education, and promoting technological entrepreneurship. Generally speaking, a technology platform could form the foundation for a unique strategy of universities that implies convergence and integration of research, innovation and teaching for the purposes of anticipatory training of professionals.

4 TECHNOLOGY PLATFORM FOR A BREAKTHROUGH IN PROFESSIONAL EDUCATION

Super dynamic non-linear and all too often chaotic changes in the global environment and stronger competition, including in geopolitics, fuel the demand for a new model of managing business development. The core of the model is preventive actions of management, its ability to respond swiftly and adequately to new circumstances, and timely preparation for upcoming change.

Managers need new competencies:

- holistic vision of trends, processes and factors that determine the emergence, development and behaviour of complex systems in the global contest;
- decision making that takes into account interdisciplinary connections between technology, reliability and safety, economy and business results;
- quick adaptation to a situation that has changed at an early stage of its emergence;
- access to global knowledge bases, rapid analysis and generalisation of knowledge;
- organisation of an active innovation process at all levels of management.

This clearly calls for anticipatory learning that the authors define as the process of building competencies and knowledge for addressing future tasks that are in line with global trends and national development programmes.

A breakthrough is needed to introduce a *new system*, as soon as possible, that will make it possible *to foresee* scientific and research, social and economic and organisational innovations with specific stochastic characteristics, and to *nurture competencies* for rapid design of new educational products and quick readjustment of the learning process in order to be able to embrace promising industry technologies and tasks in a specific profession (Fig. 5).

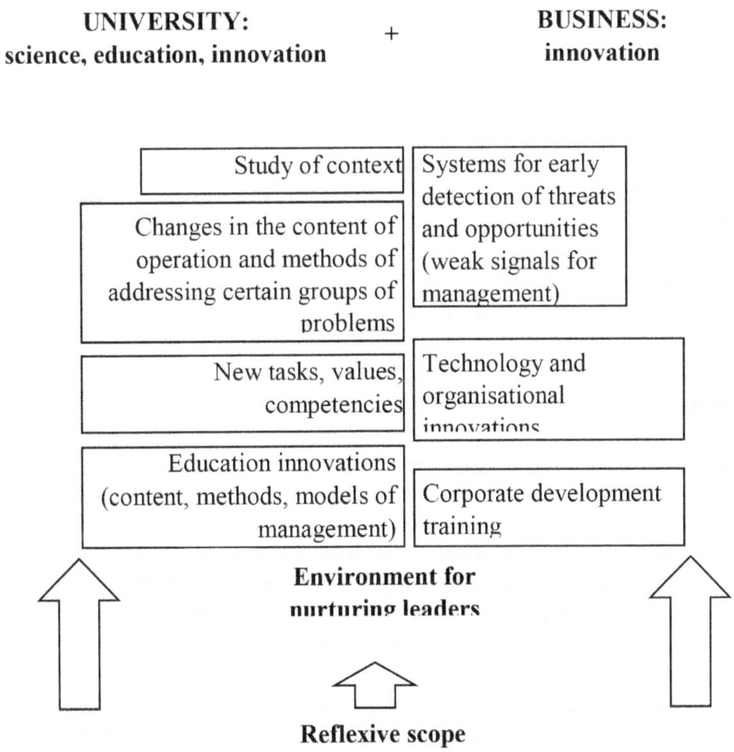

Figure 5: Conceptualisation of a breakthrough in anticipatory learning.

The only way to solve the problem is by conducting *purpose-specific studies* that will focus on foreseeing technological development, shaping a vision of the future and integrating these activities with learning and innovation (both at university and in business).

What comes to the forefront is the specific mechanism of cooperation between *science, education and business* as part of a *holistic system*. This extremely complex system is unique because a real partnership will work if each unit of the 'science – education – business' trio have their interests and intellectual resources combined into a single project on the basis of a common focus, ambitions and methodological framework:

- pre-emptive solutions and anticipatory learning that supplies management with new knowledge and relevant competences:
- staff development using innovation and personal development tools;
- attention to achievements in science and technology and promising industrial technologies as a powerful and defining driver of contemporary economic development.

The authors' experience shows that the technology platform proves to be the most effective solution to the problem. Figure 6 depicts the structure of an anticipatory learning platform that is being introduced on the initiative of the research and education centre 'ENGEC' of Ural Federal University.

The Integrated System for Research, Consulting, Training and Transformation (ISCT) [15] provides a methodological framework for the platform. Among the technologies used within the ISCT methodology on the basis of its principles and approaches are 'Conveyor for nurturing new leaders', 'We engineer the future of our company (university) with our own hands', 'Making management

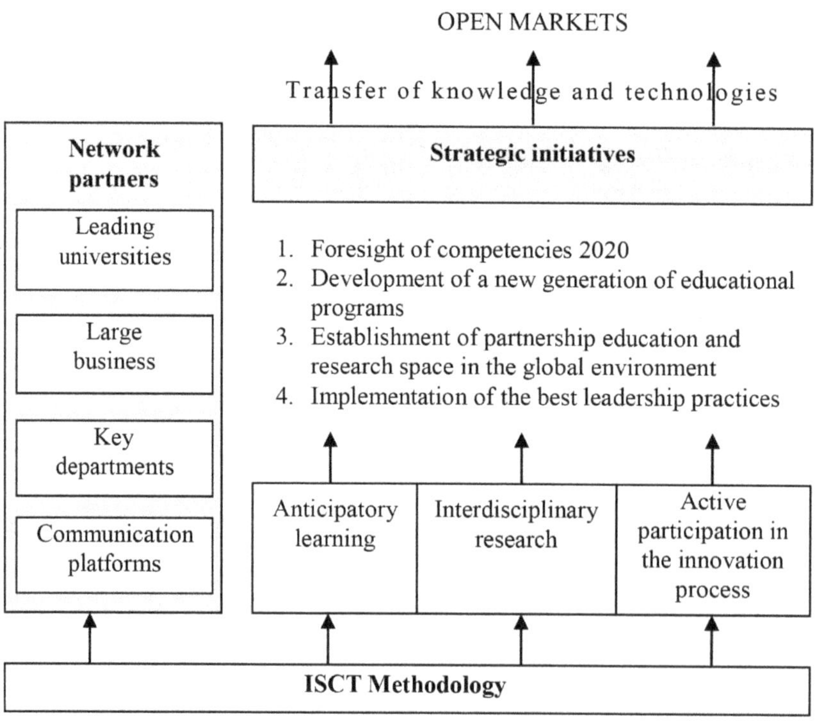

Figure 6: Structure of a technology platform for anticipatory learning.

ready for modernisation and organisational transformation', 'Corporate management incubator', 'Corporate university academy', a series of organisational activity games, strategic and project sessions 'CLIMB to excellence'.

The methodology proposes a new framework of the vision of the holistic system described above. An intelligent 'super-smart' macro-environment is being created for governing the 'smart' environment' ('the smart above the smart').

The agenda of the priority research as part of the technology platform includes:

- interdisciplinary connections between technologies, environmental safety, investment, business results and the quality of human resources;
- principles of the engineering of the future and weak signal management systems;
- foresight methods;
- methods for the study of complex systems and risk assessment amid uncertainty;
- management of a corporate environment generating new knowledge and innovations;
- development of the principles and the structure of technology-based customised learning.

5 CONCLUSION

Present-day industrial challenges dictate the need to create technology platforms that focus on innovation, transformations and associated fields, including staff training. This stems from the emergence of new technologies and competencies that will be in demand soon. As a result, universities that bring together science, professional training and the introduction of innovation in emerging industries start to play a significant and peculiar role.

A good example is German universities as most of them combine platform clusters and focus on Industry 4.0 problems. Without close cooperation with major companies, universities are unable to work out a relevant research agenda. The results of studies are subsequently integrated into sought-after educational content. This is how new courses or even branches of study appear, mechatronics, virtual prototyping, intelligent technical systems, and the internet of the things being some of the examples.

The authors' study shows that university technology platforms are an effective instrument for responding to the challenge. The platforms envisage the deployment of anticipatory learning solutions that are delivered in the global scientific and educational space that operates through networks involving leading universities, innovative businesses and various communication platforms.

ACKNOWLEDGMENTS

The work was supported by Act 211 of the Government of the Russian Federation, contract № 02.A03.21.0006.

REFERENCES

[1] Fatkhullin, B., Improving cross-company scientific, industrial and technological cooperation in the framework of technological platforms, [in Russian]. *Transport Business in Russia*, **6**, pp. 103–104, 2012.

[2] 'Smart' solutions for intelligent energy systems, [in Russian], available at https://www.pwc.ru/en/energy-utilities-mining/publications/assets/smartgrid_rus.pdf

[3] Godenov, I.S., *European Technology Platforms*, [in Russian], Tomsk State University: Tomsk, p. 72, 2011.

[4] Bebeshko, I.U., Technology platforms as a tool to promote innovative development of the Russian economy. *Theory and Practice of Social Development*, **12**, pp. 506–508, 2009.
[5] Gawer, A., Bridging differing perspectives on technological platforms: Toward an integrative framework. *Research Policy*, **7**(43), pp. 1239–1249, 2014.
[6] Osterwalder, A. & Pigneur, Y., *Business Model Generation*, Wiley & Sons: New Jersey, pp. 76–87, 2010.
[7] Shankar, V., Berry, L. & Dotzel, T., A practical guide to combining products and services, available at https://hbr.org/2009/11/a-practical-guide-to-combining-products-and-services
[8] Malik, K., Georghiou, L. & Grieve, B., Developing new technology platforms for new business models: Syngenta's partnership with the University of Manchester, available at https://www.escholar.manchester.ac.uk/api/datastream?publicationPid=uk-ac-man-scw:109173&datastreamId=POST-PEER-REVIEW-PUBLISHERS-DOCUMENT.PDF
[9] The Internationalisation of Business R&D, Evidence, Impacts and Implications, OECD, 2008, available at www.sourceoecd.org/scienceIT/9789264044043
[10] Connecting Universities to Regional Growth: A Practical Guide, available at http://ec.europa.eu/regional_policy/sources/docgener/presenta/universities2011/universities2011_en.pdf
[11] Hammadou, H., Paty, S. & Savona, M., Strategic interactions in public R&D across European countries: a spatial econometric analysis. *Research Policy*, **7**(43), pp. 1217–1226, 2014.
[12] Sendogdu, A. & Diken, A., A research on the problems encountered in the collaboration between university and industry. *Social and Behavioral Science*, **99**, pp. 966–975, 2013.
[13] Bubela, T. & Caulfield, T., Role and reality: technology transfer at Canadian universities. *Trends in Biotechnology*, **9**(28), pp. 447–451, 2010.
[14] Gitelman, L.D., Sandler, D.G., Kozhevnikov, M.V. & Tretyakov, V.S., Technology platform as a tool for transformation of university science and education activities, [in Russian]. *University Management: Practice and Analysis*, **4**(98), pp. 31–42, 2015.
[15] Gitelman, L.D., Isayev, A., Gavrilova, T., Gamburg, A., Gitelman, L.M., Yeliseyeva, K., Zorin, N., Kozhevnikov, M., Kostina, G., Rostik, O., Uglova, L., Chazov, A. & Chazova, T., *Methodology of Innovative Management Education*, Ekonomika Publishing House: Moscow, pp. 53–65, 2015.

THE NETWORK ORGANIZATIONAL CHART AS A TOOL FOR MANAGING ORGANIZATIONAL COMPLEXITY

RAFAEL AITA
University of Lima, Peru.

ABSTRACT
Companies in the twenty first century are required to continually adapt to an uncertain and ever-changing environment, which requires a more organic and flexible organization, without losing the benefits of volume production and economies of scale. The Organizational Network is a tool that allows the company to study the channels through which materials, money, and information resources flow between jobs, unlike the traditional Organizational Chart that only shows the hierarchy of power. Analyzing these links like a complex network will allow the manager to identify bottlenecks and agglomerations of power and measure the entropy generated in the organization. This gives a quantitative measure of the complexity of the business and, therefore, both its organic growth and efficient growth can be managed.
Keywords: chaos theory, complexity, entropy organization, information, strategical planning, uncertainty.

1 THE LINEAR MODEL

The Industrial Revolution introduced a series of social and economic changes that shaped the world of the nineteenth and twentieth centuries, beginning the 'industrial society'. The structure of industries in this society was consolidated thanks to the theories of Henri Fayol, Max Weber, and Frederick Taylor who laid the foundations to abandon handcrafted and empirical work.

A century later, we find ourselves again facing a new conjuncture of change due to the emergence of the new Information Technologies that has brought a significant increase in the volatility of the market and in acceleration of change. Faced with this new situation, the question arises whether we should continue to see companies in the traditional way or should we consider it as a complex system that requires non-linear tools to be managed. Management books provide a response by predicting the birth of a new management theory while they see the need for it. So says Gary Hamel [1]: 'Could the exercise of the management change in a manner just as radical during the first two or three decades of this century as it did during the early years of the twentieth century? I believe so. More than that, I think we should make it so'. Faced with this new reality, the principles of the classical and neoclassical administrative theories become obsolete, leaving the executives with poor management tools for business that need to adjust to the characteristics of a volatile market.

2 THE COMPLEXITY IN THE COMPANY

However, the question how to model the complexity in the company still remains. Although the traditional scheme that synthesizes relations in a company hierarchy is the organizational chart, this tool fails to collect the interrelationships of exchange of information, materials, and money between the different job positions. It is these relationships that are precisely the source of greater complexity. To model this emerging complexity in the company, it is intended to take the organizational chart as

DOI: 10.2495/DNE-V11-N4-593-599

a three-dimensional object which needs to be observed from a top view rather than from a front view. This new perspective allows to show the relationships of communication and interaction between each position forming a complex network instead of a hierarchical structure. According to Aldana [2], complex networks are sets of many connected nodes that interact in some way. The nodes of a network, also called vertices or elements, are represented by the symbols v_1, v_2, ... v_n, where n is the total number of nodes in the network. If a v_i node is connected to a v_j node, this connection is represented by the ordered pair (v_i, v_j). This connection is known as an 'edge' within the network. In the case of the organization, nodes are given by the job positions and the edges by the relations between these positions. The relationship between positions should not be linked arbitrarily, but among those positions where there is a real exchange of material and financial resources, or the most important resource of this era: information. Therefore, the connections must be drawn based on the connections that exist in the Process Diagrams of the Company. Process diagrams show the relationship between of exchange of information corresponding to each function of the company. In summary, the Network Diagram of the Organization is an R complex network consisting of a set of nodes $V = \{v_1, v_2, \ldots v_n\}$ and a set of ordered pairs $E = \{(v_i, v_j)\}$, which belongs to V x V and for its construction, it is powered by two diagrams: the organizational chart to draw the network nodes (positions of the Company) and the Process Diagrams to draw the edges that connect the nodes according to each link that exists in the process.

Finally, it is necessary to emphasize that the links between nodes are no communication channels. If so, all positions of the company would be connected to each other, as any person can communicate verbally or through e-mail with any other employee of the company. The links between exchange of information should only display those information flows that occur continuously as part of the production process of goods or services of the company and not from a casual or sporadic interaction. For this reason the links of interaction should be taken from the Process Diagrams of the Company and must explicitly appear in it. The complex network of the organization is a non-directed network, since for each pair (v_i, v_j) belonging to E there is also the pair (v_j, v_i) accordingly.

The advantage of focusing the organization from a Network Diagram is that it allows analyzing the company as a complex network and, therefore, helps in viewing the degree of connectivity between its elements, the level of integration between areas, and the formation of islands or people who centralize information. Figure 1 shows an example of the Organizational Network of a company of Information Technology based on the chart in Fig. 2.

3 THE CONNECTED COMPANY

To outline the organization as a network, the first thing you may notice is that not all nodes are necessarily connected, so that not all nodes have the same number of links within the network. The number k_i of neighbors of the node v_i (the number of connections of v_i) is called the v_i connectivity and the average of these connectivities in all the nodes is called the network connectivity. The connectivity index defines the degree to which the network is connected and is calculated according to the formula:

$$\text{Index of network connectivity} = \frac{\text{sum of links}}{\text{sum of nodes}} \quad (1)$$

The advantages of a network with a greater degree of connectivity are as follows:

- Increased robustness of the network: The structural strength of the network is defined by Aldana [2] as the invariance of phenotypes in the presence of internal or external disturbances to the

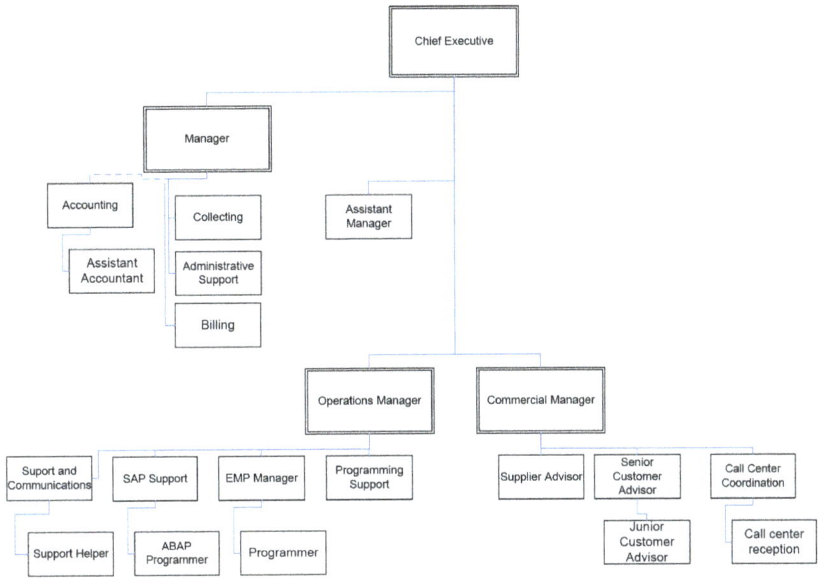

Figure 1: Organizational chart or network diagram of the organization.

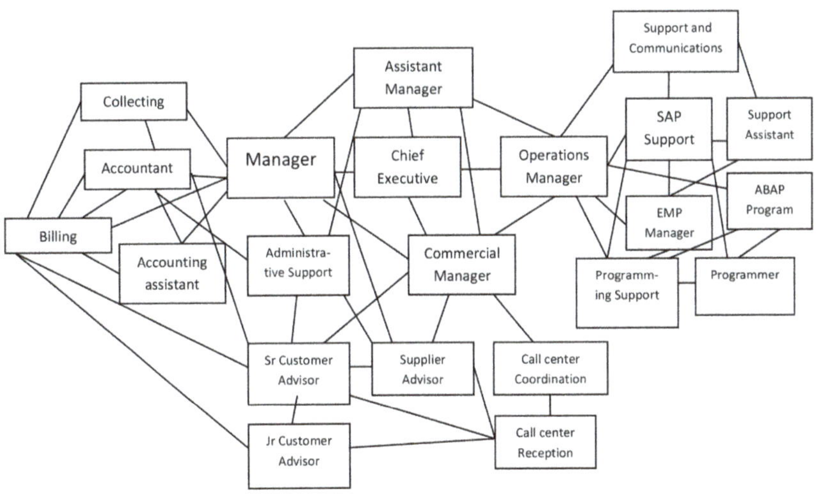

Figure 2: Organizational Chart of an IT Company.

organism. In this case it would measure the capacity of the enterprise network to re-adapt after one of the nodes (workstation) is either temporarily or indefinitely disabled. A useful analysis for the Human Resources department to anticipate the consequences of sudden resignation, absenteeism, or job redesign.

- Enhanced communicability of the network: The minimum length between two nodes is defined as the minimum number of edges that must be passed to get from one node to another node of the network. The average length of the network is the average of minimum lengths between all possible pairs of nodes of the network. A higher network connectivity offers a greater number of minimum lengths within the network and thus reduces the average length of the network. This translates as a greater horizontality in the organization, less bureaucracy in the communication, and greater closeness between the strategic apex with the rest of the company.

- Spread of information on the network: Rosvall, Trusina, Minhagen, and Sneppen *et al.* [3] proposed that a complex network structure is linked to its reliability and the speed of propagation of information, defining the entropy of the network as the degree of difficulty in seeking information on the network.
- Highest degree of network evolution: The size of the network does not end with the organization's boundary vertices, since these are connected with customers who are at the same time connected with other's customers or non-customers. This means that the network is expanding, not only internally through the incorporation of new job positions, but also externally. Therefore, the growth of the network will be studied as part of the market growth.

4 NETWORK ROBUSTNESS

Rodrigues, da F. Costa *et al.* [4] propose a method to determine the vulnerability of a network looking for its most vulnerable vertices. The vulnerability of a vertex can be defined according to the drop in network performance when the vertex, with all its edges, is removed. The vulnerability of the network would be equivalent to its most vulnerable vertex. The analysis of disturbances consists in measuring changes in the network to the non-use of their different vertices and helps to determine the robustness of the network against external shocks.

In the case of the Network flow chart in Fig. 1, the index of network connectivity will be measured before and after removing a job. The position that, at the time of being withdrawn, registers a further decline in the index of network connectivity will be the one that determines the degree of vulnerability of the network. To calculate this, Table 1 shows the indexes of the network connectivity of Fig. 1 before and after removing the designated positions.

The main conclusion that is drawn from the results of Table 1 is that there is little dispersion between the degree of variation when removing each of the nodes. This means that there is no centrality in a specific position and the network is homogenously interconnected. It also highlights that the absence of the Chief Executive position represents a low variation in network connectivity. This

Table 1: Robustness of the network.

Position to be withdrawn	Original connectivity index	Connectivity index without the position	% Variation
Manager	2.41	2.09	13.27
Chief executive	2.41	2.33	3.31
Commercial manager	2.41	2.19	9.12
Operations manager	2.41	2.14	11.2
Adminstrative support	2.41	2.28	5.39
SAP support	2.41	2.23	7.46
ERP manager	2.41	2.23	7.46

means that the senior management policy is based on delegating tasks to subordinates. An eventual absence of the Chief Executive may be supplied between the functions of the other managers without compromising the structure of the network. Finally, the vulnerability of the network lies in the position of the manager, which is the most connected in the company, lies in the position of Manager. This result shows that it is not necessary that a position belongs to the senior management to have a significance in network connectivity, which until now has been listed as an informal organization within the company, but which, with this tool, can be identified and quantified.

The index of network connectivity can be misleading, since not all generated links have the same quality. In 1976, Price proposed the model of preferential attachment. Based on the number of quotations from scientific literature, he deduced that the probability that an article is cited is directly proportional to the number of quotations it already has. The preferential attachment indicates that nodes with lots of connections will prefer to connect with other nodes that have large numbers of connections. Thus, higher quality links are generated in the network.

5 SPREAD OF INFORMATION IN THE NETWORK

The measure of information is the entropy, which led Claude Shannon and Weaver [5] to design a Mathematical Information Theory based on the measurement of entropy. In a complex network, entropy measures the difficulty of spreading information in the network and, therefore, it is an index to measure the level of bureaucracy in the organization. Rosvall, Trusina, Minhagen, and Sneppen *et al.* [3] proposed to quantify the information associated to locate a specific target in the network by measuring the probability of taking the shortest path from vertex i to vertex b in a random choice through the formula:

$$P\left[p(i,b)\right] = \frac{1}{k_i}\prod_{j\in p(i,b)} \frac{1}{k_j - 1} \qquad (2)$$

where P is the probability to take the shortest path p(i, b) from vertex i to vertex b and k_j is the degree of vertex j, and the product includes all vertices j on the path p(i, b) excluding i and b themselves. The amount of information used in the search to find the shortest path between i and b is given by the entropy S(i, b):

$$S(i,b) = -\log_2 \sum_{\{p(i,b)\}} P\left[p(i,b)\right] \qquad (3)$$

Where the sum is taken from all shortest paths p (i, b) from i to b.

In the case of the network shown in Fig. 1, we can calculate the entropy from an order that comes from the Chief Executive and goes up to the programmer searching for the probability of finding the shortest path in a random choice between these two vertices in the network.

There are three shorter paths between the Chief Executive and the programmer and they are:

1. Chief Executive - Operations manager - Programming support - Programmer
2. Chief Executive - Operations manager - SAP Support - Programmer
3. Chief Executive - Operations manager - ABAP Programming - Programmer

The probability of finding the shortest path following route 1 is = 0.892%
The probability of finding the shortest path following route 2 is = 0.714%
The probability of finding the shortest path following route 3 is = 1.785%

The entropy of the three paths equals to: - Log_2 (0.00892 + 0.00714 + 0.01785) = 4.8818 bits.

That is the amount of information lost in delivering a message from the Chief Executive to the Programmer.

The total entropy of the organization would be equivalent to the sum of the entropies of all pairs of nodes. This result depends on the index of connectivity of the company, but paradoxically, in this case, a greater connectivity of the company generates a greater entropy. The more the paths between nodes, the lower the probability of finding the shortest path in a random choice and, therefore, a greater entropy and loss of information travels over the network. That is why the nodes with greater connectivity, in this case the SAP Support, are the ones that add more entropy. This is in contradiction to the conclusion of the preceding paragraph, where a greater connectivity granted greater robustness to the network and resistance to the disqualification of a node. Is there a possibility of increasing connectivity and, thereby, the network robustness, without significantly increasing the entropy?

6 CLUSTERING

Clustering is the intermediate response between the increase of the connectivity through the number of links and the location of the optimal way to distribute information. Clustering consists in increasing the connectivity in concentrated clusters of nodes in order to strengthen their robustness, but not across the network to avoid significantly increasing the entropy thus, optimizing the likelihood of increasing the shortest path between two points practically until the end of the line where the cluster is located. It is important to highlight that, historically, the grouping of functions in departments, areas, and business units have followed this principle of clustering, demonstrating a greater organizational efficiency even beyond the company as evidenced by the industrial clusters proposed by Michael Porter. However, this principle of clustering should not necessarily be limited to business areas, but it can also be applied to functions that are close in Process Diagrams rather than the organizational chart. This is an example of the multidisciplinary work teams that so successfully brought to organizations that specialize in the development of projects.

The level of clustering in a complex network is measured through the coefficient of aggregation, which is defined as the probability that two nodes connected to a third node, are, in their turn, connected among themselves. This probability is measured through the transitivity of the network, which is defined by Newman [6] as the presence of a larger number of triangles in the network, being the triangle, the set of three vertices where each is connected to each of the others. Likewise, da F. Costa *et al.* [4] presented the formula to calculate the level of clustering C:

$$C = \frac{3\,N_\Delta}{N_3} \tag{4}$$

Where N_Δ is the number of triangles in the network and N_3 is the number of triples connected. A triangle is a set of three vertices with edges between every pair of vertices.

If k_i is the number of neighbors of the vertex i, then $N_3(i) = k_i\,(k_i - 1)/2$. In addition, if $N_\Delta(i)$ counts the number of edges between neighbors of i and the number of edges are represented between neighbors of i as L_i, then the equation to calculate the level of clustering C can be rewritten as:

$$C = \frac{2\,L_i}{k_i\left(k_i - 1\right)} \tag{5}$$

7 CONCLUSIONS

The Network Organizational Chart is a tool that uses complex networks to study the exchange of materials, money, and information between the different positions of the company. The network nodes represent the job positions and are taken from the organizational chart of the company, while the edges represent the relationships between those positions according to the flows that appear in the Process Diagrams of the company. Analyzing the organization as a complex network allows to study, among other things, the centrality of power, the robustness of the organization, and the spread of information through the company.

The companies show an informal organization that emerges from the relationship among the job positions that do not appear on the organizational chart. The Network Organizational Chart may determine the centrality of the job positions according to the network and detect the focus of power.

The robustness of the organization is determined by the invariability that the network shows after any of its nodes is withdrawn. This robustness can be increased by increasing the number of links within the network and is measured through an indicator called the index of connectivity.

The efficiency of the transmission of the information by the network is calculated based on the probability of finding at random the shortest path between two nodes and is quantified through the entropy of the network. With increased connectivity of the network, the lower the probability of finding the shortest path and, therefore, a greater entropy.

To increase the robustness of the network without significantly increasing the entropy, the connectivity in targeted clusters can be increased, which is known as clustering and it is this principle that governs the division of the organization into departments and areas. In the complex network, clustering is measured through the index of clustering, which, together with the entropy and the rate of connectivity, provides the guidelines for an optimal organizational design of the company.

REFERENCES

[1] Hamel, G. & Breen, B., *The Future of Management*, Harvard Business Press, 2013.

[2] Aldana, M., Complex networks. Taken from de available at http://www. fis. unam. mx/~ max/ English/notasredes.pdf, 2006

[3] Sneppen, K.. Trusina, A. & Rosvall, M., Measuring information networks. *Pramana*, **64**(6), pp. 1121–1125, 2005.
http://dx.doi.org/10.1007/BF02704173

[4] Costa, L.D.F., Rodrigues, F.A., Travieso, G. & Villas B.P.R., Characterization of complex networks: a survey of measurements. *Advances in Physics*, **56**(1), pp. 167–242, 2007.
http://dx.doi.org/10.1080/00018730601170527

[5] Shannon, C.E. & Weaver, W., The mathematical theory of information, 1949.

[6] Newman, M.E., The structure and function of complex networks. *SIAM Review*, **45**, pp. 167–256, 2003.
http://dx.doi.org/10.1137/S003614450342480

THE CLOSED-LOOP ECONOMY

RICHARD G. CARRANZA
Chemical Engineering Consultant, Montgomery, Texas, USA.

ABSTRACT
The economy is modeled as an engineering pumping system. At the start, no control is used; thus, the open-loop system is established. Flow, or analogously GDP, is sent from a pump to a storage tank (analogously, inventories). Control is then applied where investment (flow out of the tank) is controlled via the interest rate. Thus, the open loop becomes a closed loop. It is shown from simple example that frequent government interference in the economy causes more instability than less government interference. Dynamic systems tend toward steady-state systems and government interference should be very selective and careful. In this work, the interest rate is used to control the economy by the Federal Reserve, ceteris paribus. Simulations indicate that when the Fed attempts to control the economy using interest rates, the net effect is that economic equilibrium is delayed with respect to an economy that comes to equilibrium naturally.
Keywords: consumption, control, economics, feedback, investment, loop, macroeconomics, savings.

1 INTRODUCTION

The similarities between economics and biology are striking. Symbiotic relationships prevail everywhere in nature. Yet, the goal of the scientist is to pull this apparently disjointed universe into a coherent logical system. This work makes an attempt to do this for a dynamic macroeconomic system.

The theory of evolution asserts that, between 3 to 4 billion years ago, the atmosphere of the earth is primarily hydrogen, water, ammonia, and methane [1]. Energy obtained from sunlight and lighting then turn these simple molecules into complex structures such as carboxylic acids, amino acids, and other organic molecules. Ultimately, proteins, nucleic acids and polysaccharides follow in sequence. Inorganic elements, like phosphorus and sulfur, in the earth also help to create more complex organic materials.

Since proteins, which are made from amino acids, are polar molecules; the polar regions of the protein are attracted to the polar sides of water molecules. As proteins aggregate in this way, a type of membrane is formed. Another theory proposes that membranes are formed via lipids. Lipids are not soluble in water and easily form membranes two molecules thick. Nucleic acids slip past the membrane and life is formed.

As photosynthesis is introduced, oxygen fills the earth's atmosphere, and the first cells are classified into several categories. Autotrophs are cells that use photosynthesis for energy (the fuel source is inorganic), both aerobic and anaerobic. Heterotrophs, aerobic and anaerobic, are cells that do not use photosynthesis – they use organic matter for energy.

Evolutionary theory suggests that anerobic heterotrophs begin to absorb aerobic heterotrophs; and that both benefit because the anaerobic cell acquires the organic food for the aerobic cell, while on the other hand the aerobic cell produces energy more efficiently – energy that is shared with the

DOI: 10.2495/DNE-V11-N4-600-609

anaerobic cell. A symbiotic relationship is formed. The aerobic cell evolves into what is now known as the mitochondria.

The heterotrophs that contain mitochondria now start to absorb autotrophs. The autotrophs evolve into chloroplast. This is the foundation of plant life.

The earth, at this stage, is starting to exhibit signs of the specialization of labor. The theory of evolution teaches that cells begin to specialize in such things as leaves, roots, branches, etc. The symbiotic relationship is continuing within the plant kingdom: the cells that specialize in forming leaves help to maintain the cells that become roots and branches – and vice versa. The specialization of labor continues in the animal kingdom too. Cells start to form hearts, brains, kidneys, livers, etc.

Specialization of labor is nature's way of struggling to survive. So far in this discussion, cells cooperate to survive. Those cells that survive until they can replicate, pass on their genetic information to the next generation. Darwin calls this the survival of the fittest.

The sociologist Herbert Spencer extends this thought further. He claims that human society is rooted in the natural tendency of cellular survival. This is done by humans forming many tribes, ethnicities, cities, and countries. Specialization of labor is continuing in the form of the different branches of government and the diversity of businesses. Labor is specialized in terms of law, engineering, politics, education, construction, medicine, etc. Spencer draws analogies between specialized branches of government and human organs. He compares highways and rail lines to arteries. Ultimately, he claims that money is the lifeblood of human existence: money is what sends nutrition to all parts of the body.

Here is Spencer in his own words, "The circulating mass of nutritive materials in individual organisms and in social organisms, becoming at once better in the quality of its ingredients and more heterogeneous in composition, as the type of structure becomes higher, eventually has added to it in both cases another element, which is not itself nutritive but facilitates the processes of nutrition. We refer, in the case of the individual organism, to the blood-discs; and in the case of the social organism, to money. This analogy has been observed by Liebig, who in his Familiar Letters on Chemistry says:

'Silver and gold have to perform in the organism of the state, the same function as the blood-corpuscles in the human organism. As these round discs, without themselves taking an immediate share in the nutritive process, are the medium, the essential condition of the change of matter, of the production of the heat and of the force by which the temperature of the body is kept up, and the motions of the blood and all the juices are determined, so has gold become the medium of all activity in the life of the state.'

And blood-corpuscles being like coin in their functions, and in the fact that they are not consumed in nutrition, he further points out that the number of them which in a considerable interval flows through the great centers, is enormous when compared with their absolute number; just as the quantity of money which annually passes through the great mercantile centers, is enormous when compared with the quantity of money in the kingdom. Nor is this all. Liebig has omitted the significant circumstance that only at a certain stage of organization does this element of the circulation make its appearance. Throughout extensive divisions of the lower animals, the blood contains no corpuscles; and in societies of low civilization, there is no money." [2]

Spencer now brings the discussion full circle: a model for the behavior of a complex economic system. This discussion hints that the political economy is simply one more step in cellular evolution. Nevertheless, the microscopic enzymatic reactions that occur within cell walls are modeled using a macroscopic equation called the Michaelis–Menten equation. Analogously, the economy is modeled using macroscopic equations too, for instance aggregate demand. Aggregate demand is simply the summation of all the little demands that originate from all the individual human organisms in the society.

2 TERMS AND DEFINITIONS

The linguistic philosopher, Ludwig Wittgenstein, asserts that language is a game with rules and that confusion occurs when people do not stick to the rules: "It is the philosophers, of course, according to Wittgenstein in Philosophical Investigations, who persist to misusing language. The problems of philosophy are only linguistic problems, word puzzles which trap philosophers because they do not follow the rules of the ordinary English language game. Philosophers mistakenly look for the essence of words; they mistakenly construct ideal languages and criteria of meaning from the features of a few examples, as if such languages were more perfect than everyday langauge; they mistakenly try, as Plato and Hegel did, to synthesize all human experience and knowledge, but this, according to Wittgenstein, is total confusion, a mixing up of many different language games – art, religion, science, ethics, politics – into one nonsensical hodgepodge. Philosophers' problems are not genuine problems but only the nonsense that results from not knowing how to handle language." [3] Therefore, a few important definitions are established before proceeding.

Complex systems, definitively, are "open" – that is, the system has exchange with the environment. The closed-loop economy here does not refer to material nor energy exchange with the environment. On the contrary, raw materials come into the economy and leave as goods and services. Rather, the word closed refers to signal processing.

The terms "open" and "closed" are used in the context of traditional engineering control theory. For instance, assume you have a pot of water that you wish to heat to 100°F. If one simply places the pot on a gas burner and walks away, the temperature reaches 212°F; because there is no feedback to a "controller" that adjusts the amount of heat supplied such that the temperature of the water stays at 100°F. This is called an open-loop system – no feedback.

If a visiable temperature guage is placed in the water, and one stands at the burner and stops the flow of gas when the temperature is 100°F, feedback is established and this is called a closed loop. In other words, the individual provides a set point temperature. The thermometer provides a sample of the real temperature in the water. The individual, or the controller, makes a comparison between the set point and the actual temperatures, and adjusts the system such that the objective temperature is reached. The system comes around full circle with a process of adjustment, measurement, comparison, adjustment, etc. – a closed loop. Typically, automatic control is applied with electronics: thermocouples, controllers, actuated control valves, and other such devices.

3 OPEN-LOOP ECONOMY

A detailed analysis of the open-loop economy is given by Carranza [4]. However, a short review is given here. Figure 1 is an example of an open-loop economy as given by Carranza [4]. The ratios of mps to mpc are natural elements of the economy. Thus, in Fig. 1, the ratios are maintained by the natural flow mechanics of the system. Note that there is no friction in the system.

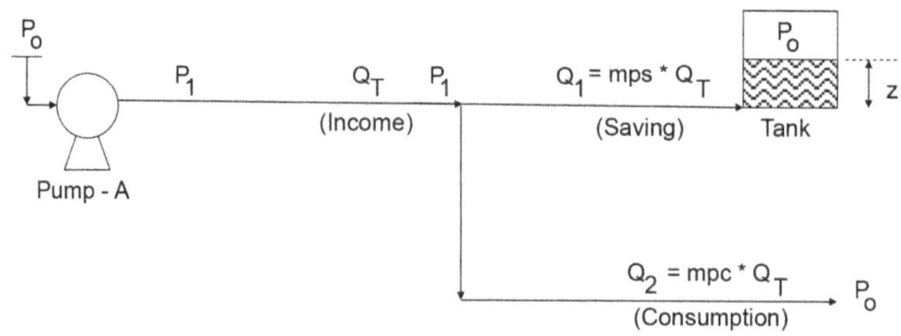

Figure 1: Basic open-loop economy.

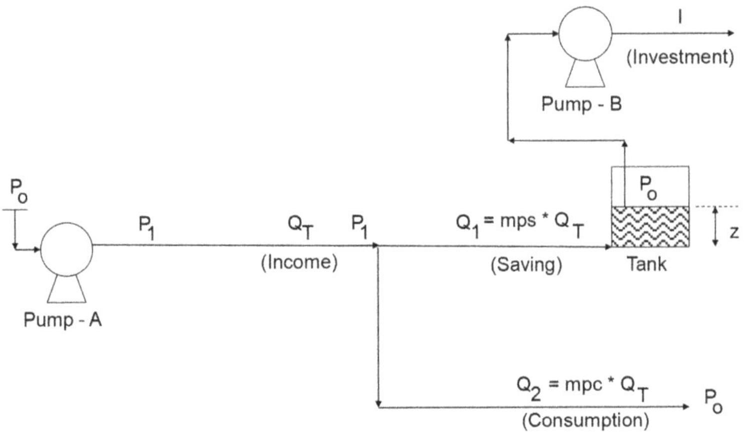

Figure 2: Open-loop economy with investment.

Figure 2 is the same system but this time pump B is added to the tank. This outflow is called investment. In economics, this is more accurately called net investment because some investment is saved as inventories. The saved inventories are represented by the liquid level in the tank.

From Bernoulli's equation, pump A discharge pressure is given as the following simple relationship (no friction, no velocity effects):

$$\frac{P_1}{\rho g} = \frac{P_o}{\rho g} + z \tag{1}$$

It must be remembered that eqn (1) is a steady–state equation. It must be converted to a dynamic equation. This is done by adding an infinitesimally small step change to the right hand side of eqn (1). The step change takes into account the amount of liquid that enters the tank for each time interval:

$$\Delta z = \frac{(Q_1 - I)\Delta t}{A} \tag{2}$$

Note the following important relationships: mps + mpc = 1.0, I = 0.1 m^3/s, A = 0.2 m^2, Δt = 0.01 s, mps = 0.2. Also note the following important eqns (observe that the pressure across pump A is essentially the height of the liquid in the tank, since there is no friction):

$$\Delta P = P_1 - P_o = \rho g h_p \tag{3}$$

$$Q_1 = mps\ Q_T \tag{4}$$

$$h_p = -Q_T^2 + 1 \Rightarrow pump\,curve \tag{5}$$

When eqns (1)–(5) are all combined, a quadratic polynomial is the result. The polynomial is then solved using the quadratic equation. The final solution is given in eqn (6). Equation (6) gives the flow through pump A for each time interval.

$$Q_{T,t+1} = \frac{\Delta t - \sqrt{\Delta t^2 + 4(1 - z_t + 5I\Delta t)}}{-2} \tag{6}$$

It can also be shown that the level in the tank is given by eqn (7).

$$z_{t+1} = z_t + \frac{(mpsQ_{T,t+1} - I)\Delta t}{A} \tag{7}$$

Equations (6) and (7) are plotted in Fig. 3 when I = 0, no investment. The way the hydraulics network is set up, the inventories in the tank bring the system to a near halt. Analogously, rising inventories slow down an economy. Without government intervention, each individual shop owner in the economy simply looks at his inventory room and curbs production as inventories start to rise. The result is a drop in GDP and unemployment. It is important to point out here that Q_T is analogous to real GDP, which is equal to C + I since at present government spending and net exports are neglected. Savings is equal to I, net investment plus inventories.

Investment is added to the economy and represented in Fig. 4. Investment is fixed at 0.1 m^3/s. Such being the case, the economy naturally comes to equilibrium when the total flow or real GDP is equal

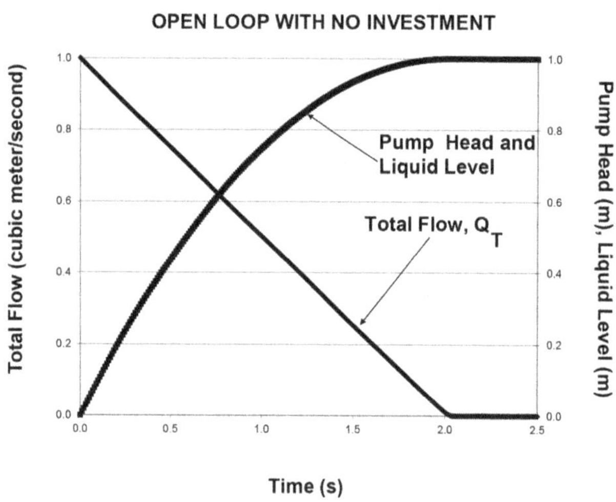

Figure 3: Open-loop economy with no investment.

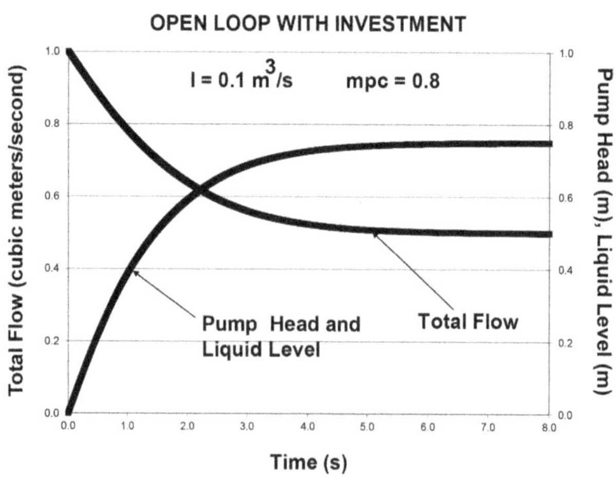

Figure 4: Open-loop economy with investment.

to 0.5 m^3/s. In other words, the rising inventory in the tank naturally forces the pump to flow less liquid. Analogously, the economy responds to rising inventories and generates less product. Liquid level and consumption are 0.75 m and 0.4 m^3/s, respectively. Note that consumption plus investment equals total GDP at equilibrium. Furthermore, since inventories are not accumulating, S = I.

4 CLOSED-LOOP ECONOMY

The close-loop economy is exemplified in Fig. 5. The level transmitter controller (LTC), represents the Federal Reserve System. Pump B drives investment and is operated by a variable speed motor. The motor is controlled by the Fed via the discount window interest rate.

A typical scenario in which the Fed attempts to bring "stability" to the economy is by closely watching the nation's inventory levels. The Fed knows that if inventories rise, then the economy starts to slow – as is shown in Fig. 3. Thus, the Fed sets an inventory target and controls investment with interest rates. In classical economics, investment and interest rates are negatively correlated; therefore, they are here defined by the following relationship, for the sake of simplicity: $I = 1/i$.

The simulation in Fig. 4 is rerun, but this time the Fed controls investment. Note that in Fig. 5 a feedback loop is established – the liquid level is observed by the LTC or the Fed, the liquid level is compared to the target level (otherwise known as the set point), interest rate is adjusted, the rate of investment is altered, the liquid level changes and again is compared to the set point, ad infinitum.

In Fig. 4, one observes that the economy naturally comes to rest when the liquid level is 0.75 m. With the Fed involved in Fig. 5, the inventory target that is set by the Fed is 0.75 m also. Equations 5 and 6 are utilized again but with the following initial specifications at t = 0.0: Δt = 0.01 s, i = 10%, liquid level = 0.0 m, set point = 0.75 m. Since there is no liquid level pushing back on Pump A, initially, the pump operates at full capacity, 1.0 m^3/s.

Using the above information, the first simulation of four involves the Fed adjusting the interest rate at every time interval. If the inventory level is below the set point, the Fed raises interest rates by 0.1%. This rise in interest rates slows investment and allows inventory to build in the tank. Likewise, if the inventory level is above the set point then the Fed will lower interest rates by 0.1%. The results of simulation 1 are given in Fig. 6.

In simulation 2, everything is kept the same except the set point. The new set point is 0.5 m. The results are given in Fig. 7. In both simulations 1 and 2, one overarching conclusion is made. The

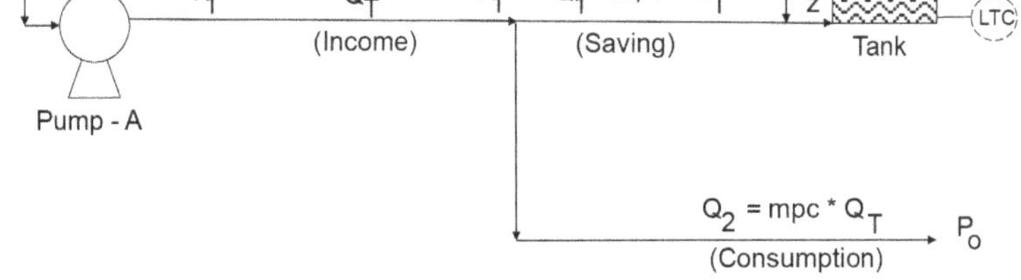

Figure 5: Closed-loop economy.

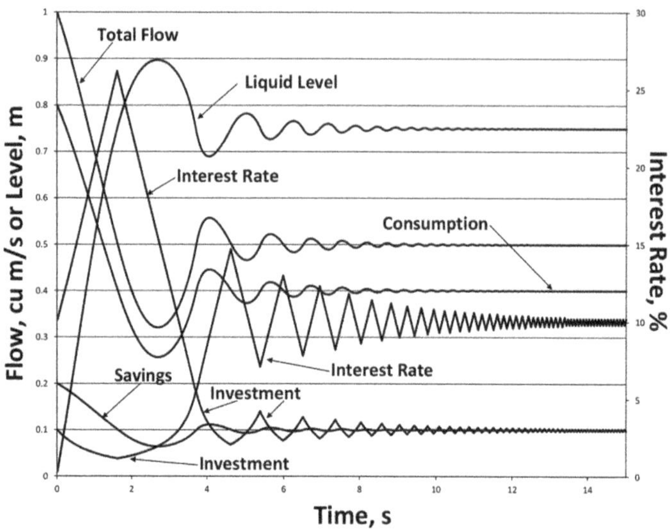

Figure 6: Simulation 1.

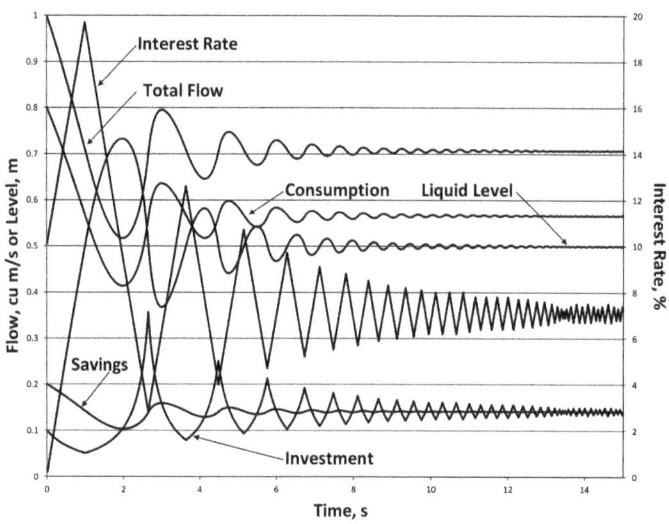

Figure 7: Simulation 2.

system is in complete chaos. It is barely being kept in control. Another way to say this is poor control loop tuning.

Simulation 3 is the same as simulation 1 but with a single difference: the Fed does not respond as fast, this time the Fed changes interest rates with every set of 10 time intervals – otherwise known as dead time, see Fig. 8.

Simulation 4 is the same as simulation 3 but with a different set point: inventories = 0.5 m, see Fig. 9. Observing both simulations 3 and 4, an interesting observation is made – less Fed involvement makes for a better economy.

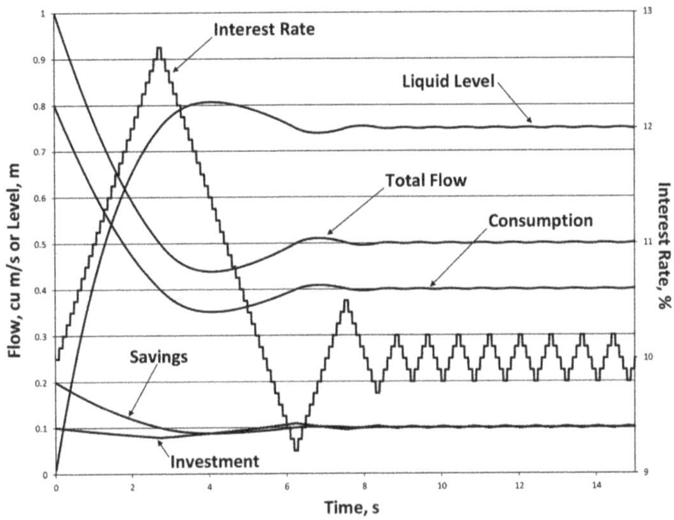

Figure 8: Simulation 3.

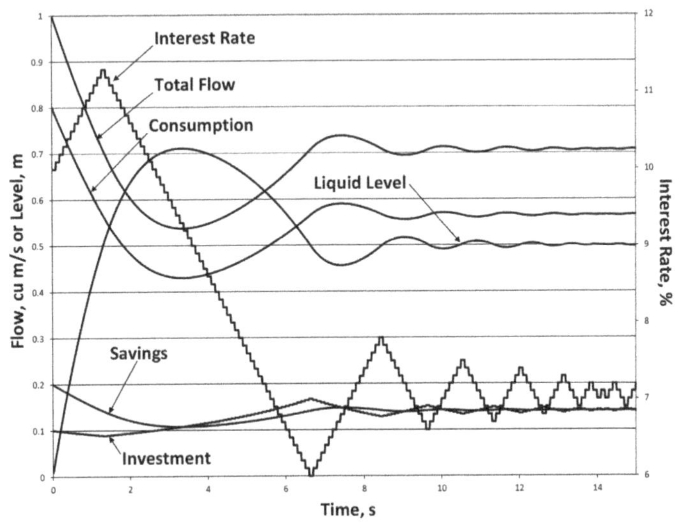

Figure 9: Simulation 4.

5 RESULTS AND CONCLUSIONS

As is shown in Figs 8 and 9, less government involvement in the economy is highly desirable. Most schools of economic thought springing forth after the time of Keynesian economics state the same principle.

"Unlike Keynesian economists, monetarists do not believe that the economy is subject to a disequilibrium that must be offset by government action. Most monetarists believe that the economy tends toward equilibrium at the level of potential national income. Their faith in the free market (price) system leads them to favor minimal government intervention . . . Monetarists often argue that government policy heightens the effects of the business cycle . . . Monetarist favor non-activist government policy because they believe that the government's attempt to make the economy better off by aiming monetary and fiscal policies at low inflation and low unemployment often make things worse." [5]

The results of this work definitely support the monetarists in their point of view. Other schools of thought that believe in minimum government involvement include new classical economics and rational expectations.

By looking at the historical data for the discount window interest rate, one sees the hopelessness of government intervention in the economy. Between 2005 and 2008, Fed chairmen Greenspan and Bernanke both raise interest rates to "cool off" the economy, but the reality is that the economy is torpedoed by sending countless adjustable rate mortgages into default. Since 2008, the Fed tries to "revive" the economy with low interest rates, with no avail.

In conclusion, human activity, past and future, is very much a function of human genetics. There is room for reason, but in conjunction with sound judgment. In the words of Ilya Prigogine, "Economic systems are also unstable, 'chaotic' systems. . . . we have to expect instability in social systems as decisions are no longer associated with some deterministic rule."

"The decision-making process introduces an essential difference between physical and social systems. . . . No longer is there a gap between the 'hard' sciences speaking of certitudes and the 'soft' sciences dealing with possibilities."

". . . . the universe is ruled both by laws and by events . . . . Therefore, we have choices, we have values. It goes far beyond the competence of a physicist to describe the origin and variety of human values. My more modest role has been to emphasize that the existence of values, and therefore also of economic values, is in line with our present description of the physical universe." [6]

NOMENCLATURE

A	area (m^2)
C	outlet volumetric stream or consumption (m^3/s or \$/time)
I	liquid draw off or investment (m^3/s or \$/time)
P	pressure (N/m^2)
Q	volumetric rate (m^3/s)
S	savings (m^3/s or \$/time)
g	gravity constant (m/s^2)
h_p	pump head (m)
i	interest rate (%)
mpc	marginal propensity to consume (-)
mps	marginal propensity to save (-)
t	time (s)
z	liquid level (m)

GREEK LETTERS

Δ	change
ρ	density (kg/m^3)

SUBSCRIPTS

T	total
o	atmospheric condition or initial time condition
t	time

REFERENCES

[1] Wolfe, S.L. & Miller, G.T., *Biology: The Foundations*, Wadsworth Publishing Company: Belmont, California, 1977.
[2] Spencer, H., The Social Organism, *The Westminster Review*, 1860.
[3] Lavine, T.Z., *From Socrates to Sartre: The Philosophic Quest*, Bantam Books: New York, 1984.
[4] Carranza, R., The open loop economy. *The Business Review, Cambridge*, **6**(2), 2008.
[5] Boyes, W. & Melvin, M., *Economics*, Houghton Mifflin Co: Boston, 1991.
[6] Prigogine, I., The rediscovery of value and the opening of economics (Chapter 2). *The Evolutionary Foundations of Economics*, ed. K. Dopfer, Cambridge University Press: Cambridge, 2005.

ECONOMIC AND TECHNOLOGICAL PRIORITIES OF COMPETITIVE DEVELOPMENT OF RUSSIAN SYSTEMS OF ENERGY COGENERATION SOURCES

A. DOMNIKOV, G. CHEBOTAREVA & L. DOMNIKOVA
Academic Department of banking and investment management, Ural Federal University named after the first President of Russia B.N. Yeltsin, Russia.

ABSTRACT

The article studies the problem of evaluation of competitive advantages of energy cogeneration sources (ECS) included in centralised and decentralised systems of energy cogeneration. For this purpose, a methodological apparatus has been developed to evaluate the effectiveness of business processes, and competitive cogeneration plants adapted to the market exploitation conditions have been chosen.

The methodological framework for conducting a comprehensive economic assessment of business processes in development of competitive advantages of cogeneration energy has been developed, which ensures improvement of the capacities structure and optimisation of fuel consumption in territorial energy generating companies, and allows choosing the most competitive options for the development of cogeneration energy sources of various types and capacities with consideration of the competitive environment on the electric and thermal energy markets. The proposed methodological approach makes it possible to reveal the economic and technological priorities for the development of ECS in regional energy systems and can be spread to the developing economies within the BRICS.

In the course of the study, it was found out, that the development of competitive advantages of energy sources cogeneration may be based on implementing the following business processes: optimisation of generating capacities and fuel consumption structure, technical re-equipment, reconstruction, expansion of existing and construction of new cogeneration energy sources.

Keywords: cogeneration systems, competitive development, energy

1 INTRODUCTION

The current market development shows that the high-tech industries set the development and ensure the competitiveness of any economy. Power generation, as its part, remains the most important and life-supporting industry. It provides a multidimensional and profound impact on socio-economic development of society and environment. This is due to the high importance and uniqueness of the sector product – electric energy and heat, which ensures operation and development of the national economy, as well as the social life, in the harsh climate of Russia. Therefore, preservation and expansion of the capacity of electric power industry, modernisation of its production facilities, as well as solution of tasks, faced by the industry, requires investments. The difficulty of attracting private capital in the power-generation sector has been considered by the authors in several publications [1–5]. However, the methodological problems related to the high level of subjectivity in risk assessment and, in general, investment attractiveness of the companies in the industry are on another development stage. The solution of this problem, aimed to increase the level of estimations objectivity, lies in the methodological development of the mathematical apparatus, allowing minimisation of the experts' opinion significance [6–9].

DOI: 10.2495/DNE-V11-N4-610-619

2 GENERAL LAWS OF ENERGY COGENERATION SYSTEMS DEVELOPMENT IN THE RUSSIAN CONTEXT

In the process of the Russian electric power industry development, the new conditions have formed, characterised primarily by a high degree of uncertainty in the development of a competitive environment. They have led to the change of former priorities to qualitative development of generation capacities due to improvement of their efficiency and reliability.

Modern trends of electric power industry development are characterised both by restructuring processes, which initiate commercial activity of energy companies, and by the increase of competitive tension in the markets of electric and heat energy, which leads to the appearance of specific forms of competitive struggle in the sphere of energy generation at the level of territories, related to the implementation of competitive advantages of generating energy sources.

A link between an electric power system and a consumer, and also the basis, providing at the territorial level such processes as electrification and district heating, are ECS, representing a joint production of electricity and thermal energy through cogeneration power plants. Depending on their type and capacity, ECSs form two system of energy cogeneration (SEC) – centralised and distributed.

2.1 Centralised system of energy

A centralised SEC includes a territorial generating company (TGC), which occupies some intermediate position between "system-wide" generation and consumption complex, formed as a result of the development of the Russian electric power industry in recent time. On the basis of the structure of TGC generating facilities, cogeneration steam turbine units (STU) are installed at combined heat-and-power stations (HPS). Combined heat-and-power plants (CHP) and gas turbine cogeneration plants (GTU) plants may be considered as the most advantageous ones.

The problem of a centralised SEC development under competitive conditions may be associated with a loss of competitive advantages and disturbance of system properties mainly due to high depreciation of equipment and due to the desire of the TGC leadership to maximise profits at the expense of intensive exploitation of its residual life. This may result in a reduced level of ECS competitiveness due to: (1) sub-optimal mode of operation of cogeneration plants; (2) increase of fuel consumption; (3) growth of production costs.

2.2 Distributed system of energy

Currently, along with the traditional centralised SEC, a new structural element appears in the regional energy sector – an independent producer, including small cogeneration units forming the basis of a distributed SEC.

ECS included in a distributed SEC can be installed in remote areas of a region, having no centralised power supply but having enough relatively cheap local fuel (e.g., biomass, coal), and are economically feasible at industrial enterprises and in housing and communal systems. Under certain circumstances, such cogeneration installations may be competitive compared to a centralised SEC. In technical terms, the development of distributed SEC is only possible at high energy and operational efficiencies of small cogeneration plants, resulting in the following: 1. Change of the structure of fuel and energy balance of the territory in favor of cheaper, than natural gas, solid fuels; 2. Restructuring of electricity supply towards distributed SEC, on the basis of energy markets liberalisation; 3. Change of the technology of biomass and solid fuels energy utilisation (with the use of modern gas technologies in cogeneration plants).

In general, the suggested directions are not contrary to the principles of competitive development of the entire SEC and do not reduce the system effects due to the spread of low-power ECS. This produces an additional link in the regional system of energy generation, which provides energy to small communities or industrial enterprises, where it is impractical for any reason to use the ECS included in a centralised SEC. Moreover, under certain circumstances, the ECS included in a distributed SEC can organically fit into the competitive environment, taking its market niche in energy generation in the region.

The rational combination of centralised and distributed systems can mutually complement each other and allow creating fairly flexible SECs on territories. Such SEC is able to reliably provide consumers with electric and thermal energy and to successfully compete with sources, conducting separate generation of energy through the implementation of competitive advantages of cogeneration.

From the point of view of system approach, SEC is an open system, providing a unity of interrelated parts – both centralised and distributed SECs, each of them can be divided into smaller elements – ECS and individual cogeneration units, altogether forming the SEC structure.

The existence and evolution of SEC as an open system, depends, on the one hand, on its internal structure, and on the other – on the interaction with the external environment. As a result, the level of ECS competitiveness may increase or decrease.

3 COMPETITIVE ADVANTAGES OF ENERGY COGENERATION SYSTEMS

The development of a competitive environment in energy generation on a certain territory requires the full realisation of competitive advantages of the SEC, revealed at comparison with alternative separate production of electricity and heat in condensing power plants (CPP) and in boiler houses. The main competitive advantages follow from: (1) fuel economy; (2) improvement of power supply reliability; (3) reduction of costs for the construction and operation of electric and heat networks; and (4) reduction of toxic emissions and greenhouse gases.

It is known that energy cogeneration is a highly efficient technology with a general EF of 60% to 80%, at that the more electric energy is produced in heat consumption (in heat extraction mode), the higher efficiency factor (EF) of the cogeneration plant is. It clearly shows the economic feasibility of cogeneration development and demonstrates its main competitive advantage, which consists in the saving of heat at energy production, and ultimately leads to a significant reduction of the largest fuel component in the cost of energy.

It should be noted, that the future development of centralised and distributed SEC largely depends on the state of the emerging competitive environment. The competition in the regional energy market forces the producers of energy, realising their competitive advantages, to improve the efficiency and reliability of ECS, which makes it possible to reduce tariffs for industrial and household consumers.

Taking into account the above, a scheme has been developed (Fig. 1), which shows that it is possible to build up competitive advantages of SEC.

4 PRACTICAL IMPLEMENTATION OF METHODICAL APPROACH

It is known that competitive advantages of ECS are represented by the price and quality characteristics of electrical and thermal energy, which favorably distinguish its manufacturer and provide him a stable position in the energy market [10].

The development of SEC competitive advantages is directly related to resolution of the following problems defining the further development of ECS:

1. Reduction of the cost of heat and electricity joint production by improving and increasing the efficiency of the structure of generating facilities in a centralised SEC;

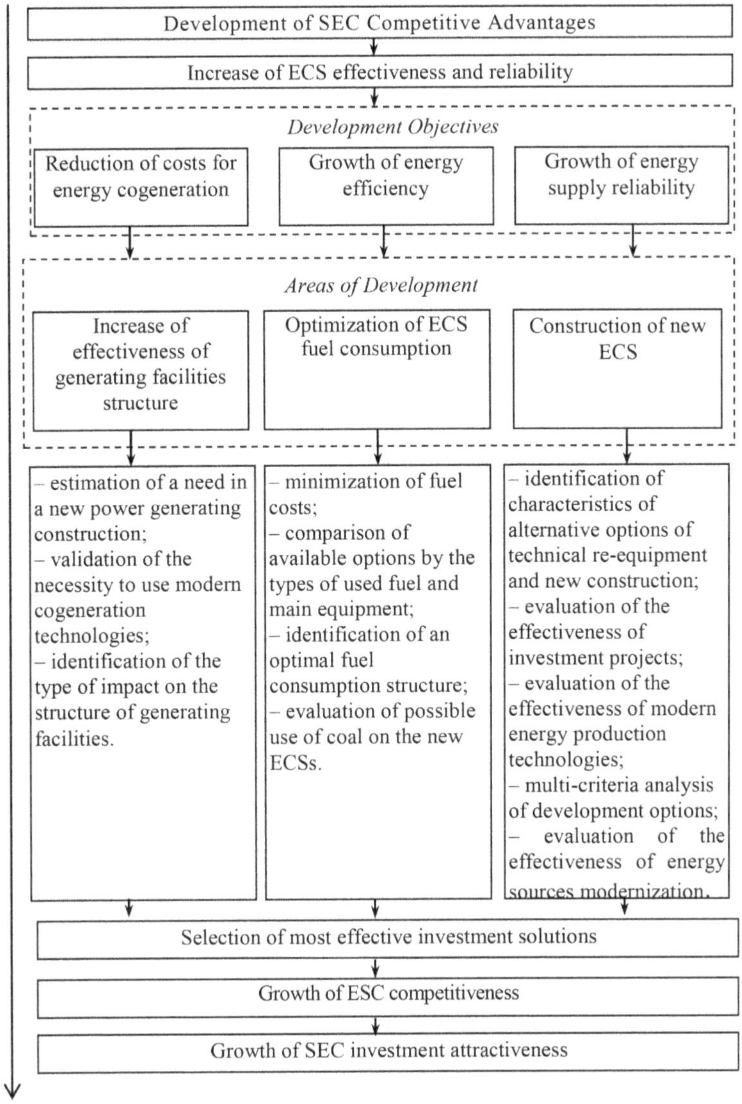

Figure 1: The scheme of building-up SEC competitive advantages.

2. Optimisation of ECS heat consumption in a centralised SEC;
3. Application of modern cogeneration plants with high reliability and efficiency.

4.1 Raising of centralised SEC competitive advantages

The structure of ECS included in the TGC has a significant impact on the realisation of competitive advantages of a centralised SEC. In this regard, the technical base of domestic cogeneration differs by an excessively large share of used inefficient steam turbine plants running on natural gas. Many of them have significant physical wear. Practical application of competitive and highly efficient cogeneration technology based on STU and gas turbines is still insignificant [11].

Improving the structure of generating facilities of TGC through its optimisation is a very important step towards the realisation of competitive advantages of a centralised SEC, since the existing steam turbine cogeneration units located at the CHP plant, in some cases, find themselves unable to compete with producers of thermal energy, mainly due to low total efficiency and reliability.

Optimisation of the structure of TGC generating facilities will help to reveal low-efficient cogeneration plants and to propose the best possible development options from the point of view of economic criteria, for which the following can be used: (a) integral costs; (b) integral effect.

The proposed criteria allow taking into account different economic interests of subjects of management of the centralised SEC. Such interests can be divided into two groups. The first group reflects the desire to define the necessary costs of production and distribution of electric and thermal energy, based on the desire to minimise them, to enhance the competitiveness of the ECS. The second group reflects the desire of power producers, as independent entities, to achieve the most favorable financial results in their activities, often to the detriment of the reliability of cogeneration plants.

The improvement of the ECS structure in TGC is achieved with the help of an optimisation object-structural model, specially developed for this purpose. This model shows the possibility of adaptation of a centralised SEC to the competitive environment, taking into account the development of ECS competitive advantages [12].

4.1.1 Analysis of estimates according to the criteria of the integral cost and integral effect

The analysis of the results of calculations, presented in Figs 2 and 3, shows the following. The optimisation of the generating facilities of TGC-9 has revealed some weaknesses in its composition and structure, which are characterised by a significant lag in the development of the energy generating company, and in the near future, they will inevitably become an obstacle to the development of industry and social sphere of the region.

By the criterion of integral costs, certain measures are required for most CHP plants on replacement of equipment with a total capacity of 2 255 MW (56%). Partial or complete dismantling of the equipment must be carried out in the amount of 520 MW (13%) due to its low technical and economic indicators, mainly due to high fuel consumption for energy production. It should be noted, that a

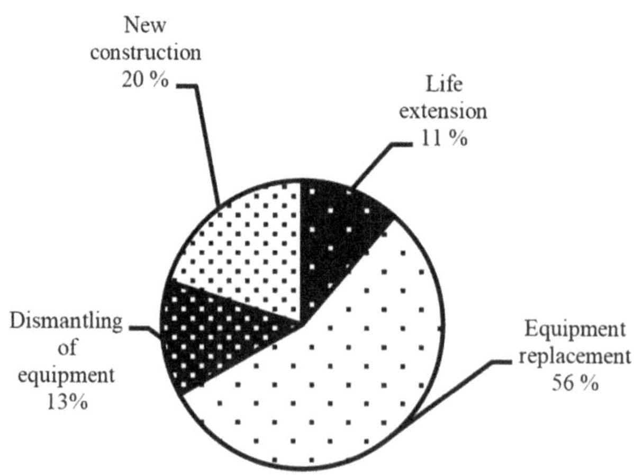

Figure 2: Optimised structure of measures at TGC-9 according to the integral costs criterion.

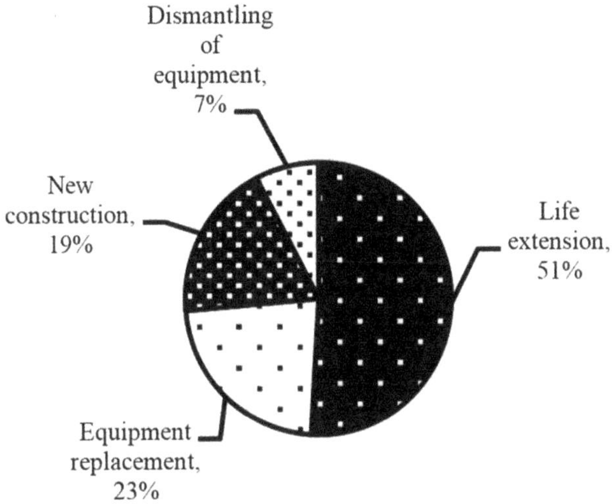

Figure 3: Optimised structure of measures at TGC-9 according to the integral effect criterion.

number of power plants of TGC-9 will need to extend the life of their equipment by upgrading (11%), with a total installed capacity of power plants of the power company increasing by 240 MW and thus comprising 3,535 MW in 2018.

According to the integral effect criterion, the largest share in the structure of measures takes the measure of generating capacity life extension by 50%, which is 1,690 MW. The activities associated with the replacement of equipment will be needed in the amount of 752 MW (23%). The dismantling of obsolete equipment having low technical and economic indicators will make 254 MW (8%), the new construction based on modern technologies will make 620 MW (19%), a large proportion of which should lie within the STU-CHP mainly because of their high efficiency that would be especially true at the increase of price for natural gas [13]. The total installed capacity of company will increase by 417 MW and will reach 3,712 MW.

In general, the optimisation results show that, depending on a selected criterion, a different structure of the types of impact on generating capacities of TGC-9 is observed. So, by the criterion of integral costs, the largest share will fall on the replacement of major equipment – 56%, and by the integral effect, only 25% will be effected by this type of impact.

According to the integral effect criterion, a large proportion of 50% falls on life extension of power equipment, whereas this type of impact makes only 11% according to the criterion of integral cost. For other types of impact, no significant change will take place according to both criteria. As it is shown by the calculations, new power-generation construction will take 20% and 19%, and dismantling – 13% and 8%, respectively, according to the criteria of integrated costs and effect. It should be noted, that if the management of TGC-9 in their decisions will adhere to the criterion of integral costs while improving the structure of its capacity, it will lead to a more rapid transition to sustainable development of the regional economy, because such a strategy focuses primarily on decommissioning of equipment with expired service life and thus has relatively low technical and economic indicators. However, it is not possible to obtain maximum revenue in the long term, primarily due to significant investments in the development of generating facilities.

In accordance with the second criterion, the commercial interest of the company management is associated with their focus on the biggest income, due to intensive exploitation of the equipment residual life and extension of its service life. However, a focus on higher yield will increase the installed capacity of TGC-9 power plants at 177 MW compared to the alternative strategy. It should be noted, that the inputs of new energy capacity in the energy company by the criterion of the integral effect will be 260 MW less, than by the criterion of integral costs [14].

In the future, the focus on maximising revenue-generating that brings rather big profits at the first stages of development to the detriment of the implementation of large-scale technical re-equipment and modernisation, is likely to cause a decrease of yield due to more frequent repairs and low-cost effectiveness of the generating facilities of the power company, and to become a serious threat to sustainable development of the economy of the region.

It is obvious, that TGC-9 leadership, in the process of decision making for further development, must proceed from the priority of sustainable and long-term development and not of quick income. The strategy developed in accordance with the criterion of integral costs fully meets this approach.

The revealed contradictions in possible developments of a centralised SEC can be resolved by using three groups of criteria. An energy group is characterised by: specific fuel consumption, availability factor, and electricity and thermal energy generation volume. An environmental group is determined by: emissions of nitrogen oxides, the cost emissions, and fee for land resources. Finally, the third group of economic criteria is formed by: specific investment costs, energy production costs, integral effect and integral costs. These groups of criteria allow the study of the development directions based on specific characteristics and development status of a certain ECS, as well as the opportunities of building-up the competitive advantages.

In the course of preliminary analysis of the effectiveness of possible ECS development options, four out of all possible options have been identified as the most promising from the point of view of efficiency increase: 1. Extension of the service life of a cogeneration unit due to modernisation of the auxiliary and main power equipment with replacement of physically worn-out elements, mainly operating in areas of high temperatures and pressures; 2. Replacement of an existing cogeneration plant to a new one preserving the former type sizes; 3. Expansion of an existing ECS due to installation of a cogeneration unit – combined-cycle gas turbine (CCGT) UNIT unit in the new main building; 4. Construction of a new CCGT UNIT-based ECS with gasification of solid fuel.

4.1.2 Additionally conducted multi-criteria analysis

The results of additionally conducted multi-criteria analysis (applying the fuzzy set theory) have shown that the most effective is an option of ECS technical re-equipment based on the new technology of electrical energy production - option 3, which provides the construction of a cogeneration CCGT UNIT. This option has the greatest degree of non-dominatedness in comparison with other alternatives, by energy and environmental criteria (with their weights equal to one). In the said conditions, the second most preferable option is installation of a new cogeneration CCGT UNIT (option 4). A lower grade of the latter, as the analysis shows, is explained: according to the energy criterion – by significantly smaller volumes of electricity generation at the initial stage (due to longer terms of the power-generation construction) and according to the environmental criterion – by a need for additional land acquisition for the new ECS. With the increased weight of the economic criterion (0.75–1.0), an option of life extension of an existing cogeneration unit (option 1) with more favorable investment characteristics becomes more effective. As a result, options 1 and 3 get into the zone of the highest efficiency. So, for a short term ECS development, it is most preferable to extend the residual life of main equipment, and for a long-term development – to expand the existing ECS with an CCGT UNIT installation [15].

An important problem in the realisation of competitive advantages of a centralised SEC is the fuel consumption, which largely determines the efficiency of ECS. It requires a development of a methodological approach to the study of the problem of diversification of types of fuel used in a centralised

SEC, which needs coordination among investors and the linking of the rate of fuel production, its transportation and consumption by a centralised SEC.

As an example, a forward-looking assessment of ECS energy efficiency in TGC-9 for 2018 was conducted according to the following two variants: the first variant – preservation of the existing level of energy efficiency; the second variant – raising the energy efficiency, by conducting of measures related to the use of modern technologies: Gas and oil-fired ECS – construction of a CCGT UNIT; pulverised coal-fired cogeneration energy sources, using circulating fluidised bed (CFB) technology.

By the results of the fuel component forecast for 2018, one can make the following conclusions. With the projected growth in gas prices, the performance of the fuel component for gas-fired ECS will increase to a much greater extent than coal, even with the broad use of modern technologies.

According to the first variant of the forecast, assuming that the current levels of efficiency at ECS will stay the same, while the fuel component indices for gas and oil will increase at 265.4%, and for coal – at 157.8% by 2018.

According to the second variant of the forecast providing for the ECS reconstruction and increase of their efficiency, the indices of the fuel component in the oil-gas units will increase at 203.3%, and on coal ones to 120.4% by 2018.

As a result, each forecast option predicts that by 2018, the fuel component indices for oil and gas ECS will be higher than for coal ones. It should be also mentioned that fuel costs for coal ECS in TGC-9 according to the second forecast variant (at transition to modern technologies of fuel combustion in CFB boilers) will be 23.4% lower than the corresponding costs according to the first forecast variant addressing the ECS work at the existing values of efficiency. Thus, the gas and oil ECS, using cogeneration CCGT UNIT, in 2018, may save 23.7% of fuel costs as compared to the costs of fuel when working on the existing steam power equipment.

4.2 Raising of distributed SEC competitive advantages

In the conditions of growing prices for certified energy resources, of deteriorating quality of fossil solid fuels and of increasing negative manmade burden on the environment, the issues of a reasonable decentralization of power supply with the use of the distributed SEC potential, as well as the issues of improvement of reliability and continuity of energy supply, when the main part of power equipment is significantly worn out, become very important. The problem of supplying electricity to remote areas can be solved through the use of distributed SEC capacity.

One of the main directions of distributed SEC development on remote territories having no central heating, taking into account the climatic features and provision with a sufficient amount of suitable fuel, is the use of biomass for production of heat and electricity in cogeneration on gas-fueled power plants with closed cycle gasification. It should be noted that under certain conditions, the gasification of biomass represents a competitive alternative to direct fuel combustion for energy generation. Therefore, autonomous ECS, operating on the basis of gas piston cogeneration units, should be recognised as a perspective direction of distributed SEC technical development. Such plants represent a quite competitive option in the field of 'mini-energy', due to the fact that they have high technical and economic indicators. Since such plants are especially useful in the presence of various solid fuels, including organic wastes (e.g. wood and agricultural waste, etc.), which increases the efficiency of distributed energy system. One of the competitive advantages of using such cogeneration plants is that they can be as close to energy consumers as possible, which does not require construction of a large network infrastructure and significantly reduces the payback period of investment projects on construction of gas generating ECS.

The analysis of feasibility of construction of gas generating ECS shows a significant economic efficiency of own cogeneration. As an option for solution of the problem of distributed SEC competitive advantages development, gas cogeneration plants, running on wastes of wood processing and agriculture, can be suggested. Another very promising option for the development of competitive advantages of distributed SEC is the modernisation of outdated energy equipment in boiler houses. As a result of such modernisation, a boiler house is transformed into ECS due to implementation of the scheme of cogeneration add-in boiler house with the gas turbine, operating with a "reset" of flue gases in the furnace of a hot water boiler. In this scheme, the functions of heat recovery, post-combustion and peak heating are combined. Before the gas is supplied into the turbine, its pressure is increased with the help of a pressurising compressor. The passing gases are cooled in an economiser before on-blast feeding to the hot water furnace. The waste gases are fed into the burner of the hot water boiler in an amount necessary for combustion with the design excess air ratio. A year-round work of the relief scheme is provided by the transfer of make-up water heating from steam boilers to hot water ones. It should be noted that the capacity of such cogeneration gas turbine must cover the year-round domestic hot water load of the city and work as much as possible in an economical mode during the inter-heating period.

The analysis of the results of calculations has confirmed the economic efficiency of cogeneration gas turbine construction, with the capacity of 58 MW at OJSC UralATI, as it provides: a relatively low-cost price of electricity (by 76%); thermal energy to the housing and utilities sector and manufacturing; fuel cost savings due to high thermal efficiency of the installation; receipt of additional income from external sales of energy. An important factor for the development of distributed SEC is an additional effect of ECS on the intensity of competition in the regional generation and appearance (under certain circumstances) of a true competition with centralised SEC. As a result, some energy consumers can shift from the category of a buyer to the category of a seller (when there are sufficient electrical power and flexibility) and supply excess electricity to the regional energy market. As shown by the economic assessment of competitive advantages of distributed SEC, its development must follow the path of increased use of modern high-performance clean solid fuel technologies, e.g. based on liquid-fueled and gas-fueled cogeneration plants, with quite high technical-economic and environmental indicators.

5 CONCLUSIONS

The assessment of SEC economic and technological development priorities showed that the ratio between its both parts may be different, depending on the state of the competitive environment. Thus, in addition to traditional STU, the priority in a centralised SEC should be given to cogeneration CCGT UNITs as the most competitive, from the point of view of load management; and in a distributed SEC, the priority should be given to cogeneration gas turbines, basically built on the basis of boilers. With this structure, the cogeneration STUs and CCGT UNITs will work in the basic part of electrical load, and the individual STUs - in semi-peak and peak zones, which will significantly improve the ECS efficiency and reliability.

The studies have shown that SEC competitive development is a process, associated with the creation of rapidly developing, cost-effective and reliable ECS, included in the centralised and distributed SEC, which are actively involved in the formation of a stable regional market of electric and thermal energy and are in permanent search of new directions and ways of realisation of their competitive advantages.

LIST OF ACRONYMS

ECS – energy cogeneration sources;
SEC – system of energy cogeneration;
TGC – territorial generating company;

STU – steam turbine units;
HPS – heat-and-power stations;
CHP – combined heat-and-power plants;
GTU – gas turbine cogeneration plants;
CPP – condensing power plants;
CCGT unit – combined-cycle gas turbine;
CFB – circulating fluidised bed;
EF – efficiency factor.

ACKNOWLEDGEMENT

The work was supported by Act 211 Government of the Russian Federation, contract № 02. A03.21.0006.

REFERENCES

[1] Domnikov, A., Chebotareva, G. & Khodorovsky, M., Evaluation of investor attractiveness of power-generating companies, given the specificity of the development risks of electric power industry. *Vestnik UrFU*, **3**, pp. 15–25, 2013.

[2] Domnikov, A., Chebotareva, G. & Khodorovsky, M., Systematic approach to diagnosis lending risks in project finance. *Audit and Finance Analyses,* **2**, pp. 114–119, 2013.

[3] Khodorovsky, M., Domnikov, A. & Chebotareva, G., Classification of investor attractiveness risks in high-precision industries: a qualitative aspects (by example of electricity). *Vestnik SUSU*, **2**, pp. 99–106, 2014.

[4] Chebotareva, G., Improving the system of power business competitive basis on an investment attractiveness risk assessment. *Modern Tendencies of Economy Development, Management and Law*, **1**, pp. 196–199, 2013.

[5] Gillingham, K. & Sweeney, J., Market failure and the structure of externalities. *Harnessing Renewable Energy in Electric Power Systems*, RFF Press, 2010.

[6] Dubrovsky, S.A., *Applied Multivariate Statistical Analysis, Finance and Statistic*, Moscow, pp. 150–152, 1982.

[7] Tatarkin, A.I., *The Complex Technique of Energy Security Diagnostics of Territorial Entities of the Russian Federation*, Economy Institute: Ekaterinburg, pp. 56–56, 2002.

[8] Laurmann, J.A., Market penetration of primary energy and its role in the green-house warming problem. *Energy,* **6**, pp. 761–775, 1985.
http://dx.doi.org/10.1016/0360-5442(85)90109-4

[9] Lance, G.N., A general theory of classificatory sorting strategies. *Computer Journal*, **9**(4), pp. 373–380, 1967.
http://dx.doi.org/10.1093/comjnl/9.4.373

[10] Gitelman, L.D. & Ratnikov, B.E., *Regional Energy*, USTU-UPI: Ekaterinburg, pp. 128–135, 2006.

[11] Gitelman, L.D. & Ratnikov, B.E., *Energy Business*, Delo: Moscow, pp. 100–110, 2006.

[12] Domnikov, A.Y., *Competitive Development of Cogeneration Systems of Energy*, USTU-UPI: Ekaterinburg, pp. 54–55, 2008.

[13] Domnikov, A., Aspects of multi-criteria analysis of technical re-equipment of power stations. *Vestnik USTU-UPI*, **1**, pp. 48–55, 2005.

[14] Domnikov, A., Khodorovsky, M. & Kosov, I., Development of optimization models for electric power industry development of the region. *Vestnik USTU-UPI*, **10**, pp. 35–42, 2004.

[15] Domnikov, A., Ryzhkov A. & Silin, V., Prospects for the creation of autonomous sources of energy based on local fuel and energy resources. *Vestnik USTU-UPI*, **3**, pp. 178–184, 2004.

EXCHANGE MODELS FOR THE EMERGENCE OF INCOME DISTRIBUTION AND ECONOMIC INEQUALITY

M.L. BERTOTTI & G. MODANESE
Faculty of Science and Technology, Free University of Bozen-Bolzano, Piazza Universitá 5, 39100 Bolzano, Italy

ABSTRACT
Also motivated by the topical problem of growing economic inequality, we propose and investigate models describing the formation process of income distribution in a closed society. Our approach fits in with a complex system perspective. We look at society as a system composed by a large number of individuals, divided into income classes. The information we derive on observable "macroscopic" features results from the interplay of a whole of interactions - money exchanges, payment of taxes, redistribution of benefits, payment of fines - occurring between individuals, i.e. at the "microscopic" level. In order to show the flexibility and the exploratory aspect of the models, we consider and compare cases characterized by different tax rates or in which means-tested welfare policies, tax evasion, taxpayer audits are present.
Keywords: complex systems, economic inequality, income distribution, income redistribution, micro to macro models, tax audit, tax evasion, taxation, welfare.

1 INTRODUCTION AND OUTLINE OF A GENERAL FRAMEWORK

Intriguing examples of complex systems include monetary exchanges in a market society and the process leading to the formation of income and wealth distribution within a given population. Understanding this process at the "microscopic" level requires considering the interactions occurring between a large number of elementary units of the systems – individuals or households. Like for many complex systems, these units are not all equal, being characterized by different behaviour and attributes. The consequence is that even in the presence of well-defined rules governing the interactions, the observable "macroscopic" regularities cannot be easily inferred.

The recent financial and economic crisis, with its ongoing negative consequences, has highlighted the need for a new approach to the modelling of economic systems. Also the ensuing widespread rise in economic inequality has led to a vivid public debate, as testified by the success of some books which are dealing with this issue [1–5].

It provides an urgent motivation to gain further insight into the fundamental mechanisms that underlie the unfolding of macro-economic processes. The ultimate goal, of course, would be to figure out which concrete actions and policies one should adopt toward the attainment of a desired trend.

The observations above show the necessity of a perspective which recognizes the role of the heterogeneity of the units and the importance of their interconnections. Such a perspective differs from the traditional view of mainstream economics, centred on the assumption of a representative rational agent; it took shape essentially during the last two decades and has been developed mainly by means of agent-based computational methods. Related references in economics are for example [6–9]. A clear discussion of complexity in a general context can be found in [10], see also [11].

Within this general conceptual framework, the technical tool we employ is that of ordinary differential equation (ODE) systems. In the traditional literature, ODEs are typically associated with

DOI: 10.2495/DNE-V11-N4-620-627

macroscopic phenomena, and accordingly, models are often described by only one or a few equations. In contrast, the model we are proposing, first developed in [12] and then variously extended and exploited e.g. in [13–16] to deal with different specific questions, is formulated by means of systems of several differential equations. These equations are nonlinear, of the kinetic discretized Boltzmann kind and involve in particular transition probabilities.

In the model, the individuals are divided into income classes and exchange money through binary and ternary interactions. The binary interactions provide a stylized representation of the payments producing income (the payment of an employer to a worker, of a patient to a physician, of a client to a lawyer and so on); the ternary interactions represent the effects of taxation and redistribution, processes which do not involve only the pair of directly interacting individuals (see [12] for the description of the mechanism). The differential equations which express the model are as many as the classes, distinguished by their average income. Let n be their number.

The j-th equation describes the variation in time of the fraction x_j of individuals belonging to the j-th class. In the basic version, it takes the form

$$\frac{dx_j}{dt} = \sum_{h=1}^{n}\sum_{k=1}^{n}\left(C_{hk}^{j} + T_{[hk]}^{j}(x)\right)x_h x_k - x_j \sum_{k=1}^{n} x_k,$$

where

- the coefficients $C_{hk}^{j} \in [0, +\infty)$, satisfying $\sum_{j=1}^{n} C_{hk}^{j} = 1$ for any fixed h and k, express the probability that an individual of the h-th class will belong to the j-th class after a direct interaction with an individual of the k-th class;

- the functions $T_{[hk]}^{j}$: $\mathbf{R}n \rightarrow \mathbf{R}$ satisfy, for any fixed h, k and $x \in \mathbf{R}^n$, $\sum_{j=1}^{n} T_{[hk]}^{j}(x) = 0$. These functions account for the indirect interactions (taxation and redistribution) and express the variation in the j-th class due to an interaction between an individual of the h-th class with one of the k-th class.

Specific expressions for the C_{hk}^{j} and $T_{[hk]}^{j}$ have been given in [12–16], also depending on the different questions therein tackled. In particular, these expressions contain parameters which can be variously tuned and represent encounter frequencies, saving propensities, average incomes, tax rates, and also tax evasion rates. In any case, the C_{hk}^{j} and $T_{[hk]}^{j}$ are constructed in such a way that the vector $(x_1(t), ..., x_n(t))$ (suitably normalized) is a discretized probability density function at time t. We point out that in the cases studied in [12–16], as also here, the total wealth is assumed to remain constant in time. Growth or variation of the wealth, however, can be considered as well and are in fact the object of ongoing investigation.

Issues like redistribution through means-tested welfare policies, the impact of tax evasion on the income distribution and the effects of taxpayer audits have been treated in [12–16]. In all these cases simulations show that after a sufficiently long time the solution $(x_1(t), ..., x_n(t))$ of the equations reaches an equilibrium state which depends on the total income and on the interaction parameters, but not on the initial distribution. The histogram of its graphical representation gives the "asymptotic" income profile. We emphasize that this equilibrium solution does not prevent single individuals to change class; what is in equilibrium is the statistical distribution. Of course, in real world the situation is not exactly stationary, because parameters are subject to changes.

2 SOME EXAMPLES OF APPLICATION OF THE OUTLINED FRAMEWORK

Our aim in this section is to show how the general framework outlined above can be used to explore the scenarios evolving from different situations. To do that, we must go into some more detail. Referring also to [12–16] for extensive motivations and comments, we fix next the following terms.

Let S denote a unit amount of money, $S = 0.1$ in numerical simulations. We define coefficients $p_{h,k}$ for $h, k = 1, \ldots, n$, expressing the probability that in an encounter between an individual of the h-th and one of the k-th income class, the one who pays is the h-individual: we take $p_{h,k} = min\{r_h, r_k\}/4r_n$, with the exception of the terms $p_{j,j} = r_j/2r_n$ for $j = 2, \ldots, n-1$, $p_{h,1} = r_1/2r_n$ for $h = 2, \ldots, n$, $p_{n,k} = r_k/2r_n$ for $k = 1, \ldots, n-1$, $p_{1,k} = 0$ for $k = 1, \ldots, n$ and $p_{h,n} = 0$ for $h = 1, \ldots, n$. Then, we define the only possibly nonzero elements among the C^j_{hk} as

$$C^i_{i+1,k} = P_{i+1,k}\frac{S(1-T_k)}{r_{i+1}-r_i},$$

$$C^i_{i+1,k} = 1 - P_{k,i}\frac{S(1-T_i)}{r_{i+1}-r_i} - P_{k,i}\frac{S(1-T_k)}{r_i - r_{i-1}},$$

$$C^i_{i-1,k} = P_{k,i-1,}\frac{S(1-T_{i-1})}{r_i - r_{i-1}},$$

where r_j represents the average income of the j-th class. In numerical simulations we take $r_j = 25j$ for $j = 1, \ldots, n$.

We also take each $T^i_{[hk]}(x)$ as a sum $T^i_{[hk]}(x) = \mathrm{U}^i_{[hk]}(x) + V^i_{[hk]}(x)$ where

$$U^i_{[hk]}(x) = \frac{p_{h,k}S_{Tk}}{\sum_{j=1}^{n} w_j x_j}\left(\frac{w_{i-1}x_{i-1}}{r_i - r_{i-1}} - \frac{w_i x_i}{r_{i+1} - r_i}\right) \tag{1}$$

and

$$V^i_{[hk]}(x) = p_{h,k}S_{Tk}\frac{\sum_{j=1}^{n-1} w_j x_j}{\sum_{j=1}^{n} w_j x_j}\left(\frac{\delta_{h,i+1}}{r_h - r_i} - \frac{\delta_{h,i}}{r_h - r_{i-1}}\right), \tag{2}$$

with $\delta_{h,k}$ denoting the *Kronecker delta*, respectively, represent the variation density corresponding to the advancement (due to the benefit of redistribution) from a class to the subsequent one and to the retrocession (due to the payment of taxes) from a class to the preceding one. The coefficients w_j in (1) and (2) denote the weights of a differently distributed welfare. For simplicity, we let them be given by

$$w_j = r_{n+1-j} + \frac{2}{n-1}\gamma\left(j - \frac{n+1}{2}\right)(r_n - r_1), \tag{3}$$

where γ is a parameter taking values in (0, 1/2). The expression in (3) implies that the w_j are decreasing in j (decreasing linearly if r_j is linear in j). By (3), one has $w_1 - w_n = (r_n - r_1)(1 - 2\gamma)$. Thus, if $\gamma = 1/2$, $w_1 = w_n$ holds true and w_j has the same value for each $j = 1, \ldots, n$; in contrast, $w_1 - w_n \to r_n - r_1$ when $\gamma \to 0$.

We take the tax rates relative to the different classes as

$$\tau_j = \tau_{min} + \frac{j-1}{n-1}(\tau_{max} - \tau_{min}),$$

for $j = 1, ..., n$, where τ_{max} and τ_{min} respectively denote the maximum and the minimum tax rate. In numerical simulations, we take $\tau_{min} = 23\%$ and $\tau_{max} = 43\%$ (just because those are the minimum and maximum IRPEF tax rate in our country, Italy). To run simulations, we take here $n = 15$.

Still, before moving to the discussion and comparison of some case studies, we point out that two indicators of interest in the economic context at hand are the *Gini index* and the *tax revenue*. The former, ranging between 0 and 1, provides a measure of the deviation from a perfectly equal distribution. The latter coincides with the amount of money collected by the government through taxation and redistributed as welfare provision. We consider here the tax revenue in the unit time, calculated as

$$TR = \sum_{h=1}^{n}\sum_{k=1}^{n}\sum_{j=1}^{n-1} p_{hk}\tau_k \frac{w_j \hat{x}_j}{\left(\sum_{i=1}^{n} w_i \hat{x}_i\right)} \hat{x}_h \hat{x}_k,$$

where $\hat{x}_i$ denotes the fraction of individuals in the i-th class at equilibrium.

We are now ready to investigate some variations of a specific fiscal policy case.

For the case, to which we refer as to the case C_0, characterized by the parameters above and by $\gamma = 0.5$ we report in the first row of the Fig. 1 the histogram of the asymptotic stationary income distribution corresponding to a given initial condition (and, accordingly, to a given value of the total income). For the asymptotic distribution of C_0 we also calculate and report in Table 1 the Gini index G and the tax revenue in the unit time TR. Then, we calculate and report in Table 1 the Gini index and the tax revenue of the asymptotic stationary income distribution corresponding to the same initial condition as for C_0, for the following partially varied situations:

1. a case C_1 differing from C_0 in the different values of τ_{min} (20% instead of 23%) and τ_{max} (65% instead of 43%);
2. a case C_2 encompassing provision of differentiated welfare for different income classes and in which, specifically, $\gamma = 0.15$ (instead of $\gamma = 0.5$). This means that the poorest class gets approximately four times what the richest class gets;
3. a case C_3 in which to some extent also tax evasion is present (specifically, half of the population is supposed to pay half of the due taxes);
4. a case C_4 in which both tax evasion and some audit procedure are postulated: specifically, one fifth of the population undergoes audit and a fine (to be paid in addition to the evaded amount) is applied to evaders, amounting to 3/4 of the evaded amount.

Our focus is on a comparison between the asymptotic scenarios, as well as the values of G and TR associated to them, relative to the cases C_1, C_2, C_3, C_4 and that of the case C_0. We underline that it is not the parameters chosen that are especially significant. Rather, what is of interest is the possibility enhanced by the framework to explore different situations and establish comparisons between their features.

Table 1: The Gini index G and the tax revenue TR in five different cases.

C0	$G = 0.379672$	$TR = 0.0175466$
C1	$G = 0.351037$	$TR = 0.022186$
C2	$G = 0.354214$	$TR = 0.0183758$
C3	$G = 0.403769$	$TR = 0.0011994$
C4	$G = 0.393106$	$TR = 0.00140869$

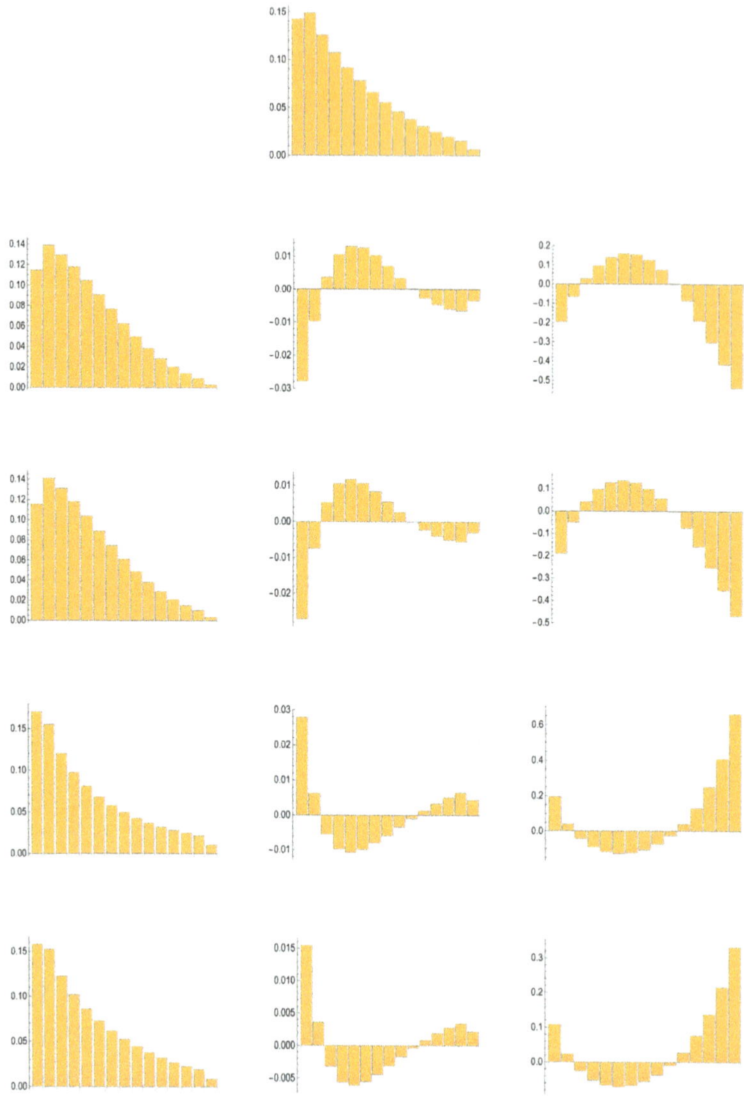

Figure 1: Asymptotic scenarios relative to five different cases.

An illustration of the different asymptotic scenarios is provided by Fig. 1. The rows from the second to the fifth in this figure respectively refer to the cases C_1, C_2, C_3, C_4. The histograms on the left represent the asymptotic stationary income distributions relative to the cases C_j for $j=1, 2, 3, 4$ and evolved from the same initial condition as the distribution in the first row (which is relative to C_0). The central histograms show the difference between the asymptotic distribution of C_j (for $j = 1, 2, 3, 4$) and that of C_0. The histograms on the right display the percentage of the difference between the asymptotic distribution of C_j and that of C_0. Notice that these histograms are differently scaled. What is evident at once is that, compared with the case C_0, both the cases C_1 and C_2 are characterized by a greater number of individuals in the middle classes and a smaller number of individuals in the poorest and the richest classes. In contrast, both for C_3 and C_4, the number of individuals in the poorest, as well as in the richest claases, is larger than in case C_0 and the contrary is true for the number of middle classes individuals. Accordingly, the Gini indices listed in Table 1 show that in the cases C_1 and C_2 economic inequality is smaller than in the case C_0, whereas in the cases C_3 and C_4 it is larger. The distinction between the features of the cases C_1 and C_2 on a side and C_3 and C_4 on another side holds true also for the values of the tax revenue: in the cases C_1 and C_2 these are larger than in the case C_0, in the cases C_3 and C_4 they are smaller. All the tax revenue values in Table 1 are pretty small, but one should keep in mind here that they just refer to a unit time.

In short, the results outlined look reasonably and qualitatively consistent with what one expects. It is not a surprise that welfare and a widening of the gap between the maximum and the minimum tax rates foster a decrease of inequality. Analogously, it is known that tax evasion is harmful to the collectivity. What our approach provides in addition is quantitative information. It may then be used, of course in combination with real data, to get insights as how to fix parameters (tax rates, fraction of audited individuals, fines, ...) to obtain some desired output.

3 CONCLUSIONS AND FURTHER RESEARCH

In this paper, we presented and compared the outputs of some versions of a model which is able to explain the emergence of aggregate-level regularities of economic processes (such as the income distribution of a population) out of a whole of interactions occurring at the individual level. The interactions we took into consideration include money exchanges, payment of taxes, redistribution of benefits, effects of means-tested welfare policies, tax evasion, taxpayer audit and payment of fines. We explored in particular the difference of long run evolutions of the income distribution taking place in the presence of different conditions. Our findings are consistent with the expectation based on empirical observations. This suggests that the model could give some insights towards the adoption of suitable policies. In fact we believe that its main value lies in its adaptability to different problems and data and in its forecasting ability.

There are now several directions, along with further investigation can be carried out. An important aspect concerns for example the introduction into the model of a network structure. We devote the following lines to a preliminary discussion of this issue.

In our discretized kinetic models, individuals are collected in classes according to their income. Likewise, in discretized kinetic models of particle systems in physics, particles are generally collected in discrete energy classes. The discretization allows to convert the integro-differential equations of the Boltzmann kind typical of continuum kinetic theory into a system of coupled ordinary differential equations which are easier to solve numerically and allow a greater degree of flexibility in the definition of the interactions. The number of classes (typically 10–30 in our numerical solutions) can be increased for better accuracy, but in fact it is not necessary to attain large values of n, because in an economic system the continuum limit is not generally needed; on the contrary, most real data on income distributions and taxation are structured according to income classes. On the other hand, it may be desirable to introduce different kinds of heterogeneity. When the model is applied to the description of tax evasion, for instance, each income class can be further subdivided into classes of individuals with different evasion behaviour (total evaders, partial evaders, honest etc.; in the previous section: partial evaders and honest). Another possibility is to consider a subdivision of each income class into "link classes", or classes of individuals with different numbers of economic connections to other individuals. This amounts to the introduction of a network structure, a procedure which has proven to be very fruitful in many dynamical models based on differential equations. It is clear that economic systems possess a network structure, with individuals having preferential interactions along certain links which can be fixed or vary in time.

A first step we propose here is to consider a fixed network structure and employ a statistical representation of the network as introduced by Boguna et al. [17] for the study of epidemics. Within this representation the link density function denoted by P (α) would represent the fraction of all individuals who have α links of economic nature to other individuals and the correlation function denoted by $P(\beta|\alpha)$ would represent the conditional probability for an individual with α links to be connected to an individual with β links. In the case of uncorrelated networks the function P $(\beta|\alpha)$ is just expressed in terms of $P(\beta)$, since $P(\beta/\alpha)=\beta P(\beta)\Big/\sum_{\gamma=1}^{N}\gamma P(\gamma)$, with N denoting the maximum number of links. Otherwise, it is an independent quantity, although the so-called network closure condition $\alpha P(\beta\,|\,\alpha)P(\alpha)=\beta P(\alpha\,|\,\beta)P(\beta), \forall\alpha,\beta=1,\ldots,N$ [18] always must hold true. Further, more technical conditions upon the kinetic equations (compare [15] for details) come from the fact that in models for economic exchanges money is conserved (while in a contagion process, unfortunately, the disease can multiply for free). Broadly speaking, the correlations can be of the assortative or disassortative kind. In the assortative case, individuals with many links are more likely to be connected to other individuals with many links; in the disassortative case, the opposite happens. In the presence of a network structure, the class population variables $x_i(t)$ which give the fraction of individuals belonging to the income class i at time t will become $x_i^{\alpha}(t)$, so that $x_i^{\alpha}(t)$ represents the fraction of individuals who have α links (fixed) and belong at time t to the income class i. In the asymptotic limit $t\to\infty$ one can expect that the income distributions corresponding to different numbers of links will be different. It makes sense, for instance, to compare the asymptotic income distribution $\{\hat{x}_1^1, \hat{x}_2^1, \ldots, \hat{x}_n^1\}$ of individuals with only 1 link to the distribution $\{\hat{x}_1^N, \hat{x}_2^N, \ldots, \hat{x}_n^N\}$ of individuals with N links. One might expect that "more links will make you richer", on the average, and therefore in the latter income distribution the richer classes will be more populated and the average income higher. This actually happens in some cases, but not always, as the outcome of the model generally depends on the structure of the network and in particular on the exponent q in the important case of scale-free networks with link density $P(\alpha) \propto \alpha^{-q}$. Carrying out this program in a practical situation is not straightforward and requires a considerable preliminary work. Indeed, one has to find reasonable values of the elements of the correlation matrix, which have to satisfy the above-mentioned technical conditions too. This will be for us the object of future investigation.

REFERENCES

[1] Wilkinson, R. & Pickett, K., *The Spirit Level. Why Equality is Better for Everyone*, Penguin Books: London, 2010.

[2] Milanovic, B., *The Haves and the Have-Nots: A Brief and Idiosyncratic History of Global Inequality*, Basic Books: New York, 2012.

[3] Stiglitz, J.E., *The Price of Inequality: How Today's Divided Society Endangers Our Future*, W.W. Norton & Company: New York, 2012.

[4] Piketty, T., *Capital in the Twenty-First Century*, Harvard University Press: Cambridge, 2014.

[5] Atkinson, A.B., *Inequality: What Can Be Done?* Harvard University Press: Cambridge, 2015.

[6] Arthur, B., Durlauf, S. & Lane, D.A., Process and the emergence in the economy. In *The Economy as an Evolving Complex System II*, eds. B. Arthur, S. Durlauf & D.A., Lane, Addison-Wesley: Reading MA, 1997.

[7] Kirman, A., *Complex Economics: Individual and Collective Rationality*, Routledge: London, 2010.

[8] Gallegati, M. & Kirman, A., Reconstructing economics: agent based models and complexity. *Complexity Economics,* **1**, pp. 5–31, 2012. http://dx.doi.org/10.7564/12-COEC2

[9] Tesfatsion, L. & Judd, K.L. (eds), *Handbook of Computational Economics, Vol. 2: Agent-based Computational Economics*, North-Holland, Amsterdam, 2006.

[10] Rzevski, G., Complexity as the defining feature of the 21st century. *International Journal of Design & Nature and Ecodynamics,* **10**, pp. 191–198, 2015.
http://dx.doi.org/10.2495/DNE-V10-N3-191-198

[11] Rzevski, G. & Brebbia, C.A. (eds), *Complex Systems: Fundamentals & Applications*, WIT Press: Accurst, 2016.

[12] Bertotti, M.L., Modelling taxation and redistribution: a discrete active particle kinetic approach. *Applied Mathematics Computation,* **217**, pp. 752–762, 2010.
http://dx.doi.org/10.1016/j.amc.2010.06.013

[13] Bertotti, M.L. & Modanese, G., Micro to macro models for income distribution in the absence and in the presence of tax evasion. *Applied Mathematics Computation,* **244**, pp. 836–846, 2014.
http://dx.doi.org/10.1016/j.amc.2014.07.055

[14] Bertotti, M.L. & Modanese, G., Microscopic models for welfare measures addressing a reduction of economic inequality. *Complexity*, in print; article first published online March 2015.
http://dx.doi.org/10.1002/cplx.21669

[15] Bertotti, M.L. & Modanese, G., Discretized kinetic theory on scale-free networks. *EPJ ST*, in print, available on arXiv:1503.03744.

[16] Bertotti. M.L. & Modanese, G., Microscopic models for the study of taxpayer audit effects. *International Journal Modern Physics C*, in print, available on arXiv:1602.08467

[17] Boguna, M., Pastor-Satorras, R. & Vespignani, A., Absence of epidemic threshold in scale-free networks with degree correlations. *Physics Review Letters,* **90**, 2003.

[18] Serrano, M.A., Boguna, M., Pastor-Satorras, R. & Vespignani, A., Correlations in complex networks. In *Large Scale Structure and Dynamics of Complex Networks: from Information Technology to Finance and Natural Sciences*, eds. G. Caldarelli & A. Vespignani, World Scientific: New Yersey, 2007.
http://dx.doi.org/10.1142/9789812771681_0004

REACHING AN ALIGNMENT OF PROGRAMMES OF SOCIAL AND ECONOMIC DEVELOPMENT OF A REGION AND ITS FUEL AND ENERGY BALANCE PLAN

V.V. DOBRODEY, L.D. GITELMAN & M.V. KOZHEVNIKOV
Department of Energy and Industrial Management Systems, Ural Federal University, Russia.

ABSTRACT

Developing methods of aligning such important documents as regional forecasts of social and economic development, and the regional fuel/energy balance plan remains a current challenge determined by the fact that the documents are drafted in isolation, as well as by different levels of detail in projections and a mismatch between target indicators. It is necessary to ensure the priority development of complex regional energy system relative to other industries.

Research indicates the absence of a clear logic of coordinating strategies at the regional level of government in the context of the lack of any formal or even expert assessment of balance between them in terms of resources and applicability of appropriate target indicators.

The paper aims to discuss the possibility of creating a methodology for comprehensive analysis of the dynamics of energy consumption in a region within the framework of modified fuel and energy balance models that would be acceptable for use when information is scarce or limited, when the energy balance plan must be in line with the aggregate indicators of the approved forecast of social and economic development, and both documents have to be monitored for compliance with law. The paper sets out the authors' methodological approach to compiling variants of the fuel and energy balance plan, possible errors and incomplete data in statistics as well as standards for forecasting accuracy.

Keywords: accuracy, dynamics, energy conservation, energy efficiency, forecast, fuel/energy balance plan, information, randomness, regional economy, trends.

1 INTRODUCTION

The International Energy Agency points to the danger posed by the lack of key data in the system of energy information [1]. The reliability of forecasts is particularly critical to such energy deficient regions as Sverdlovsk Oblast of Russia where the suggested research method was trialled. The region consumes a lot of energy, but is short of its own sources of fuel and power.

In several countries, along with Russia, law prohibits the release of statistics for energy consumption by organizations into the public domain as primary data [2]. This makes it difficult or impossible to access quantitative parameters of corporate programmes, fuel and energy consumption data for individual organizations and, as a result, for specific economic activities. Moreover, this raises a doubt about the credibility of the published data and makes it difficult to prevent the deliberate corruption of the data, including for political reasons.

Without a fuel/energy balance forecast it is impossible to substantiate energy, environmental and social and economic development forecasts [3]. The fuel and energy balance plan provides details about the consumption of fuel and energy resources, enables the analysis of energy efficiency components in gross regional product (GRP) and the key products, defines the impact of economic activities and local projects on the dynamics of energy consumption, and identifies trends in target

DOI: 10.2495/DNE-V11-N4-628-634

indicators. Taking into account the information characterizing social and economic development of a region and development programmes for the local fuel and energy sector, energy balance projections make it possible to detect potential risks to the region's energy supply.

The issue of credibility of fuel and energy balance plan as a complex information system is quite significant. The authors do not consider their proposed approach to be the best or highly satisfactory one. Nor do they aspire to find the most accurate and reliable solution to a highly unconventional task – that of designing the best method of providing a detailed elaboration of and alignment between the current and projected parameters of the region's social and economic development forecasts and its fuel and energy balance plan from a number of available options. The authors base their conclusions on their calculations and empirical framework of the fuel and energy balance plans that they drafted and adjusted between 2007 and 2013, as well as available statistics for 2004–14 and the evolving methods of forming fuel and energy balance plans that they [4] and other scholars [5–9] have used. They believe that it will be useful to discuss these research undertakings to gradually reduce the significance of information and methodological problems.

2 INFORMATION AND METHODOLOGICAL ASPECTS OF FORMING A FUEL AND ENERGY BALANCE PLAN

Available statistics do not allow for a detailed analysis of distribution of fuel and energy resources across the full list of economic uses. For example, the structure of electricity consumption is only partially described in open sources. Rounding of data in the information database of the Federal State Statistics Service (Rosstat) creates additional errors that erode value added in some economic activities to zero. As a result, a problem arises of comparing economic activities by the energy intensity of value added.

The core idea of the author's approach to forming fuel and energy balance plans is to use the Rosstat-provided detailed GRP broken down by economic activity, to analyse the dynamics of the obtained vectors between 2004 and 2013 and to forecast value added and energy intensity for each economic activity.

Given the above-mentioned issues, the method applied by the authors for decomposing GRP in the previous and current years and for making projections takes into account the following structural features of the available information base:

- estimated GRP for the current year, as computed in the regional social and economic development forecast. The available data for the past year include the total heat and electricity consumption by small enterprises, households as well as by medium-sized and large companies (for a reduced list of economic activities);
- detailed data on fuel consumption by households, medium-sized and large enterprises are available;
- information is available about the structure of turnover for basic economic activities and the characteristics of the fuel and energy balance plans for the period starting in 2007 and the indicators of GRP for the period starting in 2004;
- a volume of historical information starting from 2007 has been obtained on the trends in the output of key products, individual and aggregate energy intensity and the dynamics of the target indicators of energy efficiency and energy conservation.

We should also note fatal asymmetry in procedures of matching the parameters of fuel and energy balance and forecasts of social and economic development. The approved scenario indicators of social and economic development, in addition to the energy consumption characteristics, could not be the subject to adjustments in coordination procedures. Therefore, the reliability of approved fuel and energy balance depends on the quality of the region development forecast. In practice authors have also formed an additional fuel and energy balance embodiment, based on their own analysis and estimates of the dynamics of the certain economic activities development.

3 CALCULATION AND ANALYSIS

Given the above, the following calculations are performed.

To add structure to the formally assessed GRP in the previous and current years, historical information is analysed. Relationship is assessed between the GRP at basic prices (GRP_c) and the gross turnover (***TURN***) of medium-sized and large companies for the period between 2004 and 2013. An analysis of correlations between a time series of GDP and the turnover in the region and between value added by activity (VAL_{dcj}), $j = \overline{1,15}$ and turnover ($TURN_j$) shows that the correlation coefficient in the first case exceeds 0.87. Obviously, the correlation coefficient is partially explained by the fact that both indicators depend on a common factor – that of the positive economic environment. Regression analysis of the dependence $GRP_c = F(TURN)+\varepsilon$ confirms, to a high degree of approximation (normalized R^2 = 0.86), a relationship between the indicators: $GRP_c = Const+K*TURN+\varepsilon$, with the confidence interval containing actual and projected GRP volumes for the past 2 years. A similar correlation coefficient for individual economic activities is also quite high, so the merit of assessing the GRP and its structure on the basis of statistical relationships between the target indicators and the actual turnover data is determined by more than their definitions. It is useful to compare the confidence interval ($\underline{GRPc}$, $\overline{GRPc}$) with the projected GRP for the current year from the officially approved forecast of social and economic development.

The basic method of decomposing GRP that is based on the ‘value added-to-turnover’ ratio for the current year compared to that for the prior year and the subsequent ‘normalization’ of the indicators to ensure GRP balance is accurate enough, provided that the ratio exhibits high inertia during the last few years of the retrospective period. It is, therefore, advisable to use regression and time trend analysis for assessing the dependence of value added on turnover.

To make a conversion from basic prices to constant prices and to calculate chain-linked indices of physical volume, value-added deflators need to be computed for each type of economic activity. Information available for analysis includes actual deflators for the considered period and operational statistics. Regression analysis takes into account projected GRP at current basic prices from the officially approved forecast of social and economic development and the index of the physical volume of GRP for each scenario. A settlement procedure is implemented to determine a non-contradictory structure of the GRP.

It is impossible to create a method that would produce the most accurate prediction interval because of incomplete information. Relying on the officially approved projections of the region’s social and economic development is not entirely reasonable because of the abridged structure of GRP. Moreover, current scenarios might differ significantly from observed trends. Our calculations show, though, that statistical analysis methods can ensure the consistency and balance of estimates as well as their relative accuracy.

Breaking down GRP by economic activity makes it possible to calculate the indices of trends in energy consumption, energy conservation and energy efficiency for the base year versus the prior year. At the same time, chain-linked and aggregate deflators (against 2007) are refined for the period starting 2004.

It is advisable to apply rigorous (formal) methods for calculating GRP parameters. Assume that t is the year index (in $t \in \overline{2004; 2013}$), b is the index of the first (base) year, GRP_{tc} is the GRP at current basic prices, $\boldsymbol{GRP_{bc}}$ is the base year's GRP at current basic prices; $\boldsymbol{GRP_{ts}}$ и $\boldsymbol{GRP_{bs}}$ are similar indicators at constant prices. The following equation emerges for the baseline year $\boldsymbol{GRP_{bs} = GRP_{bc}}$. In the Rosstat data the first year (2004) is considered to be the base year and chain-linked indices of the physical volume of GRP (at prior year's prices) ***αts*** are shown. For ***t*** = ***b*** we have $a_{bs} = 1$ (or 100%); consequently, $\boldsymbol{GRP_{ts}}$ – the physical volume of GRP in the year ***t*** at constant prices of the first year (***b*** = 2004) – is defined by the equation

$$GRP_{ts} = \prod_{t=b}^{t} a_{ts} \times GRP_{bc} \tag{1}$$

or by the sequence of operations

$$GRP_{ts} = GRP_{(t-1)s} \times a_{ts}; t = \overline{2005;2013} \tag{2}$$

Chain-lined deflator indices $\boldsymbol{\beta_{th}}$ for the year 2004 are computed as follows:

$$\beta_{th} = \frac{GRP_{tc}}{GRP_{(t-1)c} / a_{ts}}; t = \overline{2005;2013} \tag{3}$$

Aggregate deflator indices $\boldsymbol{\beta_{t\,int\,b}}$ for the year 2004 are derived as:

$$\beta_{t\,int\,b} = \frac{GRP_{tc}}{GRP_{bs}}; t = \overline{2005;2013} \tag{4}$$

or by chain-linking:

$$\beta_{t\,int\,b} = \prod_{t=b}^{t} \beta_{th}; t = \overline{2005;2013} \tag{5}$$

Russian regulations for calculating energy efficiency and energy conservation accept 2007 as the base year. The formula for computing aggregate deflators to the new base year (b_1 = 2007) in our case (when aggregate deflators relative to the first year of the period are already known) is given by

$$\beta_{t\,int\,b_1} = \frac{\beta_{t\,int\,b}}{\beta_{b_1\,int\,b}}; t = \overline{2004;2013} \tag{6}$$

Similar operations are performed for each type of economic activity for the entire time period, including the current year when information is incomplete. The structural parameters of GRP by economic activity, including the evolution of the chain-linked and aggregate added value deflators adjusted to the start of the period and the baseline year of 2004, are refined in line with the scenarios. This information is essential for verifying the accuracy of calculations. More specifically, a check is needed on the GRP ratio at current and constant prices of the base year.

$$GRP_{tc} = \sum_j AddedValue_{tcj}; GRP_{ts} = \sum_j AddedValue_{tsj}; j = \overline{1;15} \tag{7}$$

where $\boldsymbol{AddedValue_{tcj}}$; $\boldsymbol{AddedValue_{tsj}}$, $j = \overline{1;15}$ is value added at current and adjusted prices by economic activity. The calculations of deflators are also verified for accuracy and error.

$$GRP_{ts} = GRP_{tc} \times \bar{\beta}_{t\,int\,b}; t = \overline{2004;2013} \tag{8}$$

$\bar{\beta}_{t\,int\,b}$ is chain-linked GRP deflator, computed as a mean for all types of economic activity.

$$\bar{\beta}_{t\,int\,b} = \sum_j \beta_{t\,int\,b_j} \times \delta_{tj}; j = \overline{1;15} \tag{9}$$

where $\delta_{tj} = AddedValue_{tcj} / GRP_{tc}$; $t = \overline{2004; 2013}$; $j = \overline{1;15}$ is the share of the economic activity in the region's GRP (at current prices).

It is essential to verify the Rosstat data to account for rounding errors in the structure of GRP by economic activity. Because of the rounded shares of economic activities in GRP, the sum total of value added at current prices does not coincide with the corresponding GRP volumes for the region. The rounding of physical volume indices for economic activities also upsets the balance of GRP reporting by economic activity and value added. The adjustment procedures for current and constant prices are identical for each year of the historical period. The following formal description of data adjustment at 2007 constant prices implies that appropriate adjustments in current prices have already been made.

We shall use the following notation: γ_{jt} are the indices of the physical volume of value added by economic activity j = 1, 2, …, 15; $t = \overline{2008,2012}$ is the year index; l_{jt} are the required adjustment factors (their values are found in a neighbourhood of 1); $l_{jt} * \gamma_{jt}$ are the target adjustment results; d_{jt} are the initial value-added volumes , $d_{jt} = d_{jt-1} * \gamma_{jt}$, for the base year t = 2007 $d_{j\,2007}$ is the known constant value at current prices.

Criteria in the procedures are written as:

$$F_t = \sum_{j=1}^{j=15} (\lambda_{jt} \times \gamma_{jt} - \gamma_{jt})^2 \rightarrow min, t = \overline{2008;2012} \tag{10}$$

Variables to be optimized are λ_{jt}, j = 1, 2, …, 15.

The restriction is that balancing conditions (11) should be met for each year in the historical period (the procedure is unnecessary for the base year):

$$\sum_{j=1}^{j=15} d_{jt} = \sum_{j=1}^{j=15} \lambda_{jt} \times \gamma_{jt} \times d_{jt-1} = GRP_{ts}, t = \overline{2008;2012} \tag{11}$$

It provided that all built-in associations among value-added measures in MS Excel calculations are observed.

The historical data are only adjusted once or when updated by Rosstat. The operations are easily automated using MS Excel tools, and the method can also be used for normalizing (balancing) the structural parameters of GRP when assessing its projected structure. The approach imitates the maximum likelihood method (the rounding of yielded results does not contradict the Rosstat data).

A general pattern for producing fuel and energy balance forecasts combines two key approaches:

- the balance plan should reflect the regional production profile and the energy intensity of the area's key products and services (production section of the balance plan);
- the balance plan should indicate the structure of energy consumption by economic activity, making it possible to take into account the sectoral scope of the balance plan.

A forecasting method based on product aggregation only proves ineffective because many areas of economic activity such as health care, education, financial services, public administration, science and culture are not included in statistics in a comprehensive enough way. Such information is fragmented and is prone to change.

Statistical estimates of added value indicators were normalized in line with known projected GRP values for the region, and correlation between formally approved scenarios and confidence interval estimates was analysed. The share of each type of economic activity in the region's GRP was subsequently determined.

Factor regression analysis or methods of time series analysis were used to produce a forecast of value added at constant prices for the year 2007 and the energy intensity of each type of economic activity. The most suitable regression equations were selected according to the approximation criterion (mostly, power laws, logarithmic dependence and autocorrelation) that did not contradict the expert logic of the region's economic development. Expert stabilization factors or exponential smoothing were applied to time series data that exhibited spikes, breaks or drops closer to the end of the historical period. The method is to be further elaborated to include, among other things, price trends. To forecast fuel and energy consumption by households, regression was used with the 'time' variable and the 'floor area per person' variable; to produce a forecast for losses of gas, power and heat in transmission and for self-consumption, regressions were used with the 'time' variable and 'distribution' variable.

Comparing forecast trends in GRP energy intensity with target values of appropriate indicators makes it possible to assess the attainability of the targets. For example, total energy consumption $\Delta \boldsymbol{EnergCost_t}$ for any year $\boldsymbol{t}$ during the forecast period and the impact of structural shifts in GRP $\delta \boldsymbol{EnergCost_t}$ may be assessed with the formulas (12) and (13).

$$\bar{\Delta} EnergCost_t = \sum_{j=\overline{1;15}} AddedValue_{tsj} \times (EnergCost_{2007j} - EnergCost_{tj}) \tag{12}$$

$$\delta EnergCost_t = \sum_{j=\overline{1;15}} \left(\delta_{tj} - \delta_{2007j}\right) \times EnergCost_{2007j}; t = \overline{2013;2017} \tag{13}$$

4 CONCLUSION

Estimates have shown that the found confidence intervals for energy consumption indicators considerably exceed their variation when observed retrospectively. That is why, the officially approved 'optimistic' forecasts of GRP and its structure can be interpreted as finding the midrange. The conditional probability of formal estimates falling within the fixed limits of projected variation depends on the degree of bias in forecasts of social and economic development.

Unlike the guidelines of the IEA, Eurostat and the UN, the blueprint for fuel and energy balance plans in the regions of Russia breaks down energy consumption in industry, agriculture, transport and housing according to products, services, processes and types of work. The fuel and energy balance concept used by the authors distinguishes a set of key products that determine the industrial profile of the region, making it possible to reflect the physical indicators of energy efficiency that are not influenced by price distortions in GRP and added value [10].

Incomplete and imprecise information is only partly to blame for increasing the uncertainty and unreliability of fuel and energy balance plans. But it is information-related problems that create fundamental obstacles to developing and improving methods that could produce reliable forecasts.

It is important that the information problems should be addressed at the legislative level. For example, data forms should be designed that would match up with the types and structure of fuel and energy balance plans in regions and municipalities; deadlines for forming fuel and energy balance plans and their publication should be agreed upon.

ACKNOWLEDGEMENT

The work was supported by Act 211 Government of the Russian Federation, contract № 02. A03.21.0006.

REFERENCES

[1] World Energy Outlook, available at http://www.iea.org/bookshop/700-World_Energy_Outlook_2015

[2] Federal law № 282, On official statistical accounting and state statistics system in the Russian Federation of 29 Nov 2007 (edited 23 Jul 2013), [in Russian], available at http://www.consultant.ru/document/cons_doc_LAW_72844/

[3] Federal law № 172, On strategic planning in the Russian Federation of 28 Jun 2014, [in Russian], available at http://www.rg.ru/2014/07/03/strategia-dok.html

[4] Danilov, N., Silin, V., Dobrodey, V. & Popov, V., Energy problems of the rational use of the economic potential of the region. *WIT Transactions on Ecology and the Environment*, **190**, pp. 419–423, 2014.
http://dx.doi.org/10.2495/EQ140401

[5] Fuel and energy balance plans, [in Russian], available at http://www.eriras.ru/data/4/rus.

[6] Bashmakov, I.A., Siniak, YV. & Makarov, A.A., *Costs and Benefits of Low-Carbon Economy and Social Transformation in Russia*, [in Russian], Centre for Effective Energy Use: Moscow, 178, 2014.

[7] Bashmakov, I. & Myshak, A., Factor analysis of evolution of russian energy efficiency: metodology and outcomes, [in Russian]. *Problems of Economics*, **10**, pp. 117–131, 2012.

[8] Brandoni, C. & Polonara, F., The role of municipal energy planning in the regional energy-planning process. *Energy*, **48**, pp. 323–338. 2012.
http://dx.doi.org/10.1016/j.energy.2012.06.061

[9] Sampaio, H.C., Dias, R.A. & Balestieri, J., Sustainable urban energy planning: the case study of a tropical city. *Applied Energy*, **104**, pp. 924–935, 2013.
http://dx.doi.org/10.1016/j.apenergy.2012.12.022

[10] *Tracking Industrial Energy Efficiency and CO_2 Emissions*, OECD/IEA: Paris, 321, 2007.

AGENT-BASED MODELLING OF OFFSHORE UPSTREAM PETROLEUM LOGISTICS

P. ENGELSETH & T.M. PETTERSEN
Department of Logistics, Molde University College, Norway.

ABSTRACT

Faced with dwindling oil prices, the petroleum industry now needs to economize its operations. A case study of upstream petroleum logistics, the management of supplies to and from petroleum platforms and rigs, is used to describe current logistics and supply chain management (SCM) challenges facing these operations in Norway. The provided case narrative reveals how an integrated planning system used to co-ordinate operations is proven relatively difficult to implement and use. As alternative research approach agent-based modelling (ABM) is applied and discussed in association with actor network theory in a SCM business-functional setting to theoretically ground use of ABM as methodology. An empirically-grounded conceptual model, the first stage of ABM methodology, is created for petroleum logistics. Findings from this first stage of inquiry also suggest how and why ABM is applicable in petroleum logistics and SCM.
Keywords: actor network theory, agent-based modelling, petroleum logistics, supply chain management.

1 THE BUSINESS CHALLENGES AND THE RESEARCH ISSUE

Due to the fall in oil prices, the Norwegian petroleum industry must economize what is deemed too costly offshore operations. Upstream offshore petroleum logistics operations secure platform operations and are associated with risk; shutting down platform oil production has severe consequences. Keeping the platform in production continuously is therefore a core aim of petroleum logistics (PL). To secure its production the platform is continuously maintained and upgraded. Each platform has, however, a unique design. This implies that service associated with maintaining and upgrading the platform needs to be engineered. PL is modelled as a network of agents in Fig. 1.

Another platform activity, drilling is vital to secure oil production and since it implies interaction with nature it is associated with uncertainty. Other constraints include platform capacity to hold people and goods. In addition, weather impacts on transport by helicopter and platform supply vessels. Different operations cannot normally be carried out simultaneously. In cases of delays or early completion of preceding services, these following services need to be re-planned. This indicates that petroleum logistics may easily be described as a complex dynamic system (CAS). To co-ordinate different offshore platform activities, the petroleum company has developed an information technology (IT)-based integrated activity planning system that has proven challenging to use. Agent-based modelling (ABM) is proposed to develop PL a case study of petroleum logistics of an offshore production platform in North-Western Norway. A conceptual model of how networking takes place embedded in various risks to secure production on an oil platform in Norway is developed. The conceptual model provides empirically-grounded understanding of operational interdependencies as perceived by the studied network agents to found later development of a computer-based model for simulation; developing understanding on how people interact to carry out PL in a complex and dynamic network context cloaked in uncertainty.

DOI: 10.2495/DNE-V11-N4-635-643

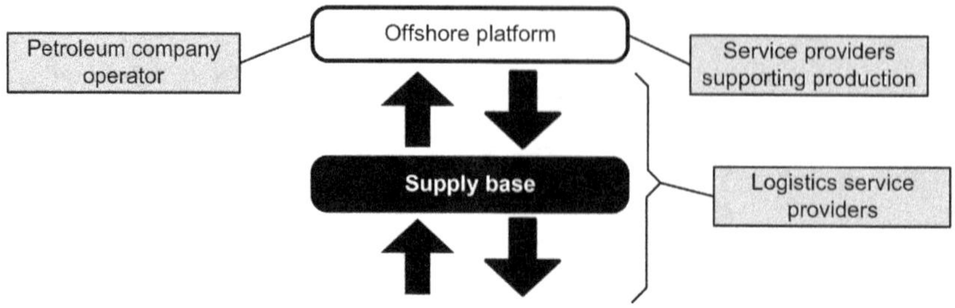

Figure 1: The petroleum logistics supply network.

2 LITERATURE REVIEW

The "as-is" mental business model scenario, including current economic challenges, triggers a wish to change supply operations. Surana *et al.* [1] state in relation to the "real world" that: "Many natural systems and increasingly many artificial systems as well, are characterized by apparently complex behaviours that arise as the result of non-linear spatio-temporal interactions among a large number of components or "subsystems". A traditional planning paradigm (mental model) will not necessarily help to improve the functioning of petroleum logistics since inherently such a system is both complex and dynamic. Loasby [2] points out with the traditional planning paradigm in mind the following: "Close attention to recipes does not ensure excellent results, and even detailed manuals often make crucial, if unconscious, assumptions about the user's skills. It is notorious that standard operating procedure, followed to the letter, produce unsatisfactory levels of performance…" A CAS is a dynamic network of many agents interacting with other agents [3]. CAS is especially fit to handle complex scenarios with a high level of uncertainty. Surana *et al.* [1] discern CAS, from other forms of complex systems as being systems that accommodate for human decision making; a decentralized and inherently uncontrollable system. Occurrences of coherent behaviour in the system arise from competition and cooperation among the agents within the system. In PL key types of agents are (1) oil company customer, (2) service providers to the platform, and (3) logistics service providers. Based on considerations regarding effects of micro-level decisions and resulting behaviour CAS is chosen as *overall* research approach [4]. According to a CAS approach, these studied systems are modelled as collections of autonomous agents, interacting with their environment and exchanging information with other agents. Such a system is not externally coordinated to produce some output; it is not driven by "plan".

CAS can be simulated using agent-based models [5]. Based on work by Boero and Squazzoni [6], Gilbert [5] discerns three levels of abstraction for agent-based models, requiring different levels of empirical validity: (1) abstract models, aiming to demonstrate theoretical properties of social processes lying behind many areas of social life; (2) middle-range models, aiming to describe characteristics of a social phenomenon sufficiently general that conclusion can be widely applied and (3) facsimile models called *case-based models* by Boero and Squazzoni [6], providing as exactly as possible a reproduction of a particular target system, often with the intention to make predictions of future states, consequences of institutional changes, etc.

Agent-based models have been applied to supply chains, for example, in studies aiming to control the bullwhip effect [7]. In agent-based models information can be more richly structured than just the value of a continuous real variable. This applies to both the information representation in the agents' "minds" and the information exchange with other agents to study the effects of information sharing in organisations; used to test the design and test production processes and information systems supporting these processes [8]. Sierhuis *et al.* [9] describe an approach based on the "Brahms" software system to simulate human behaviour in a particular organisation structure, in which information systems can be implemented and their performance and effects can be simulated and tested under different scenarios of human action. This form of approach holds promises for the development of supply chain information systems, which are hard to test in business practice involving many actors at different stages in the supply chain. Dessalles *et al.* [10] discern agents implementing different levels of cognition: (1) Reactive agents, that have no cognitive rationality but only respond to environmental events; (2) Behavioural agents, that can modify their behaviours according to observed pay-offs; (3) Epistemic agents, that are aware of their behaviour and may anticipate the effects, possibly combined with social intelligence, extending the awareness to the effect of collective action. In order to interpret and use the information, agents in simulated CASs must have cognitive capabilities when ABM is used to develop petroleum logistics.

PL is embedded in supply chain management (SCM) strategy. Building on established definitions of logistics and SCM created by the Council of Supply Chain Management Professionals (CSCMP), the supply chain is modelled by Cooper *et al.* [11] as consisting of a set of more or less integrated business processes in a complex structure. The purpose of SCM is *viewed* in line with the CSCMP SCM definition; that is, as associated with integration within and between firms in a supply-oriented network context facilitating operations coordination. Cooper *et al.* [11] model, inspired in part by Alderson's [12] marketing channels literature, takes an end-to-end perspective focusing on the logistics of goods supply. Building on this model, Lambert *et al.* [13] provide, based on Cooper *et al.* [11], a framework for SCM implementation. This framework consists of three constructs: (1) supply chain business processes, (2) SCM components, and (3) supply network structure; applicable to ABM where agents represent the focus; processes the inputs and outputs of agent decision making, and structure as.

ABM and SCM should be grounded in social theory. Terming people in supply networks as "agents" represents social ordering [14]; to comprehend and create discourse on, for example, petroleum logistics. These patterns that this social ordering creates in networks may be considered as expression of reflexive as well as self-reflexive "modernistic" activities in a local setting where production takes place [15]. Following Foucault [16], power relations are "...intentional but non-subjective. If in fact they are intelligible, this is not because they are the effect of another 'instance' that explains them, but rather because they are imbued through and through with calculation: there is no power that is exercised without a series of aims an objectives…" [16]. Typical of this line of thought is: "Truth only knows and recognizes itself in a context – a context it creates for itself" [17]. Within sociology symbolic interactionism and actor network theory (ANT) provide guidance as to societally anchoring ABM to e.g. logistics systems; that context is an un-isolated complex and dynamic *local* form. This places focus on "…that there is ordering rather than order" [18]. A main distinction between the forerunning symbolic interactionism [19] and ANT, is that this approach includes artefacts as agents. ANT describes a form of inter-objectivity where the local setting is pertinent in understanding action, an arena for framing and summing up by human agents. Within ANT thinking Law [14] states that agents may be treated as relational effects. Second agents may be considered as ordering struggles; they are dynamic. Agency is considered as an achievement. Finally, the world is considered a set of more or less related bits and pieces having no social order. Within ANT agents may be completely or in-part non-human (artefacts); in practice what we normally consider machines with decision-making properties, as well as machines assisting people in decision-making. The agent can be a conglomerate of interacting resources; agency then rendered a process itself. This ordering is contingent, there is a layer of strategy within which reflexivity and self-reflexivity takes place. This strategy may be understood as discourse. In line with Foucault [20], reconstructions of material conditions or knowledge including exhibiting power relations in a supply network. Discourses convey meaning through interaction. PL viewed as "economy" includes discourse is understood as; a social pattern susceptible to change.

Within SCM Lambert *et al.* [13] discern between (1) physical and technical management, and (2) managerial and behavioural management. ABM brings the second aspect of SCM into focus, with physical and technical components rendered contextual. Given the inherent complexity of these technical subsystems this implies a need for engineering (designing) these systems. In this approach, however, these technicalities are rendered sub-systems. There is no vison of systemic optimizing of the PL economy. This implies need for mechanisms that listen to experiences from the "factory floor", a bottom up view of strategy-making, typically found in Lean genealogy [21]. In accordance with Hammervoll [22], economy may be viewed as comprising of two forms of complementary parts; (1) production and (2) exchange. ABM places focus on the agent and therefore on "exchange" represented by information sharing supporting production. "Production" encompasses pooling resources creating deliverables; a goods, information and services mix.

The transformation of these deliverable resources is what constitutes a market offering, what Penrose [23] terms as "service" (singular form). "Service" represents in this view appreciation that the core purpose of business is to achieve customer value. Although suppliers may be clearly aware of how this may be produced, the final test is whether customers experience satisfaction. Logistics and SCM is based on a tradition that is "goods dominant [24]. A greater degree of customer orientation is called for in order to improve the competitive standing in the marketplace. Clearly production and customer value are interrelated, and the organisation of service provision as "value co-creation" [24,25] represents a gateway to improved competitive standing in the marketplace. The exchange economy can be viewed in relation to SCM; seen as embedded in management practices associated with (1) management methods, (2) power and leadership structure, (3) risk and reward structure and (4) culture and attitude [11]. These are features that are taken into consideration when applying ABM. This implies a clearly micro-level approach to these four concerns.

3 METHOD

The empirical evidence the conceptual modelling process rests upon is a case study of a petroleum logistics system. Following Miles and Huberman [26] and Yin [27], the case study research strategy is used when: (1) they make it possible to answer "how" and "why" research questions, (2) researchers cannot manipulate the behaviour involved during the research process and (3) researchers can seek a picture of the context the phenomenon is embedded in. A single case approach to evoke details at a micro-level [28], demanded in ABM. Taylor and Fearne [29], Fernie and Thorpe [30] and Holweg and Pil [31] state that this method is appropriate for describing actors, structure and agency relations taking place through social interaction. Research was carried out by an executive of the petroleum company described in the case. This involved mainly applying own experiences. To secure objectivity, and more in-depth understanding of management and processes a few informal semi-structured interviews were conducted. This study highlighted an ongoing quest of the petroleum producer to implement a computer-supported integrated activity planning system.

According to Croom *et al.* [32], the context of supply-related interaction can be described at complementary levels of analysis; as "dyads", "chain" or "network". Supply chains indicate a normative quest to integrate different firms following a linear flow of goods. Network encompasses innovation; the potential to navigate trading with different business partners as well as the feature of interacting with multiple suppliers and multiple customers simultaneously. Interaction takes place in the context of the dyad governing customer and supplier interaction [33]. Agency may accordingly be analysed in the context of (1) relationship, (2) chain or (3) network providing three distinct, but interrelated and complementary, forms of supply-related economic analysis. Supply network modelling and simulation can benefit from the following features of agent-based models. Agent-based supply network models can be built in facsimile, furthermore, modelling of cognitive capabilities, decision making and action of agents can build on empirically validated models formulated in the social sciences. The effects of cooperation and competition in networks are simulated from the bottom up with no assumption about macro-level relations have to be made. Agent-based simulations will help to improve the understanding of the processes resulting in the behaviour and performance of supply networks as systems [34]. The explicit modelling of information exchange embedded in discourse between supply chain actors enables the application of an agent-based supply network simulation as a test-bed for supply chain information. Janssen and Ostrom [35] argue that given empirical problems with data collection and the explicit inclusion of cognitive, institutional and social processes, macro-level statistical performance is not sufficient for validation of an agent-based model. Additional criteria are, according to Janssen and Ostrom [35]: (1) plausibility of the model, given the understanding of processes, (2) understanding why a model performs well, (3) better understanding of empirical observations gained through the model, and (4) stakeholder validation of model behaviour. A soft system modelling process is proposed by Checkland [36,37] that involves the following seven stages: (1) Entering the problem situation; (2) expressing the problem situation; (3) formulating root definitions of relevant systems; (4) building conceptual models of human activity systems; (5) comparing the models with the real world; (6) defining changes that are desirable and feasible; and (7) taking action to improve the real world situation. This analysis clarifies the purpose of modelling. Three basic criteria useful in every situation that is modelled regarding agency are: (1) efficacy which indicates, whether the transformation provides the intended outcome, (2) efficiency which indicates, whether the least possible amount of resources is being used to implement the transformation and (3) effectiveness which indicates whether the transformation helps to realize a more long-term goal. This helps to understand the root cause of a problem associated with the research topic. Fig. 2 describes this cyclic process of modelling, simulation and business process development:

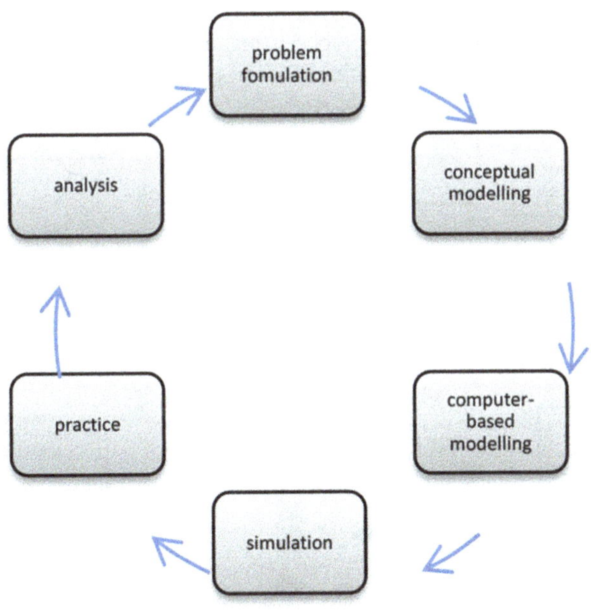

Figure 2: Agent-based modelling as a continuous cyclic development process.

This is a form of "plan-do-check-act" circle found in "total quality management" (TQM) and widespread in lean thinking [38], developed by Shewart [39] and popularised by Deming [40]. In this study, we limit investigation to first create a conceptual model.

4 CASE DESCRIPTION

Annually about 20,000 tons of goods are transported mainly by sea to the platform. The supply base on land handles these goods prior to shipment. Materials have relatively long lead times demanding. In 2014, 6,800 people were transported by helicopter to and from the studied platform. Shipping includes a reverse flow of wastes and tools. It is essential that the equipment is in place when specialists arrive at the platform. The platform has a berth capacity of 130 places. There is rift about capacity use between different operators. Activities are therefore carefully coordinated with cargo deck space, helicopter capacity, berth capacity and lifeboat capacity. Another obstacle is the unruly weather. The platform is managed from the land-based headquarters of the oil company. One of the main types of activities is the maintenance and modification operations (MMO). An integrated activity planning (IAP) system is in place to manage these MMOs. AIP aims to co-ordinate plans from different actors to create common plan applying a planning frame of 90 days. While the oil company applies a 90-day plan horizon, the MMO service provider operates its own planning system with a 6-week plan horizon creating a major obstacle in information integration. The MMO service provider internally works with several plan horizons, where project plans are broken down into work packages and "plot" into the operational plan (6 week plan). This operational plan is then fed as input to the oil company's IAP.

Conflict is generated by lack of capacity. The currently used information systems do not give an accurate picture of the planned scope of activities making it challenging to accurately visualize the oil company's own activities in the network. Sharing real-time information is also difficult. The scope of work in IAP plan is visualized through the number of hours an activity is estimated to take a start and an end date. Personnel demand is not estimated which leads not considering capacity on platforms. Information exchanged uses ERP systems with different interfaces; manual operations to gather information The oil company has developed a manual Excel document entitled "Staffing Plan": a time-consuming and error prone operation. This has created an auxiliary Heli Booking sys-

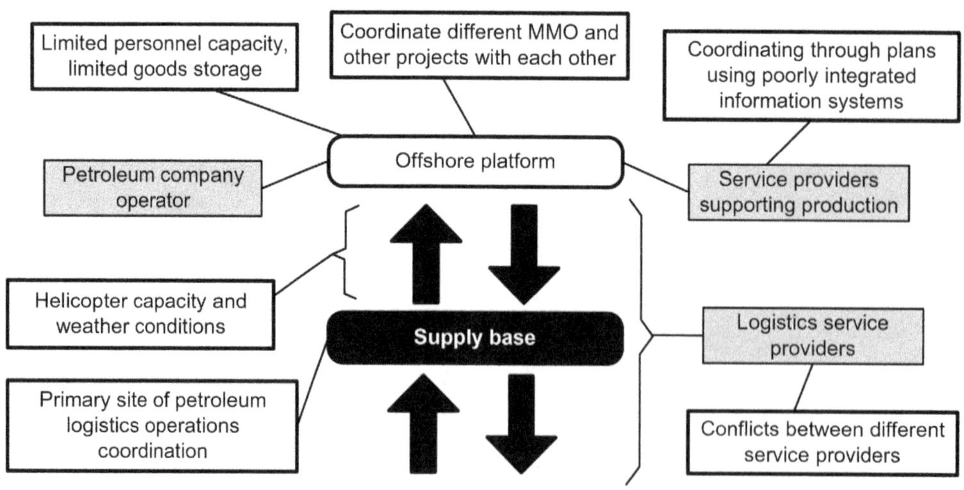

Figure 3: The preliminary conceptual model.

tem to manage helicopter transport. In practice, sorting out conflicts due to limited helicopter capacity is a task involving revealing who is "guilty" of over-booking. The IAP plan that supports the goods flow is often overridden by the management since needed information does not reach the right decision makers on-time. It is also challenging to gather all parties to jointly plan due to various ERP systems used by different companies including variation in manual routines.

5 MODELLING

Modelling takes Fig. 1 as the point of departure. The conceptual model illustrated in Fig. 3 includes the business concerns described in the case narrative:

The supply base also has a key role as the site of operations, and therefore represents a node (hierarchy) when it comes to goods flow operations management, goods identification and updating information about this flow.

The case reveals complex human-machine interaction in two aspects; (1) people managing through the use of information technology software, and (2) people operating the flows of goods to and from the platform. The case description reveals how petroleum logistics involves complex network organisation embedded in a *sea* of uncertainty. Due to the revealed complexity and the dynamics needed to handle uncertainty, ABM should provide ample reason to apply this as development strategy. This also requires parting with the quest of planning in miniscule detail. It is proposed to develop plans that provide a guiding context. It is emphasized that the quest could be pictured as to sandwich such a plan between the petroleum logistics discourse (that is itself dynamic) and the ABM provided micro-level operational structure. This implies, in line with Fig. 2, that ABM is a dynamic enacted multiplicity; agency a form of Kaizen where established lean thinking may guide practice. Value co-creation ascends as a key feature of this proposed ABM-founded petroleum logistics development system involving an ANT-influenced "inter-objective" view of interaction; a systemic multiplicity of reality, enactment and choice. With business challenges described and discussed in light of ABM thinking founded in ANT thinking, the next step in this research is to create a computer-based ABM simulation model to initiate the *Kaizen* wheel of learning founded on this ontological view.

6 CONCLUDING REMARKS

This study is limited to investigating an initial phase of ABM; generating the conceptual model. This study provides accordingly foundation for future studies. As shown in this paper, this implies active process improvement. Most immediate is the next stage to develop a computer-based model to simulate the processes described in the conceptual model. However, given the vision to establish simulation as a Kaizen wheel of learning implies implementing simulation as a lean thinking tool [38]. This involves taking into consideration how ABM best may fit into a lean organisation applying ANT. Further research is proposed following a single case research strategy using ABM and SCM founded on ANT, also facilitating computer-based simulation.

REFERENCES

[1] Surana, A., Kumara, S., Greaves, M. & Raghavan, U.N., Supply-chain networks: a complex adaptive systems perspective. *International Journal of Production Research*, **43**(20), pp. 4235–4265, 2005.

[2] Loasby, B.J., The concept of capabilities. *In Economic Organization, Capabilities and Co-ordination, Essays in Honour of G.B. Richardson*, eds. N.J. Foss & B.J. Loasby, Routledge: London, 1998.

[3] Mitchell Waldrop, M., *Complexity: The Emerging Science at the Edge of Order and Chaos*, Penguin: Harmondsworth, UK, 1994.
[4] Miller, J.H. & Page, S.E., *Complex Adaptive Systems*, Princeton University Press: Princeton, 2007.
[5] Gilbert, N., *Agent-Based Models*, London: Sage, 2007.
[6] Boero, R. & Squazzoni, F., Does empirical embeddedness matter? methodological issues on agent-based models for analytical social science. *Journal of Artificial Societies and Social Simulation*, **8**(4), p. 6, 2005.
[7] Moyaux, T., Chaib-draa, B. & D'Amours, S., Information sharing as a coordination mechanism for reducing the bullwhip effect in a supply chain. *IEEE Transactions on Systems, Man, and Cybernetics - part C: Applications and Reviews*, **37**(3), pp. 396–409, 2007. http://dx.doi.org/10.1109/TSMCC.2006.887014
[8] Clancey, W.J., Sachs, P., Sierhuis, M. & van Hoof, R., Brahms: simulating practice for work systems design. *International Journal on Human-Computer Studies*, **49**(6), pp. 831–865, 1998. http://dx.doi.org/10.1006/ijhc.1998.0229
[9] Sierhuis, M., Clancey, W.J. & van Hoof, R., Brahms: a multiagent modeling environment for simulating work processes and practices. *International Journal of Simulation and Process Modelling*, **3**(3), pp. 134–152. 2007.
http://dx.doi.org/10.1504/IJSPM.2007.015238
[10] Dessalles, J.L., Müller, J.P. & Pahn, D., Emergence in multi-agent systems: conceptual and methodological issues. *In Agent-based Modelling and Simulation in the Social and Human Sciences*, eds. D. Phanpp & F. Amblard, The Bardwell Press: Oxford, pp. 327–355, 2007.
[11] Cooper, M.C., Lambert, D.M. & Pagh, J.D., Supply chain management: more than a new name for logistics. *The International Journal of Logistics Management*, **8**(1). pp. 1–13, 1997. http://dx.doi.org/10.1108/09574099710805556
[12] Alderson, W., *Dynamic Marketing Behavior, A Functionalist Theory of Marketing*, Richard D. Irwin: Homewood IL, 1965.
[13] Lambert, D.M., Cooper, M.C. & Pagh, J.D., Supply chain management: implementation issues and research opportunities. *The International Journal of Logistics Management*, **9**(2), pp. 1–19, 1998.
http://dx.doi.org/10.1108/09574099810805807
[14] Law, J., *Organizing Modernity*, Blackwell: Oxford, 1994.
[15] Giddens, A., *Modernity and Self-Identity. Self and Society in the Late Modern Age*, Polity Press: Cambridge, 1991.
[16] Foucault, M., *The History of Sexuality: Volume 1; An Introduction*, Penguin: Harmondsworth UK, pp. 94–95, 1981.
[17] Law, J., *Organizing Modernity*, Blackwell, Oxford, p. 106, 1994.
[18] Law, J., *Organizing Modernity*, Blackwell, Oxford, p. 59, 1994.
[19] Mead, G.H., *Mind, Self and Society,* Chicago University Press, Chicago, 1934.
[20] Foucault, M., *L'Archéologie du savoir*, Editions Gallimard: Paris, 1969.
[21] Holweg M., The genealogy of lean production. *Journal of Operations Management*, **25**(2), pp. 420–437. 2007.
[22] Hammervoll, T., Service provision for co-creation of value: insights from exchange- and production economy perspectives. *International Journal of Physical Distribution and Logistics Management*, 44(½), pp. 155–168, 2014.
[23] Penrose, E., *The Theory of the Growth of the Firm*, Oxford University Press: Oxford, 1959.

[24] Vargo, S.L. & Lusch, R.F., Evolving to a new dominant logic for marketing. *Journal of Marketing*, **68**(1), pp. 1–17, 2004.
[25] Prahaladad, C.K. & Ramaswamy, V., Co-opting customer competence, *Harvard Business Review*, **78**(1), pp. 79–87, 2000.
[26] Miles, M.B. & Huberman, A.N., *Qualitative Data Analysis*, Sage: Thousand Oaks, CA, 1994.
[27] Yin, R.K., *Case Studies Research: Design and Methods,* Sage: Thousand Oaks, CA, 2009.
[28] Voss, C., Tsikriktsis, N. & Frohlich, M., Case research in operations management. *International Journal of Operations and Production Management*, **22**(2), pp. 195–219, 2002. http://dx.doi.org/10.1108/01443570210414329
[29] Taylor, D. & Fearne, A., Towards a framework for improvement in the management of agri-food SCs. *Supply Chain Management: An International Journal*, **11**(5), pp. 379–384, 2006. http://dx.doi.org/10.1108/13598540610682381
[30] Fernie, S. & Thorpe, A., Exploring change in construction: Supply chain management. *Engineering, Construction and Architectural Management*, **14**(4): pp. 319–333, 2007. http://dx.doi.org/10.1108/09699980710760649
[31] Holweg, M. & Pil, F.K., Theoretical perspectives on the coordination of supply chains. *Journal of Operations Management*, **26**(3), pp. 389–406, 2008. http://dx.doi.org/10.1016/j.jom.2007.08.003
[32] Croom, S., Romano, P. & Giannakis, M., Supply chain management: an analytical framework for critical literature review. *European Journal of Purchasing and Supply Management*, **6**(1), pp. 67–83, 2000. http://dx.doi.org/10.1016/S0969-7012(99)00030-1
[33] Zwick, D., Bonsu, S.K. & Darmody, A., Putting consumers to work 'co-creation' and new marketing govern-mentality. *Journal of Consumer Culture*, **8**(2), pp. 163–196, 2008. http://dx.doi.org/10.1177/1469540508090089
[34] Moss, S., *Relevance, Realism and Rigour: A Third Way for Social and Economic Research,* Manchester Metropolitan University, Centre for Policy Modelling Report: Manchester, 99(56), 1999.
[35] Janssen, M.A. & Ostrom, E., Empirically based, agent-based models. *Ecology and Society,* **11**(2), p. 37, 2006.
[36] Checkland, P.B., *Systems Thinking, Systems Practice*, Chichester, UK: John Wiley & Sons, 1981.
[37] Checkland, P.B., *Soft Systems Methodology in Action*, John Wiley & Sons: Chichester, UK, 2001.
[38] Womack, J.P. & Jones, D.T., *Lean Thinking: Banish Waste and Create Wealth for Your Corporation,* Simon and Schuster: New York, 2003.
[39] Shewhart, W.A., *Statistical Method from the Viewpoint of Quality Control,* Dover: New York, 1939.
[40] Deming, W.E., *Out of the Crisis*, MIT Press: Cambridge MA, 1986.

SELF-ORGANIZATION AND EMERGENCE IN GLOBAL HEALTH. INSIGHTS FROM PRACTICE, BLIND SPOTS AND POSSIBLE CONTRIBUTIONS FROM COMPLEXITY SCIENCE

E.G. SARRIOT[1], E.R. SACKS[1] & A. LARSON[2]
[1]ICF International Center for Design and Research in Sustainable Health and Human Development (CEDARS), USA
[2]National Centre for Epidemiology and Population Health, The Australian National University, Canberra, Australia

ABSTRACT

Global health as a discipline is at a moment of tremendous achievements and general optimism, but faces enduring questions, notably about the sustainability of its endeavours. The field of global health defines itself through enumerated goals, targets, and indicators to track progress of complex changes. With this comes a broad panoply of objective-driven strategies resting on assumed control over chains of cause and effect. We present areas where emergence and self-organization play an important role in determining the sustainability of positive outcomes at scale and behaviours of both the population and its service providers. The role of self-organization and emergence seem under-appreciated by the field and we consider three reasons for a possible "blind spot" of global health decision makers and influencers: (1) the culture of command and control inherent to central planning, (2) the structural limitations of large complex institutional edifices in processing information (variety), and (3) epistemological/cognitive constraints. More pro-active allowance for the conditions of positive self-organization and emergence may be required to avoid unintended and negative long-term consequences of global strategies, in spite of best intentions. Observational, analytical and computational multi-disciplinary studies are needed to bring clarity to these questions, and to inform a parallel process of dialogue, provocation, and learning to evolve the institutional culture of global health at both national and international levels.
Keywords: agent-based modelling, complex adaptive systems, computational sciences, emergence, global health, self-organization, viable system model.

1 INTRODUCTION

Global health has emerged in the last 20–30 years into its own field. Its goals are to solve the largest mortality and morbidity problems globally, and address health disparities [1]. It builds on decades of international health and international development, and brings together a wide range of actors: country institutions, government-to-government programs, United Nations, non-governmental agencies (NGOs), public–private partnerships, etc. Global health's culture emphasizes performance, effectiveness, accountability, country ownership, and evidence-based interventions [2, 3]. This has come with notable progress on reducing child, neonatal, and maternal mortality, and towards disease control and expansion of services. This evolution is part of the Millennium Development Goals (MDGs), which were partially achieved by their 2015 target, and now, the Sustainable Development Goals (SDGs) [4]. Global health has achieved great success overall and is going through a time of relative optimism about what can be achieved in the next generation [3, 4].

Some caution is however warranted for at least two reasons: First, mortality gains are due to far more than effectiveness in pursuit of defined goals [5, 6]. Progress towards the MDGs were achieved during a period of increase in gross domestic product (GDP) in developing countries, and an important part of the gains in child mortality are associated with GDP gains and social determinants of health

DOI: 10.2495/DNE-V11-N4-644-653

rather than specific health interventions supported by global health strategies [5, 7]. Secondly, most progress has been made in a context of increased development assistance [7, 8], and this raises questions about the sustainability at scale of systems, of quality, and of health outcomes [9, 10].

These questions have stimulated a growing body of literature on complexity and systems thinking in global health [11, 12], which this paper adds to, to make three points: emergence and self-organization play an important role in achieving and sustaining global health results (section 2); global health is constrained in its capacity to discern these phenomena (section 3); finally, section 4 discusses a way forward, in terms complex systems research, and in terms of our cultural evolution.

2 EMERGENCE AND SELF-ORGANIZATION IN GLOBAL HEALTH

We refer readers to the literature for a definition of the related but different concepts of self-organization and emergence [13, 14]. We use the limited literature on self-organization and emergence in global health to make the case that these phenomena play a greater role in global health than usually acknowledged, and identify examples at three levels:

2.1 Self-organization in community and women's groups

Two models for improving community health – Participatory Women's Groups (PWGs) and Care Groups (CGs) – have drawn attention recently with large scale international implementation and a substantial level of evidence [15, 16]. The models bring women together in social structures to address specific health issues. The similarities and differences in the models are beyond our scope. Promoters of these approaches have defined the models to allow replicability and building an evidence basis. But informal discussions with authors, if not in the publications themselves, suggest at least the potential for emergence and self-organization in the participatory and social interactions' elements of these models. Successful women's groups evolve, adjust behaviours, and take on new tasks or objectives over time and of their own volition. Longitudinal studies are not available, but these unanticipated self-organizing processes seem to increase with time – more tasks, new roles, complex group dynamics. Whether this supports initial goals or not, self-organization seems to be a factor in these group approaches.

2.2 Self-organization within the health workforce and between health workers and clients

We identified different examples of self-organization in the literature: health providers with formal day-time functions in the health system and informal clinical practice at night [17]; community health worker groups developing new functions (savings activities) [18]; and district managers strategies to deal with the unreliability of central funds [19].

Less documented is the use and misuse of per diems for health workers. In the context of development, external actors can sponsor activities by providing in kind or financial (per diem) compensation. This can both stimulate and disrupt planned activities [20]. But self-organization can also be used to *correct* such unintended effects. In two cases (undocumented), management teams systematically redistributed part of the per diems to incentivize staff who failed to benefit from these payments. Rather than establishing patronage, local self-organizing redistributed incentives more equitably.

Finally, Lanham [21] describes initiatives, which "took advantage of self-organization" between health workers and clients to strengthen scale-up strategies. An intervention using cellular messaging to increase compliance to treatment saw unplanned uptake of phone communication between patients and nurses, intervention group patients relaying messages to control group friends; and spontaneous formation of a support network of patients.

2.3 Scaling and sustaining global health achievements

Decentralized management and autonomy can be hard to distinguish from actual self-organizing in the literature [22, 23]. But we find examples of self-organization and emergence in publications on sustaining health outcomes: Sarriot [24] surveyed primary health care projects' implementers, and reported maintenance of some activities coming through unanticipated local processes. Program managers referred to supporting "*a local process – particularly at the community level – without being able to predict or control it... it was completely unintentional in terms of where we would go next with it . . . . It wasn't exactly a mistake, but it wasn't necessarily an intended strategy*". An assessment five-year post-project [25] describes elements of self-organization among stakeholders of an urban health project in Bangladesh. While on the surface the architecture for urban community health was as-expected, different groups' self-interest led to a working equilibrium between all parties [26].

Similar language is used for scale-up processes [27]. ExpandNet, a World Health Organization collaborative, is dedicated to advancing scale-up of evidence-based health strategies, and recognizes the existence of spontaneous scaling up, as "diffusion of the innovation without deliberate guidance" [28]. Plausible examples of self-organization can be found where actions by national, local governments and community leaders changed the course of a pre-determined scale-up strategy [29]. Another series of case studies on scaling essential life-saving interventions describes self-organizing behaviours [30]: communities spontaneously providing homes and food to community health workers; and a program to expand access to post-partum family planning methods in India finding that health facilities combined planned and self-organized peer training of nurses to achieve scale.

These phenomena echo what is described in other fields of organizational and social sciences. The issue is now which factors hinder their perception by the global health community. We suggest three reasons.

3 CULTURAL, STRUCTURAL, AND EPISTEMOLOGICAL BIASES AS BLIND SPOTS

3.1 The international development culture – inherent and natural bias for control and predictability in decision-making

Policy makers require predictability from implementers, and are held accountable for return on investment by their own boards and governments. Objectives and targets translate into command and control approaches. Self-organization, by nature, runs counter to predictability. Additionally, emergence in complex adaptive systems depends on ill-defined timelines, which are poorly reconciled with deadlines.

One of our authors participated in an aid organization workshop, convening experts to address how complexity affected their work. The day saw substantial discussions about local systems' behaviours, adaptive processes, emergence, and how the organization might need to adapt its approaches. Coming to conclude the deliberations of the day, a high level leader of the institution specifically reminded participants that while adaptation to local contexts might be important, there could not and should not be ten different ways of achieving success on global targets. At another meeting, the leader of a network of implementing organizations was describing how trust and shared space within the network had allowed different partnership configurations to emerge and to respond to different challenges (for example Ebola). While sympathetic to the accomplishments, a donor agency manager present concluded the discussion by stating "unfortunately, we cannot fund 'it depends'". These examples echo the words of Lanham [21], "*the designs of [scale-up and spread] initiatives tend to focus on reproducing interventions with total fidelity, overlooking the unique attributes of local contexts [...] failed interventions could be the result of overlooking the limits of strict adherence to an intervention design.*"

3.2 Structure – the challenge of variety engineering

Institutional hierarchies require that information be processed and reduced as it moves upstream. Global health has recognized, central, and influential nodes in its network. Harmonization of indicators allows the relative few in the relative centre to process signals coming from nearly 200 countries, and thousands of districts. The diversity of experiences must be reduced level by level in order to be managed. A balance needs to be found between the quantity of signals and information (or "variety" [31]) flowing up from one level of management to the next. Self-organization and emergent phenomena generate increased variety (complexity and signals), which naturally challenge the management capacity of each level of a control hierarchy. It becomes easier for the institutional system to retrospectively describe self-organization either as "happy accidents" (exceptional leadership, serendipity) or unintended consequences (poor coordination, corruption), essentially as surprises outside of implementation research considerations, leaving the signals received by the centre primarily about anticipated processes.

The systematic study of the success factors for countries on the "fast-track" to achieving the MDGs [5] combined data from up to 144 countries. It found that "no single configuration of factors proved necessary or sufficient" to predict success. Factors identified fit broad categories: multi-sector gains including very prominently woman's education, woman's participation in both workforce and political structures, mobilization of partners across society around evidence-based strategies, using evidence for accountability, rapid adaptation to change for course correction, and the use of guiding principles. This does not demonstrate self-organization, but it certainly suggests conditions leading to decentralized intelligence, and responsive adaptation.

3.3 Knowledge – increased complexity and epistemological challenges

Contextual complexity challenges knowledge and knowledge management in development efforts [21, 32–34] and global health itself have become more complex (through scale, number of interventions, diversity and interconnections of actors,) [12, 35–37]. This challenges how to make sense of strategies and their evaluation. Jackson [38] presents decision-making and sense-making along two axes:

1. Participants can be characterized as unitary (sharing values, beliefs and interests), pluralist (different values and interests, with consensus on discrete objectives), or coercive (interests differ and power is a driving force); and
2. Systems, which can be characterized as simple (relatively stable, limited number of structured interactions, unaffected by external influences) or complex (large number of subsystems, adaptation to each other, and a shifting environment).

Global health remains often conceptualized as unitary and simple: health is a universal value; evidence-based technologies need to be scaled up with reasonable but limited adaptation to context; priorities are established in national and global fora with correspondingly, clear and objective metrics. Global health may however be more often realized at different levels on the two axes: progress takes place in a pluralist environment, where countries have different priorities, based not only on epidemiology, but also the will of the people, the competing interests of a national economy, decentralized powers, and negotiated arrangements between sectors. The environment can also be unwittingly, coercive. Governments have to align to global decision makers who overwhelmingly control finances (i.e. challenging decisions about new vaccines versus routine immunization as just one example). Similar processes affect national to subnational negotiations. Interestingly, in the development field, the complexity matrix describes how specific problems fall on a continuum from simple to complex, and has been adapted to community health [10, 37]. As complexity accrues and the world becomes pluralist or coercive, the definition of the goal to be pursued and the metrics of progress start relying on more constructivist and participatory methods for sense-making [10, 21] to answer questions such as, can we sustain a level of collaboration required to maintain equitable health outcomes in communities?

Nurtured by medicine and epidemiology, global health has learned to rely on controlling statistically for contextual factors influencing the effectiveness of new technologies. But trying to simply control for context in evidence generation may inappropriately filter out important phenomena, thus contributing to "blind spots". Anderson [14] states: "*The constructionist hypothesis breaks down when confronted with the twin difficulties of scale and complexity. At each level of complexity entirely new properties appear. Psychology is not applied biology, nor is biology applied chemistry. We can now see that the whole becomes not merely more, but very different from the sum of its parts.*" We should continue the trail and argue that complex social and institutional interventions are not applied medicine and epidemiology. Behaving as if that they were may cause substantial distortions in our understanding.

4 DISCUSSION: WHAT PATH AHEAD

The global health community has been extremely dynamic in its speed of change over the last decades but if our argument holds, cultural, structural and epistemological biases risk depriving a professional community from lessons, which may have critical impact on the sustainability of its efforts. How can we move forward?

4.1 Possible contributions from complex systems science

There is a growing body of literature on systems thinking and complexity in health [11, 12], sustainable development, and development assistance [8, 23]. This trend is welcome and should help us move forward.

Exploratory and diagnostic efforts are needed. In our examples, we have had repeated back and forth asking, "is this example really self-organization, or is it something else?" When is local organization self-organization? Our discussions navigated along a continuum, with—on the one end—system behaviours clearly fitting a command and control strategy, others needing marginal adaptation of central strategies to local conditions with global designs still dominating the solution landscape, and finally central evidence-based solutions requiring far more than adaptation. At this point, strategies may need to create space for self-organization. Studies should explicitly examine the how, when, why and impact of self-organization. Mechanisms for decentralized management, autonomous team and peer-learning, and use of performance-based incentives would benefit from "inside the black box" studies. There are a number of other issues, which could test when our hypothesis holds or is disproved [39–44]. Complex systems specialists should play an important role in translating study data in terms of underlying complex adaptive systems behaviours [21, 45–47].

Agent-based modelling (ABM) provides another direction to study the behaviours of health workers and social groups in different settings. Agent_Zero [47], for example, provides a template primed for adaptation to our questions. The model assigns agents with deliberative/cognitive and affective/emotional components of a disposition to act. Different environmental stimuli can affect their affective component, and social ties allow disposition contagion. We cannot do justice to the model here, but adaptations of the model could examine the disposition to perform of community group members, community health workers or facility-based care providers. Possible research questions could include how the cumulative aggregate performance of agents evolves under conditions, determined by environment signals and emotional response to financial and non-financial incentives, social signalling (perceived fairness), and cognitive support mechanisms (peer support, supervision, etc.). The model could also be designed to test the long-term dynamics (sustainability) of social/health groups under different types of support. ABM would allow to explore whether patterns of system level performance emerge (regularities) based on environmental, emotional and social signalling, and possibly counter-intuitively to the intentions of the environmental signals (payment and control schemes). Interdisciplinary research could combine the knowledge of computational, behavioural, health care, and social systems scientists with implementers' experiences.

Finally, complex systems science may also offer approaches for testing, on large data sets or in simulations, probabilistic models of success, based on learning from traditional methods. For example, the Sustainability Framework [10] suggests a set of conditions, which if present at a certain level, will increase the odds of emergence of sustainability. These include: policies, specific capabilities, social capital, community capacity, macro-economic conditions, conflict, political stability, feasibility and acceptability of interventions. These factors naturally affect path dependence for how levels of sustainability can be achieved. For example, a certain set of capabilities in a central governmental health institution may be required only under a specific central-decentralized functional distribution of roles. An essential health district capacity in one context may become obsolete in another where civil society organizations are implicitly and explicitly granted public service responsibilities. Simulation models informed by solid case studies could test different configurations of conditions required to achieve success (this may fit something similar to a percolation model, where success is probabilistic and subject to a tipping point).

The implications for global health planners and researchers are important [46]. Under a linear model, the objectives are to maximize progress on the proximal determinants of success, as defined by the largest and most significant coefficients. If sustainability (or scale) follows a probabilistic model instead (for example, a percolation model) then the goal is to advance enough positive conditions while respecting principles for positive self-organization, without trying to individually predict what will be achieved on each proximal determinant.

4.2 Institutional and cultural evolution

New studies can provide information and signals, but are probably not enough to bring change in the complex adaptive system of development institutions [8]. Power, influence, trust, and network dynamics within development institutions are beyond this paper, but will influence the speed of learning [48]. Processes of dialogue, provocation, reaction and consensus, along with innovation and research will be needed to translate learning into new behaviours. Principles, such as Axelrod and Cohen's "harnessing complexity" [47] may help not only global health and development practice, but also for evolving the model and behaviours of global health institutions themselves.

The evolution of new forms of governance and the Paris Declaration for Aid Effectiveness are positive examples for how global health is able to evolve. Our argument is however not simply for more national leadership in development, even if this is a natural starting point. Our call is for recognition that non-linear phenomena based on decentralized agency affect our successes and failures. Our call for exploration and learning goes to both national and international leaders and the research community.

5 CONCLUSION

This paper has tackled and tried to relate together a complicated set of questions. In so doing, it has unavoidably been cursory in its treatment of important issues. The argument rests on three parts: (1) that self-organizing and emergent phenomena affect global health achievements, particularly their sustainability at scale; (2) that the leaders of the field of global health face inherent cultural, structural and cognitive challenges in perceiving signals about the potential importance of these phenomena and (3) that multi-disciplinary research directions, notably through contributions of complex systems science, will be necessary, but not necessarily sufficient, for the global health sector to better address these issues.

Both individual and institutional agency weighs heavily in the final outcomes that central planners are investing in. Our contention is not that self-organization is some sort of panacea. Self-organization eventually takes place in human complex adaptive systems for better or for worse. Where self-organization and emergence have a positive role to play, ignorance of their potential opens the field of global health to both opportunity costs and unintended effects. "Where, when, and how does self-organization need to play out?" are the questions we need to ask. Learning how to do this could have a fundamental importance for resolving intractable problems, establishing more resilient social and health systems, and sustaining current achievements in global health. Partnership efforts between complex systems science and global health researchers could help national and global decision makers learn to create space for probabilistic (stochastic) processes within their deterministic plans, lowering the focus on individual program success but creating conditions for greater and more lasting overall impact.

ACKNOWLEDGEMENTS

We wish to thank reviewers, Sharon Arscott-Mills, Angie Brasington, David Bishai, Ligia Paina, and Inger Scheel. Thanks to Leo Ryan and Ani Hyslop for their support.

REFERENCES

[1] Hafner, T. & Shiffman, J., The emergence of global attention to health systems strengthening. *Health Policy and Planning,* **28**, pp. 41–50, 2013. http://dx.doi.org/10.1093/heapol/czs023
[2] OECD, Aid Effectiveness 2005-10: progress in implementing the Paris declaration, OECD Publishing, pp. 1–200, 2011.
[3] USAID, Acting on the call: ending preventable child and maternal deaths report. In *United States Agency for International Development (USAID)*, pp. 1–85, 2015.
[4] United Nations General Assembly, Transforming our world: the 2030 Agenda for sustainable development. In *United Nations*, New York, NY, pp. 1–35, 2015.
[5] Kuruvilla, S., Schweitzer, J., BIshai, D., Chowdhury, S., Caramani, D., Frost, L., Cortez, R., Daelmans, B., De Francisco, A., Adam, T., Cohen, R., Alfonso, Y.N., Franz-Vasdeki, Y., Saadat, S., Pratt, B.A., Eugster, B., Bandali, S., Venkatachalam, P., Hinton, R., Murray, J., Arscott-Mills, S., Axelson, H., Maliqi, B., Sarker, I., Lakshminarayanan, R., Jacobs, T., Jacks,

S., Mason, E., Ghaffar, A., Mays, N., Presern, C. & Bustreo, F., Success factors for reducing maternal and child mortality, pp. 533–544, 2014.

[6] Fukuda-Parr, S., Ely Yamin, A. & Greenstein, J., Synthesis paper - the power of numbers: a critical review of mdg targets for human development and human rights, 2013.

[7] French, D., Did the millennium development goals change trends in child mortality? 2015.

[8] Ramalingam, B., *Aid on the Edge of Chaos. Rethinking International Cooperation in a Complex World*, Oxford University Press, 2013.

[9] Chambers, DA., Glasgow, RE. & Stange, KC., The dynamic sustainability framework: addressing the paradox of sustainment amid ongoing change, 2013.

[10] Sarriot, EG, Kouletio, M, Jahan, S, Rasul, I. & Musha, A., Advancing the application of systems thinking in health: sustainability evaluation as learning and sense-making in a complex urban health system in Northern Bangladesh. *Health Research Policy and Systems,* **12**(45), 2014. http://dx.doi.org/10.1186/1478-4505-12-45

[11] Peters David, H., The application of systems thinking in health: why use systems thinking? Commentary, 2014. http://dx.doi.org/10.1186/1478-4505-12-51

[12] Ramalingam, B., *Aid on the Edge of Chaos. Rethinking International Cooperation in a Complex World*, Oxford University Press, 2013.

[13] Miller, JH. & Page, SE., *Complex Adaptive Systems: An Introduction to Computational Models of Social Life*, Princeton University Press, 2007.

[14] Anderson, P.W., More is different. *Science*, **177**, pp. 393–396, 1972. http://dx.doi.org/10.1126/science.177.4047.393

[15] Paul, VK., Participatory women's groups: time for integration into programmes. *The Lancet Global Health*, **4**, pp. e74–e75, 2016. http://dx.doi.org/10.1016/S2214-109X(16)00010-3

[16] Perry, H., Morrow, M., Borger, S., Weiss, J., DeCoster, M., Davis, T. & Ernst, P., Care groups i: an innovative community-based strategy for improving maternal, neonatal, and child health in resource-constrained settings, pp. 358–369, 2015.

[17] Paina, L., Bennett, S., Ssangooba, F. & Peters, DH., Advancing the application of systems thinking in health: exploring dual practice and its management in Kampala, Uganda. *Health Research Policy and Systems,* **12**(41), 2014. http://dx.doi.org/10.1186/1478-4505-12-41

[18] Langston, A., Weiss, J., Landegger, J., Pullum, T., Morrow, M., Kabadege. M., Mugeni, C. & Sarriot, E., Plausible role for CHW peer support groups in increasing care-seeking in an integrated community case management project in Rwanda: a mixed methods. *Global Health Science and Practice*, **2**(3), 2014. http://dx.doi.org/10.9745/GHSP-D-14-00067

[19] Asante, A.D, Zwi, A.B., Ho, M.T., Getting by on credit: how district health managers in Ghana cope with the untimely release of funds, 2006.

[20] Aitken, J-M., Voices from the inside: managing district health services in Nepal, pp. 309–340, 1994.

[21] Lanham, H.J., Leykum Luci, K., Taylor, B.S., McCannon, C.J., Lindberg, C. & Lester, R.T., How complexity science can inform scale-up and spread in health care: understanding the role of self-organization in variation across local contexts, pp. 194–202, 2013.

[22] Bossert, T.J., Can they get along without us? sustainability of donor-supported health projects in Central America and Africa. *Social Science & Medicine,* **30**(9), pp. 1015–1023, 1990. http://dx.doi.org/10.1016/0277-9536(90)90148-L

[23] Honig, D., *Navigation by Judgment: Organizational Autonomy and Country Context in the Delivery of Foreign Aid*, Working Paper, Harvard University, 2014.
[24] Sarriot, E.G., Winch, P.J., Ryan, L.J., Edison, J., Bowie, J., Swedberg, E. & Welch, R., Qualitative research to make practical sense of sustainability in primary health care projects implemented by non-governmental organizations. *International Journal of Health Planning Management*, **19**(1), pp. 3–22, 2004.
http://dx.doi.org/10.1002/hpm.743
[25] Sarriot, E.G., Kouletio, M., Jahan, S., Rasul, I. & Musha, A., Advancing the application of systems thinking in health: sustainability evaluation as learning and sense-making in a complex urban health system in Northern Bangladesh. *Health Research Policy and Systems,* **12**(45), 2014.
[26] Sarriot, E. & Kouletio, M., Community health systems as complex adaptive systems: ontology and praxis lessons from an urban health experience with demonstrated sustainability. *Systemic Practice & Action Research,* pp. 1–18, 2014.
[27] Paina, L. & Peters, D.H., Understanding pathways for scaling up health services through the lens of complex adaptive systems. *Health Policy and Planning,* pp. 1–9, 2011.
[28] ExpandNet, Nine steps for developing a scaling-up strategy, World Health Organization: Geneva, 2010.
[29] Simmons, R., Fajans, P. & Ghiron, L., Scaling Up Health Service Delivery: From Pilot Innovations to Policies and Programmes, World Health Organization: Geneva, 2007.
[30] Larson, A., Ricca, J. & Posner, J., Scaling up maternal and child health interventions in low and middle income countries: an examination of 17 case studies. presentation to the maternal and child survival project. report presented in *Maternal and Child Survival Program (MCSP),* 2015.
[31] Beer, S., *Brain of the Firm. The Managerial Cybernetcis of Organization*, John Wiley & Sons: Chichester, UK, 1981.
[32] Jackson, D., Brady, W. & Stein, I., Towards (re)conciliation: (re)constructing relationships between indigenous health workers and nurses. *Journal Advanced Nursing*, **29**(1), pp. 97–103, 1999.
http://dx.doi.org/10.1046/j.1365-2648.1999.00866.x
[33] Patton, M.Q., *Qualitative Evaluation and Research Methods*, Sage Publications, Inc., 1990.
[34] Williams, B. & Imam, I., *Systems Concepts in Evaluation*, EdgePress of Inverness: Point Reyes, 2007.
[35] Rihani, S., *Complex Systems Theory and Development Practice: Understanding Non-Linear Realities*, Zed Books, London, 2002.
[36] McCoy, D., Storeng, G., Filippi, V., Ronsmans, C., Osrin, D., Matthias, B., campbell, O.M., Wolfe, R., Prost, A., Hill, Z., Costello, A., Azad, K., Mwansambo, C., Manandhar, D.S., Maternal, neonataland child health interventions and services: moving from knowledge of what works to systems that deliver, pp. 87–98, 2010.
[37] Geyer, R. & Rihani, S., *Complexity and Public Policy: A New Approach to 21st Century Politics, Policy and Society*, Routledge, 2010.
[38] Jackson, M., Critical systems thinking and practice. *European Journal of Operational Research,* **128**(2), pp. 233–244, 2001.
http://dx.doi.org/10.1016/S0377-2217(00)00067-9
[39] Bisits Bullen, P.A., The positive deviance/hearth approach to reducing child malnutrition: systematic review, pp. 1354–1366, 2011.

[40] Underwood, C. & Boulay, M., Community capacity as means to improved health practices and an end it itself: evidence from a multi-stage study, pp. 105–127, 2013.

[41] Kar, S.B, Pacual, C.A. & Chickering, K.L., Empowerment of women for health promotion: a meta-analysis, pp. 1431–1460, 1999.

[42] Bjorkman Nyqvist, M. & Svensson, J., Power to the people: evidence from a randomized field experiment of a community-based monitoring project in uganda, 2007.

[43] Jagosh, J., Bush, P.L., Salsberg, J., Macaulay, A.C., Greenhalgh, T., Wong, G., Cargo, M., Green, L.W., Herbert, C.P. & Pluye, P., A realist evaluation of community-based participatory research: partnership synergy, trust building and related ripple effects, pp. 1–11, 2015.

[44] LeBan, K., How social capital in community systems strengthens health systems: people, structures, processes. In *USAID; Core Group; MCHIP,Maternal and Child Health Integrated Program*, pp. 1–22, 2011.

[45] Norman, C.D., Health promotion as a systems science and practice, pp. 868–872, 2009.

[46] USAID, Local Systems: a framework for supporting sustained development, Washington DC, 2014.

[47] Axelrod, R. & Cohen, M.D., *Harnessing Complexity: Organizational Implications of a Scientific Frontier*, Simon and Schuster, 2001.

[48] Shiffman, J., Knowledge, moral claims and the exercise of power in global health, pp. 297–299, 2014.

EXCESS OF SOCIAL ACTIVITY REDUCES THE RESPONSIVENESS OF SWARMS

DAVID MATEO & ROLAND BOUFFANAIS
Singapore University of Technology and Design, Somapah Road, Singapore, Singapore.

ABSTRACT
Social interaction increases significantly the collective action and operation of swarming systems. However, a growing body of empirical results reveals limitations in the number of social links in natural swarms, thereby suggesting possibly detrimental effects associated with excessive social activity. We study the consequences that such excess has on the responsiveness of a swarm facing dynamic perturbations. Using a classical model of collective motion, we show that the capacity of a swarm to perform actions critical to its survival peaks for a number of social links much smaller than the maximum possible number. This relationship between social activity, responsiveness and predator avoidance capacity presents a plausible picture of how natural selection could favor organisms that self-limit their sociality, as is common in many taxa. These results have important implications for the design of swarm robotics systems.
Keywords: collective behaviors, consensus dynamics, sociality, swarming systems.

1 INTRODUCTION

Animal groups provide paradigmatic examples of collective phenomena in which repeated interactions among individuals produce dynamic patterns and responses on a scale larger than individuals themselves. Some of the examples around us include the coordinated movements of fish and birds in a school or a flock respectively, the chemotactic aggregation of amoebae, the formation of lanes in densely-packed human crowds, the generation of vortices in bacterial colonies, the synchronized march of wingless locusts and the synchronized flashing of fireflies. Many more examples can also be found inside all of us: the firing of neurons in our brains, the clustering of differentiated cells to construct our organs both during embryonic development and wound healing, and the targeted response of neutrophils as part of the initial immune response to a bacterial infection. This non-exhaustive list of collective behaviors of unicellular and multicellular organisms is revealing of the pervasiveness of swarming in the natural world [1].

Swarming agents are interconnected organisms or agents. A key benefit of being connected is access to information. Such dynamic interconnectivity serves the purpose of channeling information exchanges, which are critical to the effectiveness in swarming. Indeed, it is well known that collective animal behavior is dependent on the existence of communication channels enabling information exchange between individuals [1, 2]. For instance, the collective surveillance against oncoming threats of a flock of birds provides a higher level of vigilance only if the information obtained by each pair of eyes is shared among the flock.

Swarm dynamics is governed by inter-agent interactions and also interactions between individual agents and their environment. Through all these interactions a social transmission of information occurs, eventually leading to a wide range of self-organizing natural behaviors. Let us come back to the example of the evasive maneuver performed by a flock of birds confronting a threat. The predator

DOI: 10.2495/DNE-V11-N4-654-662

is external to the flock and can be said to be part of the environment. The detection of its presence by birds at a close range amounts to an external signal or stimuli – a.k.a. sensory information. The detecting agents swerve to move away from this stimuli, thereby creating a signal internal to the flock – behavioral information. This behavioral signal contains information and propagates through the flock at high speed and with little distortion [3, 4]. This latter information transmission process can be said to be responsible for the overall effectiveness of the flock in dodging the predator's attack. For instance, birds detecting an environmental threat give utmost priority to sensory information, thereby discarding behavioral signals from neighboring conspecifics. In short, information is a crucial currency for animals from a behavioral standpoint. This is certainly also true for artificial swarms.

Tracing the flow of information and quantifying informational exchanges are keys to gaining insight into the functioning of swarms. It also plays a central role when designing swarming systems. From the engineering standpoint, such effective information transfers highlight the existence of an underlying communication channel that takes the form of the swarm signaling network (SSN) [5] as is schematically shown in Fig. 1. Animal collectives use this SSN to effectively respond to changes in the surroundings: e.g. coordinated evasive maneuvers upon detection of a predator or collision avoidance. The structure and dynamics of SSNs are quite unique, being temporal and adaptive networks [6] with a dynamics deeply interwoven with the agents' motion dynamics embedded in the physical space. For instance, in the particular case of bird flocks governed by a topological interaction [7], the SSN has been found to be a small-world, homogeneous clustered network whose connectedness is key to yielding resilient swarming behaviors [5]. The knowledge of and access to the structural properties of the SSN revealed the high dynamic controllability of swarms [5, 8] – where few agents are capable of driving the dynamics of the swarm as a whole – as well as very effective consensus reaching processes [9].

Increasing the amount of social links is usually known to yield an improved collective operation. From a network-theoretic standpoint, this amounts to increasing the degree distribution of the signaling network. However, there is a growing body of evidence showing that most collective behaviors

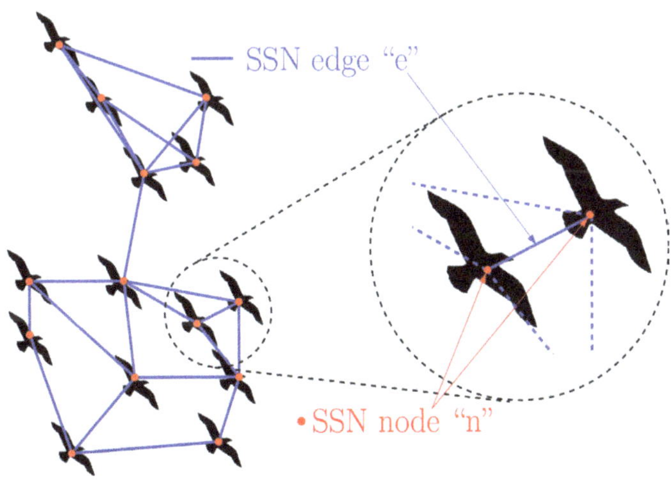

Figure 1: Schematic of a networked flock of birds with the associated information transmission channel in the form of the SSN. Edges represent an interaction between two agents. Nodes are the agents themselves, which act as routers for the behavioral information.

naturally occurring involve a limited number of social links – i.e. the SSN has a very specific degree distribution. A recent landmark study from the STARFLAG group imaged and tracked wild flocks of starlings numbering in the thousands and revealed that the neighborhood of interaction actually depends on a topological distance [7, 10]. More precisely, Ballerini *et al.* [7] discovered that each bird interacts on average with a fixed number of nearest neighbors (six to eight). This apparent limitation in the number of social links in flocks of starlings was further observed with swarms of midges [11] governed by a metric interaction distance and in which a regulation of the nearest neighbor distance, as a function of the size of the swarm, takes place. Similar limitations in the number of social links have been observed with other taxa, including insects: e.g. the ants (*Lasius fuliginosus*) regulate their rate of social encounters [12]. The important fact here is that this limitation in the number of social links established appears to be a behavioral feature and not a direct result of physical limitations associated with their sensing and/or signaling apparatuses. These findings suggest that an excess in social connectivity could have detrimental consequences for the swarm dynamics, when the latter is confronted with changing circumstances.

Here, we investigate the consequences of establishing an excessive amount of social links on the collective dynamics of a swarming system confronted with a threat. Understanding these effects has far-reaching consequences on the design of artificial swarms [13] and for the study of spreading of behaviors in online communities [14]. Previous studies of this problem [9, 15] were focused on system reaching consensus and have shown that, above a certain point, increasing the number of social links yields only marginal benefit to the collective. However, a systematic study of the responsiveness of swarms as a function of the number of social connections is lacking. To this aim, we use a canonical model of collective motion, in which agents move by adjusting their direction of travel to that of their neighbors in both two and three-dimensional setups. We characterize the responsiveness of the swarm by simulating a predator attack and measuring the survival rate of agents depending on their level of social activity measured by their degree distribution of their signaling network.

2 SELF-PROPELLED PARTICLES MODEL OF COLLECTIVE MOTION

In the past two decades, a wealth of swarming models have been developed and introduced. Most of them are aimed at generating consensus behaviors [1, 5, 8, 9, 16, 17], often in the form of group alignment or polarization. In our framework, such consensus behaviors amount to swarming that refers to a circumstance in which multiple adaptive agents – be them living creatures or artificial ones – create a certain level of spatiotemporal order characterized by one or more macro-level properties.

2.1 Base model

Following the work by Komareji & Bouffanais [5], we consider a collective of N identical and topologically interacting agents moving at the same speed. Each individual group member, at any given instant t, is assumed to be fully characterized by a given state variable $\Psi_i(t)$. Such a generic state variable may represent widely different characteristics depending on the nature of the group considered, but here it simply reduces to the direction of travel and the achievement of swarm consensus therefore yields an alignment of all the agents, in other words a polarized swarm. Hence, from a formal standpoint, by reaching a consensus, we mean asymptotically converging to a one-dimensional agreement in space characterized by $\Psi_1 = \Psi_2 = \cdots = \Psi_N$ [18].

In the dynamical model considered, the adaptive and interacting swarming agents are modeled as self-propelled particles (SPPs) for which the biological details of the internal origin of an agent's thrust is considered to be irrelevant. Such SPP-based models are a good representation of collective animal behaviors [16]. Here, these SPPs are moving about a two-dimensional plane with constant speed v_0 and subject to a topological neighborhood of interaction. As mentioned previously, each agent i is fully characterized by its direction of travel—in other words, here $\Psi_i(t) = \Psi_i(t)$—related to the agent's velocity through **V**i = $v_0 \cos \theta_i \hat{x} + v_0 \sin \theta_i \hat{y}$. The local synchronization protocol, based on relative states, is strictly equivalent to a local linear alignment rule which mathematically can be stated as:

$$\dot{\theta}_i(t) = \frac{1}{|N_i|} \sum_{j \in N_i} w_{ij}(\theta_j(t) - \theta_i(t)), \tag{1}$$

where N_i is the set of out degree neighbors in the topological neighborhood of interaction of agent i, with cardinal number $|N_i|$, and w_{ij} is the binary weight of the $i - j$ communication link. Note that in some models, wij can take a more complicated form than our binary choice. Using the k-nearest neighbor rule to define the topological distance, we have $|N_i| = k$ for the case $k = 7$) and the following dynamical equation for each individual agent i in the swarm:

$$\begin{aligned} \dot{\theta}_i &= \frac{1}{k}\left[\left(\theta_j - \theta_i\right) + \left(\theta_{j+1} - \theta_i\right) + \cdots + \left(\theta_{j+k-1} - \theta_i\right)\right] \\ &= \frac{1}{k}\left(-k\theta_i + \theta_j + \theta_{j+1} + \cdots + \theta_{j+k-1}\right), \end{aligned} \tag{2}$$

where $\theta_j, \cdots, \theta_j$+$k$-1 are its k-nearest neighbors' velocity directions.

2.2 Incorporation of noise

Noise can generally be assumed to be random fluctuations with a normal distribution [19]. In the sequel, the background noise is considered to have a normal distribution fully characterized by its noise level, η. Specifically, the presence of noise modifies the equation governing the dynamics of agent i which, using a discrete-time approach reads

$$\theta(t+1) = \theta_i(t) + \frac{1}{k}\left[\left(\theta_j(t) - \theta_i(t)\right) + \cdots + \left(\theta_{j+k-1}(t) - \theta_i(t)\right)\right] + \eta\xi_i(t) \tag{3}$$

where $\eta\xi_i(t)$ is a Gaussian white noise of magnitude η since $\xi_i(t) \in [-\pi, \pi]$.

2.3 Predator attack

The predator is introduced as an agent that does not participate in the collective motion governed by eqn (3). Instead, it is afforded predatory capabilities associated with a motion 40% faster than swarming agents, and systematically in the direction pointing to the closest one. When the predator succeeds in catching an agent, the latter is removed from the simulation. An agent can only detect the presence of the threat when it is located at a distance smaller than a fixed radius RD; as soon as the agent detects it, an evasive maneuver is initiated with the agent moving away in the direction opposite to the predator. We have set RD to be constant throughout the simulations and independent of the established number of social links. The fleeing behavior takes precedence over the collective motion of a particular agent for as long as the predator lies inside the ball of radius RD and is associated with

$$\vec{v}_i(t) = v_0 \frac{\vec{x}_i(t) - \ \vec{x}_i P(t)}{|\vec{x}_i(t) - \vec{x}P(t)|} \quad \text{if } |\vec{x}_i(t) - \vec{x}P(t)| < R_D, \tag{4}$$

where $\vec{x}$ denotes the predator's position.

The mean avoidance time shown in the next Section is obtained by computing the swarm dynamics in the presence of a single predator for 500 different runs of 5, 000 iterations each. The reason for computing several runs instead of running the calculation for longer times is that the results depend on the density of agents in the swarm, and the repeated removal of agents by the predator can cause significant changes in the density after long times.

3 SWARM RESPONSIVENESS – PREDATOR AVOIDANCE

In order to study the far-from-consensus dynamics of a swarm, we have simulated a predator attack as detailed in the previous Section, and measured the survival rate of agents with different levels of social interaction—corresponding to different degree distributions of the SSN, different space dimensions—two- and three-dimensional computational domains, and different levels of ambient noise η. The ubiquitous presence of noise is known to have very specific disruptive effects on the collective behavior. Intuitively, it is clear that noise corresponds to external perturbations hindering the process of self-organization.

The emergent collective avoidance maneuver is shown in Fig. 2 for two selected snapshots of a predator attack. A careful local analysis at different time instants reveals that agents outside the detection area ($r > R_D$) are collectively reacting to the threat thanks to the social information transmitted through the swarm, subsequently leading to a global evasive maneuver [20]. The simulations show that the social information transfer takes place strikingly fast even though the agents themselves are considerably slower than the predator. These results are consistent with recent empirical observations of collective turns in flocks of starlings [3] and startled schools of fish [4]. Specifically, Attanasi *et al.* [3] have highlighted the spatially-localized origin of the triggering behavioral signal in collective turns of starlings. In addition, for a specific event involving 176 starlings, it took approximately less than 0.6 s for the information to flow through the whole flock, corresponding to an amazingly fast speed of information transfer of the order of 20–40 m s^{-1} (3 times faster than the flock itself).

The social alignment consensus alone is sufficient for the vast majority of the agents to avoid the predator without ever detecting it. This phenomenon of augmented perceptual range in swarms due to social transfer of information is well known in flocks of birds [3] and schools of fish [21].

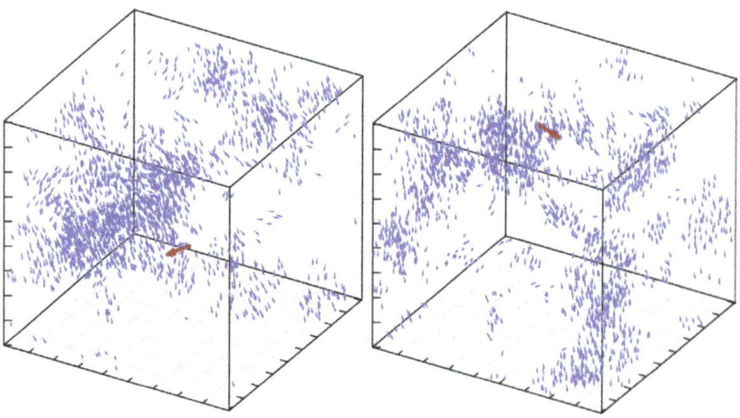

Figure 2: Snapshots of a collective evasive action induced by a predator attack at two different time instants. The $N = 500$ SPP agents are in blue while the predator appears in red.

The characteristic avoidance time for the swarm, defined as the average time elapsed between two consecutive catches by the predator, is shown in Figures 3 and 4 as a function of the number k of social links between agents. Starting from a non-interacting collective ($k = 0$), the avoidance time grows with the amount of social interaction up to a maximum value about 40% larger than the non-interacting time. From that optimal point at approximately 20 neighbors, the avoidance time monotonously decreases with increasing sociality, all the way down back to the value obtained for a non-interacting collective.

While the amount of interactionquantified here as the number of neighbors or the degree distribution of the SSN – certainly has a big impact on the ability of the swarm to perform collective predator avoidance, the properties of the dynamic interaction network itself also play a fundamental role in the swarm behavior and its efficiency. One natural first step in determining how different topologies of interaction network will affect the swarm's performance is to compare the predator avoidance using the original Vicsek model [17] when the agents are embedded in a two-dimensional or a three-dimensional space (see Fig. 3). In both cases, we observe a maximum in the mean avoidance time for a finite amount of interaction. However, in the three-dimensional case the optimal number of neighbors is considerably lower than in the two-dimensional case, from $k^*_{2D} = 18$ to $k^*_{3D} = 6$, consistent with the experimental observations that flocking starlings typically interact with 6 to 7 conspecifics [7]. It would be interesting to obtain similar experimental evidence for terrestrial swarms that essentially move in a two-dimensional space.

In order to be able to compare both sets of results between them, we have chosen: (i) to use a topological interaction instead of a metric one, as it reduces the number of dimensional parameters to consider, and (ii) to fix the density and speed of the agents such that the mean avoidance time in the limit for $k \rightarrow N$ is the same in both the 2D and 3D cases.

Vicsek's model [17] includes a noise term in the equations of motion to represent the agents imperfect sensing and imperfect motor capabilities. While this noise adds interesting phenomenology to the model itself (see Sec. 2.2), like a phase transition from order to disorder, it does not critically affect how the avoidance time depends on the amount of interaction. Higher noise will in

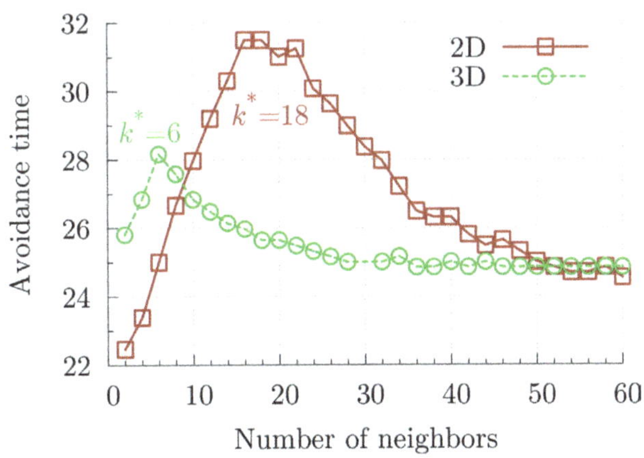

Figure 3: Efficiency in predator avoidance: Mean avoidance time for bi-dimensional and three-dimensional swarms. Mean number of iterations between two consecutive predator kills (avoidance time) as a function of the number of neighbors k for two-dimensional (□) and three-dimensional (O) swarms. The number of neighbors that maximizes avoidance time in each case is shown as k^* for a topological interaction.

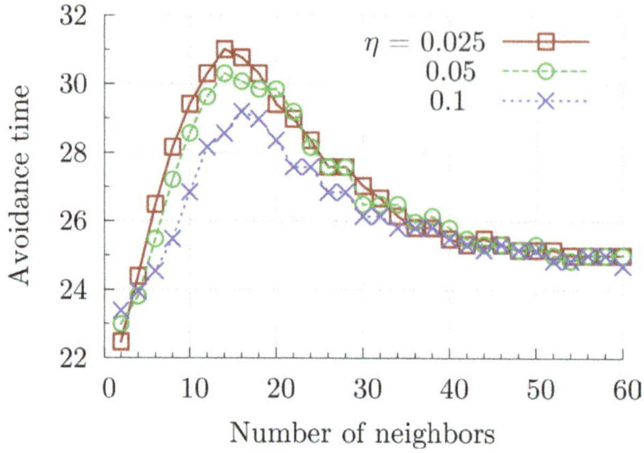

Figure 4: Influence of the ambient noise level (η) on the effectiveness in avoiding the predator. Mean number of iterations between two consecutive predator kills (avoidance time) as a function of the number of neighbors k for two-dimensional swarms for a topological interaction.

general lower the peak of avoidance time but does not affect much the position of the peak itself nor change drastically the avoidance time for low and large k. Therefore, from the results obtained, we conclude that in this particular case the topology of the interaction network (SSN) seems to play a bigger role in determining the efficiency of the swarm in avoiding a predator than the details of the consensus protocol the swarm is following.

4 CONCLUSION

In our analysis, the responsiveness of swarms reveals that an excess of social activity has detrimental effects on the collective response of swarms confronted with threats and dangers. Specifically, we have shown that for a system of self-propelling agents – subjected to a consensus protocol to align their velocities – in the presence of predator. The improvement in predator avoidance with increased social activity (ascending branch in Fig. 3) reflects the expected benefits of swarming in terms of anti-predator vigilance [2, 22]. However, the existence of a maximum in the avoidance time implies that an excess of social interaction has a negative effect on the capacity of the swarm to respond to a threat associated with a dilution of behavioral information [20].

Interestingly, the avoidance time in the limit of large sociality is approximately the same as in the non-interacting swarm ($k = 0$ in Figs 3 and 4), showing that the oversaturation in the information each agent receives (i.e. the very high degree of each node of the SSN, see Fig. 1) effectively negates any benefit of social activity on the dynamical collective response. It is important noting that our results for dynamical responsiveness complement previous studies associated with global properties, such as the robustness of the interaction network [15] or the consensus speed [9]. In these studies, increasing the amount of interaction eventually yields diminishing returns – i.e. less gain per neighbor, but never an actual reduction in the property of interest. Diminishing returns can only justify the preference for a finite number of connections if the cost for establishing links between agents is significant. However, quantifying such costs is close to impossible given the complexity associated with sensory and neurological requirements in biological swarms [9, 15]. In contrast, the present study on the dynamical responsiveness of the swarm shows an absolute reduction in swarming effectiveness when the number of neighbors is increased above a certain level.

ACKNOWLEDGMENTS

This work was supported by a grant #SMIG15003 from the Singapore–MIT Alliance for Research and Technology (SMART) under CENSAM (R.B.), and by a grant from the Temasek Lab (TL@ SUTD) under the STARS project (D.M.).

REFERENCES

[1] Bouffanais, R., *Design and Control of Swarm Dynamics*, Springer: Heidelberg, 2016. http://dx.doi.org/10.1007/978-981-287-751-2

[2] Krause, J. & Ruxton, G.D., *Living in Groups. Oxford Series in Ecology and Evolution*, Oxford University Press: Oxford, U.K., 2002.

[3] Attanasi, A., Cavagna, A., Del Castello, L., Giardina, I., Grigera, T.S., Jeli´c, A., Melillo, S., Parisi, L., Pohl, O., Shen, E. & Viale, M., Information transfer and behavioural inertia in starling flocks. *Nature Physics*, **10**, pp. 691–696, 2014. http://dx.doi.org/10.1038/nphys3035

[4] Herbert-Read, J.E., Buhl, J., Hu, F., Ward, A.J.W. & Sumpter, D.J.T., Initiation and spread of escape waves within animal groups. *Royal Society Open Science*, **2**, p. 140355, 2015. http://dx.doi.org/10.1098/rsos.140355

[5] Komareji, M. & Bouffanais, R., Resilience and controllability of dynamic collective behaviors. *PLoS ONE*, **8**, p. e82578, 2013. http://dx.doi.org/10.1371/journal.pone.0082578

[6] Holme, P. & Saramaki, J., Temporal networks. *Physics Report*, **519**, pp. 97–125, 2012. http://dx.doi.org/10.1016/j.physrep.2012.03.001

[7] Ballerini, M., Cabibbo, N., Candelier, R., Cavagna, A., Cisbani, E., Giardina, I., Lecomte, V., Orlandi, A., Parisi, G., Procaccini, A., Viale, M. & Zdravkovic, V., Interaction ruling animal collective behavior depends on topological rather than metric distance: evidence from a field study. *Proceedings of the National Academy of Science USA*, **105**, pp. 1232–1237, 2008. http://dx.doi.org/10.1073/pnas.0711437105

[8] Komareji, M. & Bouffanais, R., Controllability of a swarm of topologically interacting autonomous agents. *International Journal Complex Systems in Science*, **3**, pp. 11–19, 2013.

[9] Shang, Y. & Bouffanais, R., Influence of the number of topologically interacting neighbors on swarm dynamics. *Scientific Report*, **4**, p. 4184, 2014. http://dx.doi.org/10.1038/srep04184

[10] Ballerini, M., Cabibbo, N., Candelier, R., Cavagna, A., Cisban, E., Gia rdina, I., Orlandi, A., Parisi, G., Procaccini, A., Viale, M. & Zdravkovic, V., An empirical study of large, naturally occurring starling flocks: a benchmark in collective animal behaviour. *Animal Behaviour*, **76**, pp. 201–215, 2008. http://dx.doi.org/10.1016/j.anbehav.2008.02.004

[11] Attanasi, A., Cavagna, A., Del Castello, L., Giardina, I., Melillo, S., Parisi, L., Pohl, O., Rossaro, B., Shen, E., Silvestri, E. & Viale, M., Finite-size scaling as a way to probe near-criticality in natural swarms. *Physical Review Letters*, **113**, p. 238102, 2014. http://dx.doi.org/10.1103/PhysRevLett.113.238102

[12] Gordon, D.M., Paul, R.E. & Thorpe, K., What is the function of encounter patterns in ant colonies? *Animal Behaviour*, **45**, pp. 1083–1100, 1993. http://dx.doi.org/10.1006/anbe.1993.1134

[13] Kawashima, H. & Egerstedt, M., Manipulability of leader–follower networks with the rigid-link approximation. *Automatica*, **50**, pp. 695–706, 2014. http://dx.doi.org/10.1016/j.automatica.2013.11.041

[14] Centola, D., The spread of behavior in an online social network experiment. *Science*, **329**(5996), pp. 1194–1197, 2010.
http://dx.doi.org/10.1126/science.1185231
[15] Young, G.F., Scardovi, L., Cavagna, A., Giardina, I. & Leonard, N.E., Starling flock networks manage uncertainty in consensus at low cost. *PLoS Computational Biology*, **9**(1), p. e1002894, 2013.
http://dx.doi.org/10.1371/journal.pcbi.1002894
[16] Vicsek, T. & Zafeiris, A., Collective motion. *Physics Reports*, **517**, pp. 71–140, 2012.
http://dx.doi.org/10.1016/j.physrep.2012.03.004
[17] Vicsek, T., Cziro´k, A., Ben-Jacob, E., Cohen, I. & Shochet, O., Novel type of phase-transition in a system of self-driven particles. *Physical Review Letters*, **75**, pp. 1226–1229, 1995.
http://dx.doi.org/10.1103/PhysRevLett.75.1226
[18] Olfati-Saber, R., Fax, J.A. & Murray, R.M., Consensus and cooperation in networked multi-agent systems. *Proceedings of the IEEE*, **95**(1), pp. 215–233, 2007.
http://dx.doi.org/10.1109/JPROC.2006.887293
[19] Dusenbery, D.B., *Sensory Ecology: How Organisms Acquire and Respondto Information*. W.H. Freeman and Co.: New York, 1992.
[20] Mateo, D., Kuan, Y.K. & Bouffanais, R., Excess of social behavior reduces the capacity to respond to perturbations, 2015, arXiv:1509.08157 [nlin.AO].
[21] Sumpter, D., Buhl, J., Biro, D. & Couzin, I., Information transfer in moving animal groups. *Theory in Biosciences*, **127**, pp. 177–186, 2008.
http://dx.doi.org/10.1007/s12064-008-0040-1
[22] Sumpter, D.J.T., *Collective Animal Behavior*, Princeton University Press: Princeton, NJ, 2010.
http://dx.doi.org/10.1515/9781400837106

HEALTH RESEARCH, TEACHING AND PROVISION OF CARE: APPLYING A NEW APPROACH BASED ON COMPLEX SYSTEMS AND A KNOWLEDGE TRANSLATION COMPLEXITY NETWORK MODEL

A.H BROOK[1,2], H.M LIVERSIDGE[2], D WILSON[3], Z JORDAN[4], G HARVEY[5,6], R.J MARSHALL[5] & A.L KITSON[5,7]
[1]School of Dentistry, University of Adelaide
[2]Institute of Dentistry, Queen Mary University of London
[3]School of Medicine, University of Adelaide
[4]The Joanna Briggs Institute, University of Adelaide
[5]School of Nursing, University of Adelaide
[6]Manchester Business School, University of Manchester
[7]Green Templeton College, University of Oxford

ABSTRACT

Despite increased emphasis on the translation of research-based knowledge into practice, studies in the U.S.A. and Australia have found that up to 50 per cent of health care delivered does not accord with evidence-based guidelines. Health research, teaching and practice have traditionally emphasised defined inputs to produce specific, linear outputs and changes in teaching and practice may suffer delays in implementation when required to overcome barriers around spheres of interest. We are exploring a new approach based on the principles of complex systems and networks. In this paper, we used a successful knowledge translation project and a case study of a natural disaster, to model the effective application of these principles to a new health knowledge translation model, the Knowledge Translation Complexity Network. Following the Indian Ocean Tsunami of 2004, there were major challenges in identifying many of the dead. Research identified that Dental Age could be used to estimate the chronological age of unidentified victims up to 20 years of age. However, at the time the existing data were insufficient for this purpose and one author (HL) undertook to lead a knowledge creation and synthesis project. The research was evaluated by peer review, published in a leading journal and was subsequently implemented into practice as an identification tool in both paper and electronic forms. Subsequently the data charts and instructions have been translated into 18 languages and are used internationally in university teaching courses as well as in disaster identifications, with feedback evaluation from users providing further refinement. In conclusion, the development of the dentitions showed the characteristics of a complex adaptive system; of emergence, self-organisation, dynamic interactions, robustness and co-evolution. Further, the Dental Atlas incorporated elements of the key sub-networks of the new Knowledge Translation Complexity Network of problem identification (PI), knowledge creation (KC), knowledge synthesis (KS), implementation (I) and evaluation (E). Investigating real-world examples in this way can both highlight key aspects for future planning and identify gaps for development.
Keywords: complex systems, complexity network, health research, knowledge translation.

1 INTRODUCTION

Despite increased emphasis on the translation of research-based knowledge into practice, a landmark study in the USA found that up to 50 per cent of health care delivered does not accord with evidence-based guidelines [1]. Among the reasons for this maybe that much health research, teaching and practice has been guided by a traditional, linear process paradigm that defined inputs produce specifically predictable linear outputs [2]. The assumptions underlying this paradigm may be flawed

DOI: 10.2495/DNE-V11-N4-663-669

and so limit changes in health [2]. Rather organisations and communities are complex social systems and systems thinking is needed to understand and increase effectiveness [3].

We are exploring a new approach based on complex systems and network thinking. Our argument is that translation of new, research-derived health knowledge occurs within a complex adaptive system of many components, which can interact in a non-linear manner. The outcomes of those interactions are not fully predictable and can be greater than the sum of the inputs. The aim of this paper is to investigate a knowledge translation project in relation to a case study of a response to a natural disaster and map the effective application of these principles on a new health knowledge translation model, the Knowledge Translation Complexity Network.

2 THE KNOWLEDGE TRANSLATION COMPLEXITY NETWORK

This model has been developed by members of the Translational Science Task Force Team of the Faculty of Health Sciences, University of Adelaide to identify the movement of knowledge between those who generate it and those who use it [4].

We propose that Knowledge Translation is a complex network composed of five, sub-networks or 'core clusters' (problem identification, knowledge creation, knowledge synthesis, implementation and evaluation) that have multiple interactions. The pattern of these interactions varies at different tiems across the organisational systems of community, health, government, education and research. The components of this network are modelled in Fig. 1 and explained in the legend.

To optimise Knowledge Translation, participants in any of the five core clusters need to interact between the clusters. They also need to interact across the five organisational systems at the centre of the model which control funding for research, teaching and implementation. To progress the interactive nature of Knowledge Translation in the real world, it must be recognised and accounted for in health care systems.

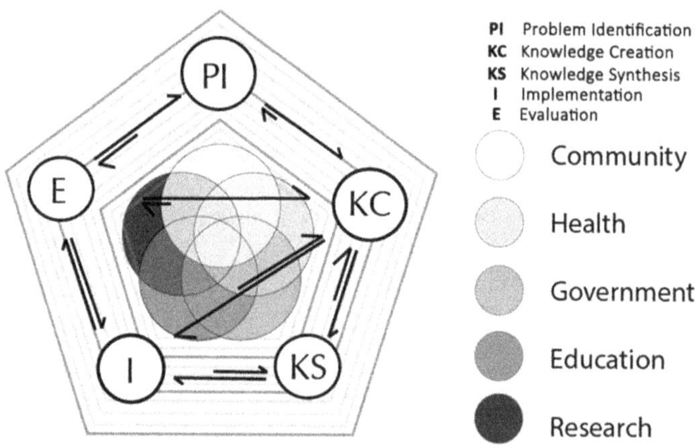

Figure 1: The KT Complexity Network Central to this knowledge translation model are five complex adaptive systems; Research, Education, Health, Government, Community. The outer five clusters of the network; problem identification (PI), knowledge creation, (KC), knowledge synthesis (KS), implementation (I), evaluation (E), function dynamically in space and time in the overall network. Each system and cluster may be weighted differently (as indicated by the size of the circles) and interact more or less frequently (as indicated by the presence/length of the arrows) depending on the needs of a given knowledge translation goal.

3 PROJECT OUTLINE OF THE LONDON ATLAS OF DENTAL DEVELOPMENT AND ERUPTION

Following the Indian Ocean tsunami of 2004, there were major problems in identifying the dead. While dental age assessed from the stage of development of the teeth would be a valuable tool to estimate the chronological age of unidentified victims up to 20 years of age, at the disaster sites there was uncertainty about how this should be undertaken and what standards to use.

A leading forensic odontologist, the late Prof Jules Kieser, was a member of one of the disaster volunteer identification teams in Thailand. He contacted Dr Helen Liversidge (an author of this paper), for advice on which data and methods were the most appropriate and accurate for estimation of human age, (this represents Problem Identification (PI)). Dr Liversidge, who at that time had recently undertaken a review of three different methods for dental age, recognised the limitations of the existing databases, charts and methodologies [5]. Dr Liversidge felt unable to recommend any one method in an evidence-based way, but undertook to lead a knowledge creation study (KC) to improve the accuracy of dentition-based age identification and knowledge synthesis project (KS), aimed at developing tools and methodology that would enable implementation (I). The initial aim of KS was to develop a new Atlas that was freely available online so that Disaster Volunteer Investigation teams could access it when needed.

The initial step in knowledge creation (KC) was evaluating (E) the existing research through review to identifying the knowledge gaps and the limitations of the existing databases, atlases and methods of measurement. Dr Liversidge had previously undertaken extensive research both of historical skeletal material and of clinical radiographs of patients [5–7] enabling her to accelerate the rate of KS. She now also contacted national and international colleagues (across health, education and research fields) to evaluate her processes. Having evaluated (E) the existing methods and identified the gaps in the available data, the problem identification (PI), knowledge creation (KC) and knowledge synthesis (KS) questions were further refined.

At this stage, the aim was described as creating a new dental Atlas with illustrations of dentition in 1-year age categories. Discussions and evaluation with colleagues in Paediatric Dentistry, Forensic Anthropology and Palaeodemography established that a uniform age distribution was necessary with equal numbers of males and females in each age group. Following evaluation and input from a North American dental anthropologist, the median tooth stage and eruption level were selected as representing development in each age category. A postgraduate student began gathering data (KC) from dental radiographs and constructing digital illustrations (KS) of age-dependent dentition. Further evaluation (E) was sourced from a several anthropologists with considerable experience in aging skeletal and dental material and from an academic forensic odontologist. From this, the methodology was further modified.

The initial project was completed and presented for evaluation (E) to the British Association of Forensic Odontology (an organisation which spans research, education, government and health). Following evaluation (E) feedback led to additional refinements and expansion of the project. Incorporating collections of teeth from the Royal College of Surgeons of England (Health) and the Natural History Museum in London (Government and Community) allowed the categorical age range to be expanded to cover almost all the entire developing dentition from 28 weeks in utero to 23 years of age. Testing an early version of the Atlas on archaeological skeletal remains demonstrated that 1-year categories were not suitable for early development. The resultant illustrations (KS) now represented prenatal tooth formation by one month categories, the first post-natal year by 3-month categories and the remaining ages by 1-year categories (Fig. 2).

An important move at this stage was to establish world copyright to the Atlas, called The London Atlas, so that this as a knowledge synthesis tool (KS) could be made freely available without cost

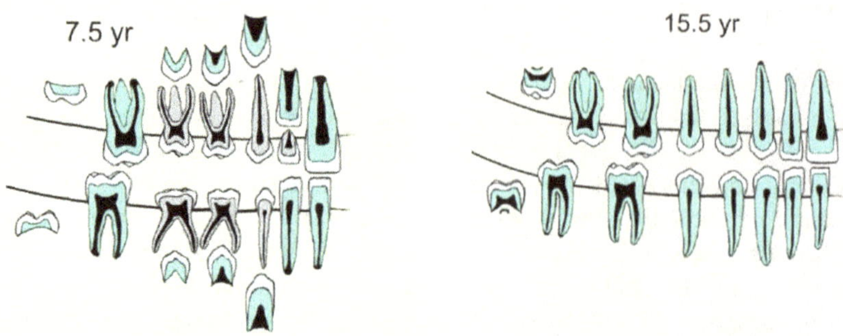

Figure 2: An example diagram from the atlas. Each tooth and eruption level represents the median tooth stage from children aged 7 and 15, respectively.

once completed. This insightful step of obtaining world copy right precluded barriers to timely implementation (I) This was followed by an initial publication in the foremost anthropological journal, The American Journal of Physical Anthropology [8].

Next, the accuracy of The London Atlas was evaluated (E) by comparison with two existing atlases enabling the final synthesis of the new knowledge product (KS) in context. Estimated age was compared to known real age for skeletal remains or radiographs for 1,514 individuals. Excellent reproducibility was observed for all three atlases, but The London Atlas correctly estimate more cases and showed no bias while the other atlases consistently underestimated chronological age. To independently evaluate (E) the implementation (I) of the application of The London Atlas, 90 third-year dental students were randomised into three groups. Each group used one of the three atlases to estimate the radiographic age and complete a questionnaire focussed on the design, clarity, simplicity and self-explanation of the atlas. The London Atlas was preferred in all quality measures tested [9]. These findings on accuracy were also published in The American Journal of Physical Anthropology [10]. To enhance implementation (I) of the new atlas, internet and mobile app versions of the Atlas were developed. The software was sent to forensic odontologists in three countries to pilot its use (E).

Throughout the duration of the project, Dr Liversidge continued to investigate ethnic differences in tooth formation in the UK, South Africa, West and East Africa, Malaysia, Japan, Australia (Aboriginal), New Zealand (Maori), Pacific Islands and North America (Inuit and Native Americans). The timing of tooth development was similar in several groups [11]. Further work has shown that malnutrition has no significant effect [12]. Both findings underpin the validity of the use of dental age in international settings and underscore the unanticipated research and educational value that continuous evaluation (E) has provided.

The London Atlas was used more widely to estimate the ages of victims in the air crash in Libya in May 2010 and the earthquake in New Zealand in February 2011. Professor Jules Kieser commented "…my team used the (London) Atlas when we did the identification of the victims of the tragic Christchurch Earthquake. The Atlas enabled even inexperienced volunteers to understand dental aging and possibly most importantly, it was hugely useful when presentation were made to the Coroner's Board". The American Society for Forensic Odontology now list The London Atlas in their guidelines as one of the methods. In addition, The London Atlas is used in teaching forensic anthropology, forensic dentistry, dental development and archaeology. It is also used in estimating disputed age of asylum seekers and age assessment of individual archaeological skeletons.

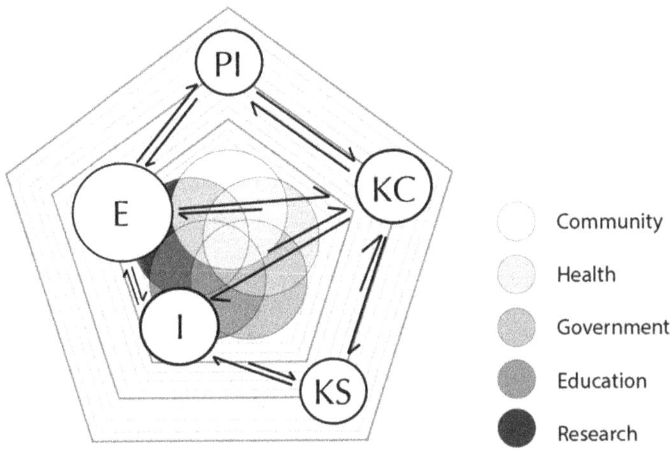

Figure 3: The London Atlas Project, identification of human age based on dentition. Problem Identification (PI) began with a goal to identify the age at death based on the dentition. Evaluation (E) of existing knowledge identified a deficiency in accuracy leading to the development of a Knowledge Creation (KC) research and interaction across the central systems of community, health, government, education and research coupled with evaluation led to an improved project goal and more defined and accurate categorical age determination. The research was developed as a tool. Knowledge Synthesis (KS). The London Atlas for chronological human dentition and further developed into an online app and a paper based illustration toolkit. Implementation of KS tools enable identification of human age. Evaluation throughout the process accelerated the rate of development.

A well-executed dissemination strategy has been key to enabling widespread use. Free access on the internet has resulted on an average 130 hits a day on the website [13]. The Atlas has been translated into 18 languages and is freely available online [14]. The app is also freely available and has been downloaded many thousands of times. A training video is also available [15].

4 DISCUSSION

Plotting the events of the successful project against the Knowledge Translation Complexity Network model allows both identification of the steps that enabled the highly positive outcome and also the gaps and problems that have impacted on the project. In this way, future developments for this work and for the application of the model can be highlighted.

A key factor has been the constant evaluation throughout the project that enabled its continued refinement even redefining the scope of the problem identification. The interaction with the community, represented by the end users, and the funding bodies, represented by government and universities, throughout the project. Each key sub-network of the model PI, KC, KS, I, E has been involved (Fig. 3). Problem identification occurred from the aftermath of the 2004 tsunami and was clarified by interactions with end users and professional colleagues, The Knowledge Creation built on existing knowledge that was also extended by interactive evaluation and feedback during the stages of the experimental work and development of The London Atlas and development of its distribution format. It was also farsighted to consider the implementation and obtain copyright at this stage to enable later open, free access to the final Atlas. It is important to highlight that repeated evaluation was embedded in the process e.g., during knowledge creation when the initial results were present to the British Association of Forensic Odontology, and the feedback was incorporated in the research. The Knowledge Synthesis, in which the research outcomes were reviewed in the light of existing data and evidence incorporated that ethnic differences and malnutrition had only limited effect on dental age, extending the application of The London Atlas as did the early establishment of world copyright.

Implementation involved the development of Internet and mobile app versions and continues with translation into different languages. Important cross links occurred with the central components of the model (Fig. 1 and 3) initially and during the project with the Saudi Arabian government funding one of the researchers (Dr AlQahtani) and Queen Mary University of London supporting implementation. The gaps were particularly limitations in funding for further applications such as an Apple version and the initial lack of involvement by the Scientific Committee of Forensic Odontology of Interpol (SCFOI). Recent discussions with SCFOI have moved this forward, and they now recommend it as one of the methods. These gaps relate to the important place of the 'government' complex within the overall network (Fig. 1 and 3) and illustrate the considerable effort required in these areas.

5 CONCLUSIONS

This project provided further validation for the use of dental development as a paradigm for general development. The careful and collaborative approach taken in initiating and carrying through of the project resulted in considerable impact, not only in providing a valuable tool in disaster investigation fulfilling the original need by also allowing emergence of multiple, multidisciplinary interactions. Mapping this completed project against the Knowledge Translation Complexity Network model suggests some initial validity for the model and its value in panning future work to enhance the translation of research into improved health outcomes. Using the model highlights the importance of targeted work on groups such as governments to promote the movement of new knowledge across systems.

REFERENCES

[1] Schuster, M., McGlynn, E. & Brook, R., How good is the quality of health care in the United States? *The Milbank Quarterly*, **76**, pp. 517–563, 1998. http://dx.doi.org/10.1111/1468-0009.00105

[2] Resnicow, K. & Page, S.E., Embracing chaos and complexity: a quantum change for public health. *American Journal Public Health*, **98**, pp. 1382–1389, 2008. http://dx.doi.org/10.2105/AJPH.2007.129460

[3] Sadia, R., The relationship between employee health, quality culture and organisational effectiveness: findings from the literature. *International Journal of Design & Nature and Ecodynamics,* **11**, pp. 1–9, 2016. http://dx.doi.org/10.2495/DNE-V11-N1-1-9

[4] Kitson, A.L., Jordan, Z., Wilson, D., Harvey, G., Marshall, R. & Brook, A.H., Complexity, chaos and the changing discourse of knowledge translation. University of Adelaide, Faculty of Health Science, Working Paper, 2016.

[5] Liversidge, H.M., Lyons, F. & Hector, M.P., The accuracy of three methods of age estimation using radiographic measurements of developing teeth. *Forensic Science International,* **131**, pp. 22–29, 2003. http://dx.doi.org/10.1016/S0379-0738(02)00373-0

[6] Liversidge, H.M., Accuracy of age estimation from developing teeth in a population of known age (0-5.4 years). *International Journal of Osteoarchaeology*, **4**, pp. 37–45, 1994. http://dx.doi.org/10.1002/oa.1390040107
[7] Liversidge, H.M. & Molleson, T.I., Developing permanent tooth length as an estimate of age. *Journal of Forensic Sciences*, **44**, pp. 917–920, 1999. http://dx.doi.org/10.1520/JFS12016J
[8] AlQahatani, S.J., Hector, M.P. & Liversidge, H.M., Brief communication: the London Atlas of human tooth development and eruption. *American Journal of Physical Anthropology*, **142**, pp. 481–490, 2010, available at http://www.ncbi.nlm.nih.gov/pubmed/20310064 http://dx.doi.org/10.1002/ajpa.21258
[9] AlQahtani, SJ., The London Atlas: developing an atlas of tooth development and testing its quality and performance measures. Queen Mary University of London 2012; PhD Thesis, available at https://qmro.qmul.ac.uk/xmlui/handle/123456789/8364
[10] AlQahtani, S.J., Hector, M.P. & Liversidge, H.M., Accuracy of dental age estimation charts Schour and Massler, Ubelaker and the London Atlas. *American Journal of Physical Anthropology*, **154**, pp. 70–78, 2014. http://dx.doi.org/10.1002/ajpa.22473
[11] Liversidge, H.M., Similarity in dental maturation in two ethnic groups of London children. *Annals of Human Biology*, **38**, pp. 702–715, 2011. http://dx.doi.org/10.3109/03014460.2011.609565
[12] Elamin, F. & Liversidge, H.M., Malnutrition has no effect on the timing of human tooth formation. *PLoS ONE*, **8**, e72274, 2013. http://dx.doi.org/10.1371/journal.pone.0072274
[13] Webpage www.atlas.dentistry.qmul.ac.uk
[14] Downloadable app www.qappsonline.com/apps/atlas/
[15] Training video www.youtube.com/watch?v=FSyQvZaiXGo&feature=c4-overview&list=UU7JSE8G0aOmtL1oYWGueSVg

THE MATURE STOMATOGNATHIC SYSTEM IS A COMPLEX ADAPTIVE SYSTEM

J.A. KAIDONIS[1], S. RANJITKAR[1], G.C. TOWNSEND[1], K.S.B. KOH[1], V.K.L. TOH[1] & A.H. BROOK[1,2]
[1]School of Dentistry, University of Adelaide, Australia
[2]School of Dentistry, Queen Mary University of London, UK

ABSTRACT

During development, the teeth, jaws and facial structures emerge in a process that has the characteristics of a complex adaptive system (CAS). The outcome is the mature stomatognathic system which is an interactive network of teeth, their occlusion and supporting mechanisms, the upper and lower jaws, the temporomandibular (jaw) joints, the muscles, the blood and nerve supplies, and the salivary glands. The aim of this paper is to explore whether this mature network also demonstrates the characteristics of a CAS. This network is a carefully balanced mechanism with functions required throughout the lifespan of the individual. It is exposed to a series of mechanical and chemical environmental challenges. Tooth wear is an example of the impact of such challenges, resulting in loss of surface enamel and dentine. The response is spread through the network: reparative secondary dentine occurs on the inner pulpal walls of the tooth; secondary cementum is formed and bone remodelling occurs in the suspensory mechanism; and the masticatory pattern of dental occlusion adapt under neuromuscular control. These adaptive changes arise from interactions at molecular and cellular levels. They are evidence of ongoing emergence and robustness in the network as it functions. Considering these changes to be positive responses that enable both maintenance of function and adaptation to age changes in the component parts of the network, represents a paradigm shift. It has major implications for oral health teaching and clinical practice. Attempting to restore teeth to the size and detailed cuspal shape, they showed when they first erupted is not only unnecessary for late adult function but can also place excessive loads on other subnetworks such as on the jaw joints and muscles. Thus, appreciating that both the developmental and mature stomatognathic networks are CAS establishes their value as exemplars for biological systems in general and has important implications for enhancing health care.
Keywords: caries, craniofacial, dental, dentine, enamel, erosion, oral health, phenomics, tooth wear

1 INTRODUCTION

Prenatal and postnatal growth and development of all craniofacial structures represent a complex adaptive system (CAS) where genetic, epigenetic and environmental influences occur on tissues at different levels with varying degrees of intensity and duration [1]. It is an interactive, carefully balanced network of components that remain dynamic throughout life.

Prenatal growth and development has a very strong genetic component that is moderated by epigenetic factors via maternal environmental exposure [1]. These interactions occur at a cellular and molecular level, however, from the commencement of postnatal growth and development and throughout life, there are continual physical environmental challenges that mechanically load the system at a ‘macro level’ influencing the craniofacial structures. The teeth, both individually and collectively as occluding opposing arches act as a ‘portal’ through which physical forces act on the network of underlying tissues. The intensity and the duration of the mechanical load acting via the occlusion is responsible for adaptive changes that can be further explained at cellular and molecular levels within the tissues. Such adaptations are uniquely evolved physiological mechanisms neces-

DOI: 10.2495/DNE-V11-N4-670-675

sary for tolerating environmental changes, and that they are necessary for survival. They are an indication to the robustness of the network. They are also a measure of physiological fitness of the individual, the population and indeed the species.

Coupled with these influences, there are also other dietary effects where food of low pH can potentially demineralise tooth surfaces through chemical action. This is naturally prevented by the symbiotic relationship developed between the human and the oral micro-biome [2].

The complexity model effectively explains craniofacial phenotypic variation, observed within and between populations, as always being in a state of flux, reflecting the intensity and duration of environmental stress. The model also emphasises that anatomical structures such as muscles and in particular bone, are physiologically 'plastic' and can change in accordance with the directional physical forces. It is when the individual cannot adapt, or is too slow to adapt in response to environmental stress that pathological inflammatory responses occur.

2 AIMS

The aims of this paper are to describe how the mature human stomatognathic system represents a CAS, and then to consider how this concept has major implications for research and teaching in oral health and clinical dental practice.

3 THE OVERALL NETWORK AND ITS INTERACTIONS WITHIN THE STOMATOGNATHIC SYSTEM

Figure 1 summarises the interactions that occur between the various components of the stomatognathic system, which is a part of the overall craniofacial complex. Moss' functional matrix hypothesis states that function determines the morphology of the craniofacial complex as an adaptive response to mechanotransduction [3]. For example, the influence of air passage through the nasal cavity when breathing can have a direct effect on the plasticity of bone and associated soft tissues to ensure the development of patent airways [4, 5]. In this paper, we discuss various subnetworks in the stomatognathic system involving multiple interactions between environmental factors and teeth that trigger various forms of adaptive responses (Fig. 1). The system is best described in terms of environmental factors, such as tooth wear from abrasive diet and acids as well as dental caries (tooth decay) from

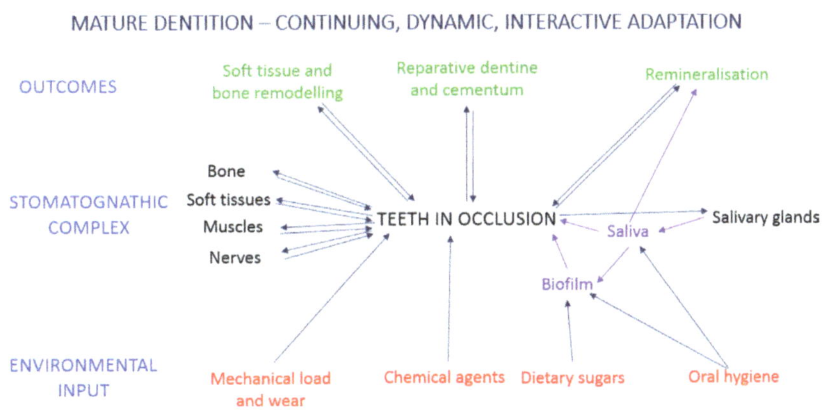

Figure 1: The mature stomatognathic system as a complex adaptive system in which reciprocal, recurrent interactions with environmental factors determine the outcome for health and disease.

dietary sugars and patient factors (i.e. poor oral hygiene and lack of protective biofilms), that interact within subnetworks to produce adaptations upon the already established stomatognathic components. The whole system is dynamic where the various inter-related components adapt in response to the environmental factors. Thus, within the overall network, one subnetwork is that between the biofilm, saliva and tooth surfaces (Fig. 1).

3.1 Environmental influence on the neuromascular network

3.1.1 Mechanical loading and wear

Once the teeth emerge and an occlusion is established, the diet becomes one of the most predominant postnatal environmental factors in the CAS. The diet helps to establish loading between the occlusal surfaces of opposing (i.e. upper and lower) teeth, establishing the masticatory cycle. The physical forces acting on the teeth stimulate and shape both immediate oral components such as the underlying periodontal structures, as well as other craniofacial structures distant to and not directly associated with the oral cavity [6]. The coarseness of the diet and the intensity of masticatory activity are responsible for both the occlusal and interproximal wear of teeth.

3.1.2 Chemical acid effects

Chemical acid effects on teeth are referred to as dental erosion (Figs 1 and 2). It can be caused by either strong intrinsic acids (e.g. gastric acids from vomiting and regurgitation, commonly known as "gastric reflux") or extrinsic acids (e.g. soft drinks, citrus fruits and alcoholic beverages) at around pH 3.0 or less [7]. Dental erosion results in immediate surface loss and softening after acid contact [8, 9], and it is a growing concern because of increasing prevalence in recent decades [10]. Repeated exposure to such agents results in the clinically apparent loss of tooth structure. Severe wear that has extended well into dentine has the potential to allow bacterial fluids and toxins from oral environment to communicate with the pulp tissue through open dentinal tubules.

3.1.3 Dietary sugars, oral hygiene and dental caries

Dental caries (decay) is caused by acidic demineralisation of teeth but at a higher acidic environment (around pH 4.0 to 5.0) compared with dental erosion (at around pH 3.0 or less). It is caused by the

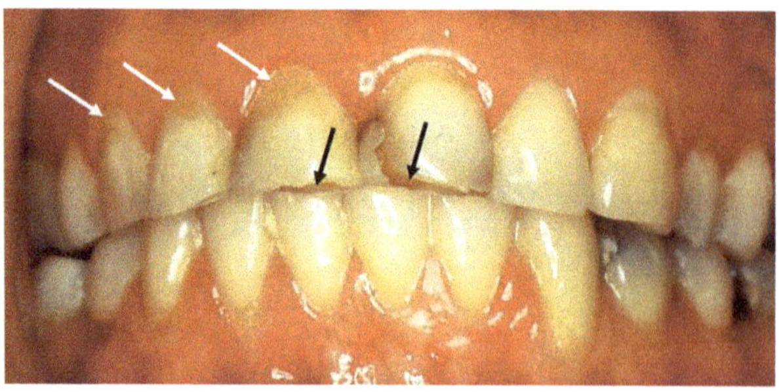

Figure 2: Tooth wear caused by a combination of both tooth grinding and dental erosion (wine tasting). Tooth grinding produces Facetted surfaces (black arrows) from contact during lateral movement of the lower jaw and dental erosion causes loss of surface features (such as surface grooves) and scooping (white arrows). [Courtesy of Dr D Hunt and Dr J McIntyre].

presence of fermented sugars (carbohydrates) that are fermented by oral bacterial to produce mild lactic acid. In initial stages, the acid causes sub-surface demineralisation as redeposition of dissolved crystals occurs within the surface layers. As the process continues, this is overcome and cavitation occurs.

Both extensive forms of erosive wear and dental caries can result in exposure of the pulp chamber, allowing free passage of bacteria and their products (toxins) to elicit cellular response (inflammation) in the pulpal tissues. Both abrasive and erosive wear can result in loss of vertical dimension (i.e. bite overclosure), whereas interproximal wear, combined with mesial tooth migration [11], is responsible for a reduction in dental arch length and width.

3.2 Adaptive response to environmental influences

The masticatory cycle can be described as a rhythmic muscular action that is centrally controlled via proprioceptive feedback loops originating from craniofacial structures. Over time, the functional forces acting on teeth result in tooth wear, the degree of which is determined by the consistency of the diet. The changing tooth morphology due to wear does not reduce masticatory efficiency due to differential wear between enamel and dentine [12, 13], and the nature of the masticatory cycle changes in tandem with the wear. The periodontal ligament, the muscle fibres and the temporomandibular joints provide proprioceptive information and are a part of a feedback loop that changes the masticatory cycle via the central ‘masticatory /grinding centre’ in the central nervous system. As the cusps of the teeth are reduced by wear, the masticatory pattern becomes broader. Collectively, the neuromuscular system is designed to perpetually change proportionally to masticatory load, and to perpetually change tooth anatomy.

The adaptive response to mechanical (abrasive) wear includes a cellular response by pulpal cells (i.e. odontonblasts) to lay reparative dentine (i.e. a calcified bridge at the dentine boundary around the pulp chamber) and isolate the pulp from the oral environment. In relation to the loss of vertical dimension of the face (from reduced tooth crown height from severe wear), the occlusal forces acting on the periodontal ligament stimulate cementoblasts on the root surface of the tooth to produce cementum. This causes the teeth to continually erupt to compensate for the occlusal wear. This is another example of how the physical forces, on the one hand, can produce wear and, on the other hand, compensate for tooth wear, thereby allowing the teeth to remain functional.

The degree of masticatory muscle development is also a cellular response to the diet. Masticatory muscle hypertrophy is not uncommon in populations living in environments where vigorous mastication is required to process fibrous, hard food. In tandem with this, vigorous masticatory forces acting on teeth also stimulates alveolar bone growth, aligns trabecular patterns within the bone and therefore is responsible for bi-maxillary protrusion observed in pre-contemporary hunter-gatherer populations. In summary, although the genetic makeup of the individual underlies the direction of general development, it can be argued that the environmental forces acting throughout life alters the genetic phenotype accordingly. Overall, this to some degree, not only explains craniofacial differences observed within populations but explains craniofacial differences between populations.

In a healthy stomatognathic system, a good balance is maintained in this dynamic subnetwork between risk factors (e.g. poor diet and poor oral hygiene for dental caries, and gastric and/or strong dietary acids for dental erosion) and host factors (e.g. protection from saliva and oral biofilm made of commensural bacteria) (Fig. 1). While biofilm is a physical barrier against acids, it also acts as a reservoir of calcium, phosphate and fluoride ions that are necessary for remineralising (repairing) demineralised tooth structure. At the tooth enamel surface, there is a constant demineralisation/

remineralisation interaction, with the healthy balance favouring remineralisation. Chewing stimulates salivary flow and an increase in its buffering capacity.

4 DISCUSSION

The application of the concept of the CAS in craniofacial growth and development highlights the dynamic nature of the craniofacial system throughout life. Static concepts of the past, such as Bonwill's Triangle, the Curve of Spee, Monson's Curves etc., are teleological concepts that do not cater for dynamic change and variability [14]. The implications for various types of dental treatment need to be explored. For example:

1. Restoring teeth to their original newly erupted anatomy in a generally worn dentition may reduce functionality, as steep cuspal inclines could lock the bite. Such iatrogenic change may be significant and akin to sudden (unnatural) environmental change where the body is unable to adapt fast enough, thereby resulting in clinical symptoms such as tenderness to biting (i.e. occlusal trauma) or even temporomandibular joint pain.
2. The use of implants in a dynamic system also needs to be explored. Implants by their nature are osseo-integrated without a periodontal ligament. However, the periodontium with its cellular, fibrous and fluid contents allows adaptation to high masticatory forces. Lack of periodontal movement and tooth migration of implants may affect the adaptive responses in the craniofacial system.
3. The physical forces on teeth also have orthodontic implications in association with relapse of the dentition after orthodontic treatment. Understanding how the force vectors change with tooth repositioning could give some insight towards avoiding failures.
4. There is a dynamic subnetwork within the overall system influencing the integrity of the tooth surface. The balances within this determine whether ongoing demineralisation will lead to caries or whether remineralisation and lesion healing will predominate. This subnetwork is incorporated in Fig. 1.

5 CONCLUSIONS

Application of the concepts of CASs to craniofacial growth and development has many positive outcomes, including an appreciation of the overview of the interactions that occur within subnetworks in the mature stomatognathic system. This approach enables appreciation of the nature of the patterns in the functional dentition, with teeth acting as a 'portal' through which adaptive changes occur. Such knowledge has implications in dental education and clinical practice in relation to restoration of functionality of teeth. Furthermore, this approach opens new directions for future research in dental phenomics that has the potential to correlate large-scale genomic and phenotypic data and to improve health outcomes [15, 16].

REFERENCES

[1] Brook, A., O'Donnell, M.B., Hone, A., Hart, E., Hughes, T., Smith, R. & Townsend, G.C., General and craniofacial development are complex adaptive processes influenced by diversity. *Australian Dental Journal*, **59**(1 Suppl), pp. 13–22, 2014. http://dx.doi.org/10.1111/adj.12158

[2] Kaidonis, J.A. & Townsend, G.C., The 'Sialo-Microbial Dental Complex' in health and disease. *Annals of Anatomy* (in press), doi:10.1016/j.aanat.2015.02.002 [EPub ahead of print). http://dx.doi.org/10.1016/j.aanat.2015.02.002

[3] Moss, M.L., The theoretical analysis of the functional matrix. *Acta Biotheoretica*, **18**, pp. 195–202, 1968.
http://dx.doi.org/10.1007/BF01556727
[4] Moss, M.L., The functional matrix hypothesis revisited. 2. The role of an osseous connected cellular network. *American Journal of Orthodontics and Dentofacial Orthopaedics*, **112**, pp. 221–226, 1997.
http://dx.doi.org/10.1016/S0889-5406(97)70249-X
[5] Moss, M.L., The functional matrix hypothesis revisited. 4. The epigenetic antithesis and the resolving synthesis. *American Journal of Orthodontics and Dentofacial Orthopedics,* **112**, pp. 410–417, 1997.
http://dx.doi.org/10.1016/S0889-5406(97)70049-0
[6] Ascenzi, A., The micromechanics versus the macromechanics of cortical bone – a comprehensive presentation. *Journal of Biomechanical Engineering*, **110**, pp. 357–363, 1998.
http://dx.doi.org/10.1115/1.3108454
[7] Ranjitkar, S., Smales, R. & Kaidonis, J.A., Oral manifestations of gastroesophageal reflux disease. *Journal of Gastroenterology and Hepatology*, **27**, pp. 21–27, 2012.
http://dx.doi.org/10.1111/j.1440-1746.2011.06945.x
[8] Mann, C., Ranjitkar, S., Lekkas, D., Hall, C., Kaidonis, J.A., Townsend, G.C. & Brook, A.H., Three-dimensional profilometric assessment of early enamel erosion simulating gastric regurgitation. *Journal of Dentistry*, **42**, pp. 1411–1421, 2014.
http://dx.doi.org/10.1016/j.jdent.2014.06.011
[9] Kwek, S., Mian, M., Hall, C., Xie, Z., Yong, R., Kaidonis, J.A., Townsend, G.C. & Ranjitkar, S., Nanoscratch testing for the assessment of enamel erosion under conditions simulating wine tasting. *Australian Dental Journal*, **60**, pp. 12–17. 2015.
http://dx.doi.org/10.1111/adj.12277
[10] Lussi, A. & Carvalho, T.S., Erosive tooth wear: a multifactorial condition of growing concern and increasing knowledge. *Monographs of Oral Science*, **25**, pp. 1–15, 2014.
http://dx.doi.org/10.1159/000360380
[11] Kaidonis, J.A., Townsend, G.C. & Richards, L.C., Interproximal tooth wear - a new observation. *American Journal of Physical Anthropology*, **88**, pp. 105–107, 1992.
http://dx.doi.org/10.1002/ajpa.1330880109
[12] Kaidonis, J.A., Townsend, G.C. & Richards, L.C., Abrasion: an evolutionary and clinical view. *Australian Prosthodontic Journal*, **6**, pp. 9–16, 1992.
[13] Kaidonis, J.A., Tooth wear: the view of the anthropologist. *Clinical Oral Investigations*, **12**(Suppl 1), pp. S21–S26, 2008.
http://dx.doi.org/10.1007/s00784-007-0154-8
[14] Kaidonis, J.A., Ranjitkar, S., Lekkas, D., Brook, A. & Townsend, G., Functional dental occlusion: an anthropological perspective and implications for practice. *Australian Dental Journal*, **59**(1 Suppl), pp. 167–173, 2014.
http://dx.doi.org/10.1111/adj.12133
[15] Yong, R., Ranjitkar, S., Townsend, G., Smith, R., Evans, A., Hughes, T., Lekkas, D. & Brook, A.H., Dental phenomics: advancing genotype to phenotype correlations in craniofacial research. *Australian Dental Journal*, **59**(1 Suppl), pp. 34–47, 2014.
http://dx.doi.org/10.1111/adj.12156
[16] Houle, D., Govindaraju, D.R. & Omholt, S., Phenomics: the next challenge. *Nature Reviews Genetics,* **11**, pp. 855–866, 2010.
http://dx.doi.org/10.1038/nrg2897

COMPLEX SYSTEMS IN HUMAN DEVELOPMENT: SEXUAL DIMORPHISM IN TEETH AND FINGERPRINTS OF AUSTRALIAN TWINS

R.J.O. TADURAN[1], S. RANJITKAR[1], T. HUGHES[1], G. TOWNSEND[1] & A.H. BROOK[1,2]
[1]Craniofacial Biology Research Group, School of Dentistry, University of Adelaide, Australia.
[2]School of Dentistry, Queen Mary University of London, UK.

ABSTRACT

Human teeth and fingerprints have similar embryological origin from epithelial-mesenchymal interactions. This study aims to determine the nature and extent of sexual dimorphism in fingerprints and teeth of twins; investigate the influences of genetic, epigenetic and environmental factors on observed variation; identify possible developmental associations between the phenotypes; and explore whether both systems display the features of complex adaptive systems. Mesiodistal (MD) measurements from both primary and permanent teeth and ridge density (RD) from three different finger areas, namely ulnar (U), radial (R), and proximal (P), from fingerprints of the same set of monozygotic and dizygotic Australian twins (28 males and 31 females aged 8 to 10 years, and aged 13 to 16 years, respectively) were collected and analysed. Sexual dimorphism was observed in both the primary and permanent dentitions, with the latter showing greater magnitude of differences than the former. There was no observed sexual dimorphism in the fingerprints of the 8 to 10 year cohort, but a few finger areas (left index U, right index R, left little R, and left little P) of the 13- to 16-year cohort exhibited significant differences, showing that friction ridges expand over time. It was concluded that both dentition and dermatoglyphics display characteristics of complex adaptive systems.

Keywords: complex adaptive system, dentition, tooth size, dermatoglyphics, fingerprints, human development, mesiodistal, ridge density, sexual dimorphism.

1 INTRODUCTION

Sexual dimorphism is defined as the phenotypic or observable difference between males and females of the same biological species. A number of studies have been conducted on sexual dimorphism in the human dentition. In general, males have larger crown diameters than females [1,2], and sexual dimorphism is greater in permanent than in primary dentition [2,3]. Meanwhile, studies on sexual dimorphism in adult human dermatoglyphs reveal that males have fewer ridges than females [4,5].

The development of the human dentition and of dermatoglyphs has similar embryological origin from epithelial-mesenchymal interactions [6]. During embryonic growth, patterning, or the establishment of groups of cells in the proper relationship to each other and to surrounding tissues, occurs. Patterning is a longitudinal event that eventually leads to differentiation of cells to assume specialised functions and shapes. Primary teeth start to develop around 4 to 6 weeks in utero [6], while ridged skin begins to form around 10 to 16 weeks in utero [7].

Human development in general is a complex adaptive process that is influenced by genetic, epigenetic and environmental factors [8]. The genetic factors interact with epigenetic and environmental elements at the molecular level and form complex networks within the cells, and from these dynamics arise the higher level tissues. Sexual dimorphism has been suggested by some researchers to be

DOI: 10.2495/DNE-V11-N4-676-685

governed by sex chromosomes [9,10] but there have been others who have suggested that hormones are also important [11,12]. Dental and dermatoglyphic patterns develop in utero, and once stabilised, their unique and persistent morphology makes them valuable models in studying sexual dimorphism. This study aimed to determine the nature and extent of sexual dimorphism in mesiodistal crown measurements of teeth and ridge density counts of fingerprints of twins; to investigate the influences of genetic, epigenetic and environmental factors; and to identify possible developmental associations and covariance of the studied phenotypes.

2 MATERIALS AND METHODS

Twin samples were obtained from the ongoing longitudinal studies of the Craniofacial Biology Research Group in the School of Dentistry at the University of Adelaide [13], which is one of the four most extensive investigation of its type in the world [14]. Serial dental casts of primary and permanent dentitions, and rolled ink prints of fingers of individuals aged 8 to 10 years and 13 to 16 years of the same cohort of monozygotic and dizygotic Australian twins (28 males and 31 females) were collected and analysed. Dental casts showing wear, caries, or restorations and ten-prints with smudge ink and scarred patterns in any of the fingerprints were excluded.

Mesiodistal crown diameter (MD) was measured as the distance between the mesial and distal contact points of the tooth crown [1,15] by using a 2D imaging system. Dental casts were oriented using an adjustable stage to obtain the correct plane or angle before taking images and a calibrated Image J [16] software was used to digitise landmarks (Fig. 1). Measurements were obtained for central incisors (I1), lateral incisors (I2), canines (C), first molars (M1) and second molars (M2) of primary and permanent teeth.

Ridge density (RD) was measured by counting friction ridges diagonally on a square measuring 5 mm × 5 mm to isolate ridges within a well-defined area [4,17]. Measurements were obtained in three different finger ridge locations, namely, distal ulnar (U), distal radial (R), and proximal (P), of each of the ten fingerprints in both cohorts. The three areas were located by dividing the rolled prints into four sectors, with two perpendicular axes that cross two ridges above the centre [17]. Fingers

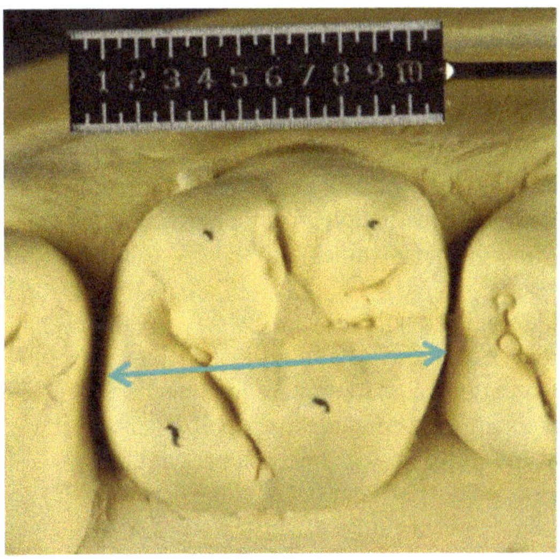

Figure 1: Mesiodistal (MD) measurement on a permanent upper first molar from the occlusal view.

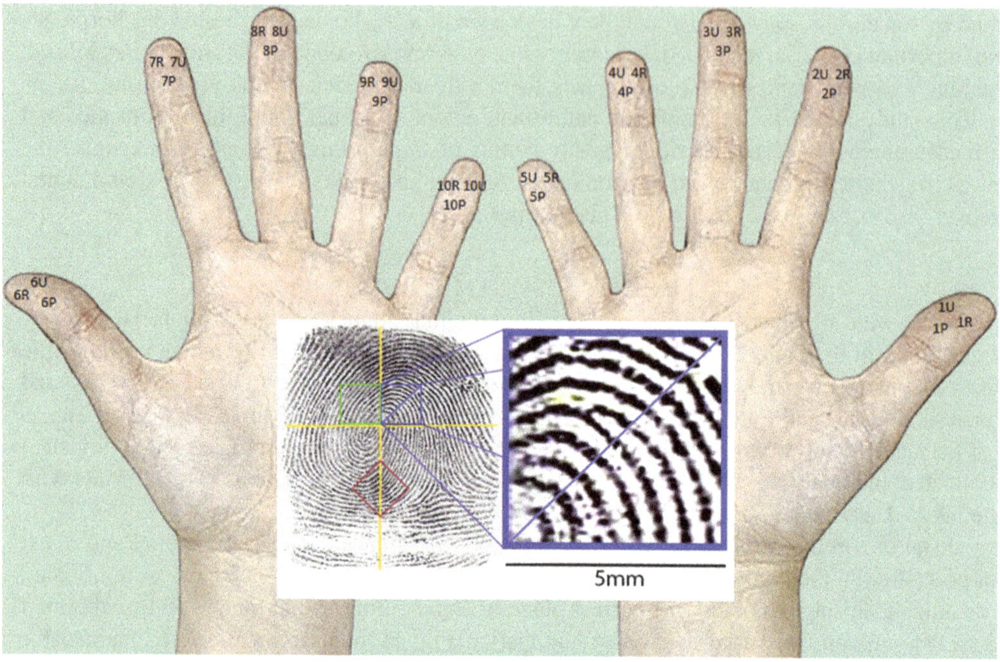

Figure 2: Locations of fingerprint areas – distal ulnar (U), distal radial (R) and proximal (P) – on each finger, and an enlarged 5 mm × 5 mm area to facilitate ridge density (RD) count.

were assigned with a numerical order 1–10, with finger 1 being the right thumb and finger 10 being the left little finger (Fig. 2).

Data were statistically analysed using XLSTAT statistical software. Descriptive statistics including means, standard deviations (SD) and coefficients of variation (CV) were computed for MD and RD. Differences between sexes and sides were calculated using student's unpaired t-test. Differences among fingers and finger areas were examined with analysis of variance (ANOVA). Finger ridge differences between age groups were compared with paired t-test. Pearson's coefficient was calculated to examine correlations between the variables.

3 RESULTS

MD measurements and RD counts were normally distributed, and results of intra and inter-operator repeatability tests showed that errors in methodological measurements were negligible and not likely to bias data. Shown in Table 1 are the mean values, SD and CV of mesiodistal (MD) measurements of primary and permanent teeth.

Highlighted in yellow background and bold text are the sexually dimorphic dental measurements, where mean values are different between sexes at $p < 0.05$. Mean values of MD crown dimensions of males were consistently greater compared to females in all teeth. Permanent dentitions displayed greater sexual dimorphism compared to primary dentitions. There were no left-right differences observed in MD measurements of all primary and permanent dentitions.

Shown in Table 2 are the mean values, SD and CV of ridge density (RD) counts of fingerprints of 8 to 10 years old cohort and 13 to 16 years old cohort.

All mean values of RD based on finger type and finger area were statistically different to each other at $p < 0.05$. Highlighted in blue background and italics are the RD counts that were found to

Table 1: Descriptive statistics for mesiodistal (MD) measurements of primary and permanent teeth in male and female twins.

	Males								Females							
	Right				Left				Right				Left			
	n	Mean	SD	CV	n	Mean	SD	CV	n	Mean	SD	CV	n	Mean	SD	CV
Primary																
Maxillary																
il	28	6.34	0.37	0.06	28	**6.32**	0.42	0.07	31	6.23	0.39	0.06	31	**6.22**	0.41	0.07
c	28	**6.87**	0.37	0.05	28	**6.84**	0.36	0.05	31	**6.70**	0.39	0.06	31	**6.69**	0.37	0.06
ml	28	**7.05**	0.44	0.06	28	**7.12**	0.41	0.06	31	**6.83**	0.38	0.06	31	**6.90**	0.40	0.06
m2	28	**8.86**	0.47	0.05	28	**8.83**	0.43	0.05	31	**8.62**	0.40	0.05	31	**8.63**	0.41	0.05
Mandibular																
il	28	**4.05**	0.31	0.08	28	**4.03**	0.30	0.07	31	**3.92**	0.27	0.07	31	**3.90**	0.28	0.07
i2	28	**4.61**	0.35	0.08	28	**4.62**	0.32	0.07	31	**4.50**	0.31	0.07	31	**4.49**	0.31	0.07
c	28	**5.89**	0.33	0.06	28	**5.90**	0.32	0.05	31	**5.73**	0.30	0.05	31	**5.74**	0.32	0.06
ml	28	**7.87**	0.49	0.06	28	**7.91**	0.44	0.06	31	**7.67**	0.38	0.05	31	**7.65**	0.40	0.05
m2	28	**10.07**	0.47	0.05	28	**10.03**	0.46	0.05	31	**9.81**	0.41	0.04	31	**9.77**	0.41	0.04
Permanent																
Maxillary																
Il	28	**8.70**	0.49	0.06	28	**8.64**	0.48	0.05	31	**8.43**	0.53	0.06	31	**8.43**	0.48	0.06
C	28	**8.13**	0.41	0.05	28	**8.18**	0.45	0.06	31	**7.71**	0.46	0.06	31	**7.66**	0.41	0.05
Ml	28	**10.45**	0.52	0.05	28	**10.46**	0.52	0.05	31	**10.09**	0.51	0.05	31	**10.11**	0.48	0.05
M2	28	**10.35**	0.58	0.06	28	**10.46**	0.64	0.06	31	**9.86**	0.53	0.05	31	**9.81**	0.56	0.06
Mandibular																
Il	28	5.42	0.33	0.06	28	**5.46**	0.32	0.06	31	5.30	0.33	0.06	31	**5.29**	0.33	0.06
12	28	6.00	0.37	0.06	28	**5.98**	0.34	0.06	31	5.80	0.37	0.06	31	**5.78**	0.37	0.06
C	28	**7.19**	0.42	0.06	28	**7.17**	0.44	0.06	31	**6.66**	0.40	0.06	31	**6.65**	0.41	0.06
Ml	28	**11.31**	0.63	0.06	28	**11.33**	0.60	0.05	31	**10.83**	0.63	0.06	31	**10.85**	0.62	0.06
M2	28	**10.95**	0.71	0.06	28	**11.11**	0.79	0.07	31	**10.21**	0.55	0.05	31	**10.42**	0.67	0.06

Table 2: Descriptive statistics for ridge density (RD) counts of fingerprints and finger areas in male and female twins.

	Males								Females							
	Right				Left				Right				Left			
	n	Mean	SD	CV	n	Mean	SD	CV	n	Mean	SD	CV	n	Mean	SD	CV
Ulnar																
8–10 years old																
Thumb	28	15.9	1.7	0.1	28	16.1	1.9	0.1	31	*15.7*	2.1	0.1	31	*16.3*	2.0	0.1
Index	28	16.4	1.7	0.1	28	16.5	2.1	0.1	31	16.3	1.8	0.1	31	16.6	2.0	0.1
Middle	28	17.1	1.7	0.1	28	17.2	2.3	0.1	31	16.9	2.1	0.1	31	17.4	2.3	0.1
Ring	28	*17.7*	2.1	0.1	28	*18.5*	2.3	0.1	31	*17.5*	2.3	0.1	31	*18.6*	2.1	0.1
Little	28	17.2	2.0	0.1	28	17.5	2.0	0.1	31	*17.0*	1.9	0.1	31	*17.7*	*1.7*	0.1
13–16 years old																
Thumb	28	14.3	1.7	0.1	28	14.0	2.1	0.1	31	14.8	2.0	0.1	31	14.5	1.8	0.1
Index	28	14.5	1.9	0.1	28	**13.8**	1.7	0.1	31	15.3	2.1	0.1	31	**15.0**	2.2	0.1
Middle	28	15.8	1.9	0.1	28	16.4	1.8	0.1	31	15.7	2.2	0.1	31	16.6	1.9	0.1
Ring	28	16.5	1.9	0.1	28	17.0	2.0	0.1	31	16.6	2.1	0.1	31	17.5	2.1	0.1
Little	28	15.9	1.9	0.1	28	15.6	1.5	0.1	31	15.4	1.9	0.1	31	16.2	1.8	0.1
Radial																
8–10 years old																
Thumb	28	17.8	1.9	0.1	28	17.3	2.1	0.1	31	17.5	1.8	0.1	31	17.2	2.4	0.1
Index	28	17.0	1.8	0.1	28	17.2	2.0	0.1	31	17.3	1.8	0.1	31	17.4	2.0	0.1
Middle	28	18.3	2.2	0.1	28	18.1	2.3	0.1	31	18.2	1.9	0.1	31	18.0	2.3	0.1
Ring	28	*19.0*	2.1	0.1	28	*18.4*	2.2	0.1	31	18.9	2.2	0.1	31	18.4	2.3	0.1
Little	28	18.2	2.1	0.1	28	17.6	2.1	0.1	31	18.3	2.0	0.1	31	17.8	2.3	0.1

							13–16 years old									
Thumb	28	15.5	2.2	0.1	28	15.7	2.2	0.1	31	16.3	1.9	0.1	31	15.5	2.9	0.2
Index	28	**15.0**	2.0	0.1	28	15.4	1.7	0.1	31	**16.3**	1.8	0.1	31	15.8	1.8	0.1
Middle	28	16.4	1.7	0.1	28	16.8	1.7	0.1	31	16.9	2.2	0.1	31	17.1	2.0	0.1
Ring	28	*17.3*	*1.7*	0.1	28	*16.3*	1.8	0.1	31	17.5	2.6	0.2	31	17.1	1.9	0.1
Little	28	16.8	2.0	0.1	28	**16.1**	1.6	0.1	31	16.9	2.8	0.2	31	**17.1**	2.1	0.1
							Proximal									
							8–10 years old									
Thumb	28	14.9	1.7	0.1	28	14.6	1.7	0.1	31	*15.1*	*1.7*	0.1	31	*14.5*	1.9	0.1
Index	28	14.2	1.9	0.1	28	13.9	2.2	0.2	31	14.3	1.9	0.1	31	13.9	2.0	0.1
Middle	28	14.3	2.1	0.1	28	14.1	1.8	0.1	31	*14.3*	1.8	0.1	31	*13.8*	*1.7*	0.1
Ring	28	14.7	2.0	0.1	28	14.3	2.1	0.1	31	14.6	1.8	0.1	31	14.1	2.1	0.2
Little	28	14.1	2.4	0.2	28	13.8	2.0	0.1	31	13.6	1.9	0.1	31	13.5	2.0	0.1
							13–16 years old									
Thumb	28	13.2	1.8	0.1	28	13.2	1.8	0.1	31	13.7	2.4	0.2	31	13.2	1.7	0.1
Index	28	12.9	1.7	0.1	28	12.7	1.5	0.1	31	13.4	1.6	0.1	31	13.0	1.6	0.1
Middle	28	12.5	2.1	0.2	28	12.6	1.2	0.1	31	13.1	1.4	0.1	31	12.8	1.5	0.1
Ring	28	13.1	1.9	0.1	28	12.8	1.6	0.1	31	13.6	1.8	0.1	31	13.4	2.2	0.2
Little	28	12.3	1.9	0.2	28	**12.1**	2.0	0.2	31	12.6	1.7	0.1	31	**13.4**	2.4	0.2

Table 3: Pearson correlation coefficients of the dental (mesiodistal width) and dermatoglyphic (ridge density) traits.

	Males	**Females**	**Males**	**Females**
Finger Area	**Primary**		**Permanent**	
		Maxillary MD		
Ulnar	**0.20**	**0.16**	**0.16**	**0.13**
Radial	**0.12**	**0.11**	0.11	0.08
Proximal	−0.03	**−0.12**	−0.04	0.00
		Mandibular MD		
Ulnar	**0.27**	**0.25**	**0.38**	**0.30**
Radial	**0.13**	**0.16**	**0.24**	**0.23**
Proximal	**−0.07**	**−0.16**	−0.11	−0.01

be statistically different on both sides. More differences were observed in the younger (8–10 years old) cohort. The ulnar area (U) was the most irregular of all finger ridge areas in terms of left-right discrepancies, yet it was observed to have smaller RD in the right side, which indicates thicker finger ridges. Most of the thicker ridges in the radial (R) and proximal (P) areas were observed in the left fingers.

Highlighted in yellow background and bold text are the sexually dimorphic dermatoglyphic measurements, where mean values are different between sexes at $p < 0.05$. Only a few sexually dimorphic finger ridge areas were observed to be greater from the older cohort (13–16 years old), with smaller mean values for RD in males, which indicates thicker and fewer friction ridges within the 5 mm^2 square. Based on paired t-test, all mean values of RD are different between age groups at $p < 0.05$, with the older cohort showing smaller RD values compared to the younger group.

Pearson's coefficients (r) between teeth and fingerprints are presented in Table 3. Highlighted in yellow background and bold text are the significant correlations between dental trait MD and dermatoglyphic characteristic RD (ulnar, radial and proximal) at $p < 0.05$. Overall, the correlations between teeth and fingerprints are low, but the RD in the ulnar area emerged with the highest coefficients with MD diameter of the maxillary and mandibular dentition in both sexes.

Correlation coefficients were calculated within groups of dental and dermatoglyphic variables at $p < 0.05$. All MD diameters taken from different tooth types were positively correlated to each other in the primary teeth (0.31–0.95). Only some MD measurements (71 of 231 in males, 119 of 231 in females) were positively correlated with each other in the permanent teeth (0.40–0.84), and more significant values were observed in females than males. Meanwhile, only some RD counts from different fingers and areas were positively correlated to each other (312 of 435 in young males and 252 of 435 in young females, 147 of 435 in old males and 112 of 435 in old females), with more significant values in males and the young cohort. Greater r values were observed in the old cohort (0.35 to 0.75) compared to the young cohort (0.19 to 0.59).

4 DISCUSSION

The degree and patterning of sexual dimorphism in the dentition varies according to tooth type. Our observation of the permanent dentition showing more pronounced sexual dimorphism than primary dentition agrees with previous findings [1,2]. The permanent lower canines displayed the largest sexual dimorphism in MD measurements, similar to the results of Garn *et al.* [3] and Ribeiro *et al.* [12] who pointed out that dental development might occur under fairly high levels of testosterone influence, and this could explain the differences in sexual dimorphism between primary and permanent teeth of same individuals.

The degree and patterning of sexual dimorphism in the dermatoglyph varies according to the finger area and finger type. In this study, there was no observed sexual dimorphism in the 8 to 10 year old group, while fingerprints of the 13 to 16 year old group displayed sexual dimorphism in the ulnar and radial areas of the index finger, and radial and proximal areas of the little finger. Few studies have investigated subadult fingerprints, and our results could be preliminary empirical evidence that friction ridges expand as individuals grow and develop, and possibly more so in males than females. It seems that sexual dimorphism in dermatoglyphic development commences during puberty, when a testosterone surge occurs in males [18].

There are three surges of testosterone that occur in normal male development. The first surge begins at around the 7th to 9th week of pregnancy, following testicular differentiation, and the testosterone level is at its highest around the 14th week [19,20]. The second surge occurs after birth due to the reduction of oestrogen produced by the placenta [18]. The third surge, as previously mentioned, occurs during puberty.

Primary dentition starts to develop at around 4 to 6 weeks in utero [6] and continues until around one year after birth. Permanent dentition begins to form 14 weeks in utero and continues to develop until at around 14 years of age [21]. Meanwhile, primary ridge formation begins at around 10 to 16 weeks and ends on the 17th week, then secondary ridges form until the 24th week in utero [7]. Our results suggest that the first two testosterone surges have a critical role in the sexual dimorphism of both the primary and permanent dentitions, while the third testosterone surge influences the sexual dimorphism of the fingerprints.

Human development is a complex adaptive process [8] and the human body is a complex adaptive system. This study has shown that both teeth and fingerprints are interconnected, yet they still have a degree of autonomy. They share a similar embryological origin and epithelial-mesenchymal interactions [6], yet they develop and interact with epigenetic and environmental factors differently. The interactions may be unpredictable, with no central control, but they are not random, as regularities and patterns emerge to find the best fit with the environment.

This research is the first to study both human dental and dermatoglyphic traits. Although a number of studies have been conducted on human dentition and dermatoglyphs separately, no attempt has been made previously to explore possible correlations between the two. This research furthers the investigation on the complex mechanisms and interactions occurring during dental, dermatoglyphic and general development with mesiodistal (MD) diameters of the teeth and ridge density (RD) counts of the fingerprints. In the future, more dental and dermatoglyphic traits could be studied together.

ACKNOWLEDGEMENTS

The authors wish to thank The Australian Dental Research Foundation (ADRF), the NHMRC of Australia, Australian Twin Registry and Australian Multiple Birth Association.

REFERENCES

[1] Moorrees, C.F.A., Thomsen, S.O., Jensen, E. & Yen, P.K., Mesiodistal crown diameters of the deciduous and permanent teeth in individuals. *Journal of Dental Research*, **36**, pp. 39–47, 1957. http://dx.doi.org/10.1177/00220345570360011501

[2] Ribeiro, D., Sampson, W., Hughes, T., Brook, A. & Townsend, G., Sexual dimorphism in the primary and permanent dentitions of twins: an approach to clarifying the role of hormonal factors (Chapter 5). *New Directions in Dental Anthropology: Paradigms, Methodologies and Outcomes*, eds. G. Townsend, E. Kanazawa & H. Takayama, University of Adelaide Press: South Australia, pp. 46–64, 2012.

[3] Garn, S.M., Lewis, A.B., Swindler, D.R. & Kerewsky, R.S., Genetic control of sexual dimorphism in tooth size. *Journal of Dental Research*, **46**, pp. 963–972, 1967. http://dx.doi.org/10.1177/00220345670460055801

[4] Acree, M.A., Is there a gender difference in fingerprint ridge density? *Forensic Science International*, **102**, pp. 35–44, 1999. http://dx.doi.org/10.1016/S0379-0738(99)00037-7

[5] Taduran, R.J.O., Tadeo, A.K.V., Escalona, N.A.C. & Townsend, G.C., Sex determination from fingerprint ridge density and white line counts in Filipinos. *HOMO - Journal of Comparative Human Biology*, 2015. http://dx.doi.org/10.1016/j.jchb.2015.11.001

[6] Nanci, A., *Ten Cate's Oral Histology: Development, Structure, and Function,* 7th edn., Elsevier Health Sciences: Missouri, pp. 16–107, 2008.

[7] Kücken, M. & Newell, A.C., Fingerprint formation. *Journal of Theoretical Biology*, **235**, pp. 71–83, 2005. http://dx.doi.org/10.1016/j.jtbi.2004.12.020

[8] Brook, A.H., Brook O'Donnell, M., Hone, A., Hart, E., Hughes, T.E., Smith, R.N. & Townsend, G.C., General and craniofacial development are complex adaptive processes influenced by diversity. *Australian Dental Journal*, **59S**, pp. 13–22, 2014. http://dx.doi.org/10.1111/adj.12158

[9] Guatelli-Steinberg, D., Sciulli, P.W. & Betsinger, T.K., Dental crown size and sex hormone concentrations: another look at the development of sexual dimorphism. *American Journal of Physical Anthropology*, **137**, pp. 324–333, 2008. http://dx.doi.org/10.1002/ajpa.20878

[10] Alvesalo, L., Human sex chromosomes in oral and craniofacial growth. *Archives of Oral Biology*, **54S**, pp. 18–24, 2009. http://dx.doi.org/10.1016/j.archoralbio.2008.06.004

[11] Dempsey, P.J., Townsend, G.C. & Richards, L.C., Increased tooth crown size in females with twin brothers: evidence for hormonal diffusion between human twins in utero. *American Journal of Human Genetics*, **11**, pp. 577–586, 1999.

[12] Ribeiro, D.C., Brook, A.H., Hughes, T.E., Sampson, W.J. & Townsend, G.C., Intrauterine hormone effects on tooth dimensions. *Journal of Dental Research*, **92**, pp. 425–431, 2013. http://dx.doi.org/10.1177/0022034513484934

[13] Townsend, G., Bockmann, M., Hughes, T., Mihailidis, S., Seow, K.W. & Brook, A., New approaches to dental anthropology based on the study of twins (Chapter 2). *New Directions in Dental Anthropology: Paradigms, Methodologies and Outcomes*, eds. G. Townsend, E. Kanazawa & H. Takayama, University of Adelaide Press: South Australia, pp. 10–21, 2012.

[14] Hughes, T.E., Townsend, G.C., Pinkerton, S.K., Bockmann, M.R., Seow, W.K., Brook, A.H., Richards, L.C., Mihailidis, S., Ranjitkar, S. & Lekkas, D., The teeth and faces of twins: providing insights into dentofacial development and oral health for practicing oral health professionals. *Australian Dental Journal*, **59S**, pp. 101–116, 2014. http://dx.doi.org/10.1111/adj.12101

[15] Brook, A.H., Smith, R.N., Elcock, C., al-Sharood, M.H., Shah, A.A., Khalaf, F., Robinson, D.L., Lath, D.L. & Karmo, M., The measurement of tooth morphology: validation of an image analysis system. *13th International Symposium of Dental Morphology*. ed. E. Zadzinska, University of Lodz Press: Lodz, pp. 475–482, 2005.

[16] Schneider, C.A., Rasband, W.S. & Eliceiri, K.W., NIH Image to ImageJ: 25 years of image analysis. *Nature Methods*, **9**, pp. 671–675, 2012. http://dx.doi.org/10.1038/nmeth.2089

[17] Gutiérrez-Redomero, E., Alonso, C., Romero, E. & Galera, V., Variability of fingerprint ridge density in a sample of Spanish Caucasians and its application to sex determination. *Forensic Science International*, **180**, pp. 17–22, 2008. http://dx.doi.org/10.1016/j.forsciint.2008.06.014

[18] Griffin, J.E. & Wilson, J.D., Disorders of the testes and the male reproductive tract (Chapter 18). *Williams Textbook of Endocrinology,* 10th edn., eds. P.R. Larsen, H.M. Kronemberg, S. Melmed & K.S. Polonsky, W.B. Saunders Company: Philadelphia, pp. 709–770, 2003.

[19] Reyes, F.I., Boroditsky, R.S., Winter, J.S.D. & Fairman, C., Studies on human sexual development. II. Fetal and maternal serum gonadotropin and sex steroid concentrations. *Journal of Clinical Endocrinology and Metabolism*, **38**, pp. 612–617, 1974. http://dx.doi.org/10.1210/jcem-38-4-612

[20] Knickmeyer, R.C. & Baron-Cohen, S., Fetal testosterone and sex differences. *Early Human Development*, **82**, pp. 755–760, 2006. http://dx.doi.org/10.1016/j.earlhumdev.2006.09.014

[21] AlQahtani, S.J., Hector, M.P. & Liversidge, H.M., Brief communication: the London atlas of human tooth development and eruption. *American Journal of Physical Anthropology*, **142**, pp. 481–490, 2010. http://dx.doi.org/10.1002/ajpa.21258

A COMPLEX ADAPTIVE SYSTEM IN WHICH ENVIRONMENTAL STRESS AFFECTS GENE EXPRESSION DURING DEVELOPMENT

K.S.B. KOH[1], V.K.L. TOH[1], M. BROOK O'DONNELL[2], S. RANJITKAR[1] & A. H. BROOK[1,3]
[1]School of Dentistry, University of Adelaide, Australia.
[2]Annenberg School of Communication, University of Pennsylvania.
[3]Institute of Dentistry, Queen Mary University of London, England.

ABSTRACT

Dental development is a complex adaptive system influenced by genetic-epigenetic-environmental factors. The mature dentition is a paradigm for general development and is an accessible, permanent record of interactions affecting development from 6 weeks in-utero to 20 years of age. Published research on the dentition of a large group of Romano-Britons has indicated that three major environmental insults acting throughout the developmental period are associated with high frequencies of dental variations. Based on those studies and the literature, the aetiological network incorporates the effects of excessive lead ingestion, poor nutrition and infections on the endocrine and immune systems, epigenome and gene expression. The environmental insults act independently and synergistically on both the endocrine and immune systems, which in turn have reciprocal interactions with each other, as well as on the epigenome. Thus, a complex subnetwork influences the epigenome and modifies one or more of three mechanisms by which it controls gene expression, that is, DNA methylation, post-translational histone modification and microRNA expression. The environmental-epigenetic-genetic interactions behave as a complex adaptive system that can be illustrated in a network diagram. The epigenetic modifications can have long-term health effects in the individual and some may also be heritable.
Keywords: complex adaptive system, dental development, environmental stress, epigenetics, gene expression, general development, Romano-Britons.

1 INTRODUCTION

Environmental stress is associated with significant health problems. In this paper, we explore its impact on the development of dentition which is a valuable paradigm for general development. The aim is to advance previous work by exploring further the actions of the environmental factors identified as having major impact in a group of Romano-Britons, and to model a complex network involving the interactions between environmental factors, body systems, epigenetic mechanisms and gene expression.

1.1 Complex adaptive system

Dental development is a complex adaptive system (CAS), and is characterised by dynamic behaviours that arise from non-linear spatial-temporal interactions among multiple component systems at different levels of organisation [1]. The ecosystem, social networks and financial systems are other examples of complex systems.

The exploration of dental development as a complex adaptive system began with Brook & Brook-O'Donnell [2]. Then, Brook *et al.* [3] demonstrated that the CAS characteristics of self-organisation,

DOI: 10.2495/DNE-V11-N4-686-695

bottom-up emergence, multitasking, self-adaptation, variation, tipping points, critical phases and robustness are present in the development of dental tissues to form an integrated functional system. The interactions among multiple agents at lower molecular and cellular levels, which result in the emergence of the dentition at a higher level, are initiated by genetic, epigenetic and environmental factors. The dentition is able to adapt and evolve in changing environments during its development and as a mature system.

1.2 Dental development as a paradigm for general development

The formation of teeth begins with a sequential and reciprocal series of inductive signals transmitted between the epithelium and mesenchyme-derived neural crest cells [4]. Then, the dental epithelium thickens to form the dental lamina, which subsequently develops a specific number of higher-level organs, called the soft tissue tooth germs [3]. The tooth germs continue to develop morphologically into incisors, canines, premolars and molars at specific locations within the dental arches. Then, specialised cells called ameloblasts and odontoblasts differentiate to control the secretion of enamel and dentine matrices respectively. The mineralised mature teeth erupt and form a dynamic system that continues to adapt during function [5,6]. Dental development is a valuable paradigm for general development because teeth are accessible, permanent records of development that extends from 6 weeks in-utero to 20 years of age. Dental development is therefore a good model for the study of normal and variations in general development and disease.

1.3 Genetic-epigenetic-environmental interactions

Dental development is initiated by genetic, epigenetic and environmental interactions. Recent research in epigenetics has changed the way we think about normal development, disease states and expression of the genome. Environmental factors act on the genome by affecting epigenetic mechanisms, such as DNA methylation, histone modification and regulation of microRNAs. Epigenetic modifications influence gene expression and can result in an altered phenotype while the genomic structure remains unchanged. Interestingly, some epigenetic changes are potentially reversible, but others persist throughout life. Children born during the period of the Dutch hunger winter from 1944 to 1945 had increased rates of adult-onset diseases such as coronary heart disease and obesity when there was maternal exposure to famine during pregnancy [7]. Some epigenetic modifications have also been shown to be heritable or transgenerational. In another Dutch hunger winter study, results showed that first generation famine exposure in-utero was associated with increased second generation adiposity and poor health in later life [8]. The epigenome changes over the human lifespan. However, it is most vulnerable to environmental factors during embryogenesis when the DNA synthesis rate is high, and the elaborate DNA methylation patterning required for normal tissue development is established [9]. In-utero and early-life events can significantly modulate the epigenome and lead to altered phenotypes and disease susceptibility throughout the life course [10].

1.4 Environmental stress in early life and dental variation: Romano-British population

The numerous studies on the substantial skeletal material of the Romano-Britons from Poundbury, Dorset in the 200–400 A.D. provide much evidence concerning lifestyle and medical conditions [11]. Major environmental insults known to affect development in the Poundbury population were high levels of lead ingestion, poor general nutrition with identified vitamin and mineral deficiencies and recurrent infections [5]. The teeth and bones were found to have high lead content [12,13]. Possible sources of lead were lead kitchenware, and utilisation of lead for preparation of drinks in some Roman recipes [11]. Besides that, an agricultural lifestyle with cereal as the predominant element in the diet contributed to their poor nutrition [11]. High incidence of rickets and scurvy were likely to have occurred following deficiency of micronutrients, such as Vitamin C and D, in addition to poor maternal health during gestation [14]. Furthermore, gastrointestinal infections aggravated their nutritional status and health [15]. The findings from Brook *et al.* [5] were that both Romano-Britons and Modern Britons had the same general dental characteristics and types of variation. However, there were substantial differences in the dimensions of teeth and prevalence of variation between the two groups. It was suggested that the long-term environmental insults during dental development influenced this complex adaptive system to produce the greater number of phenotypic variations seen in the dentition of the Romano-Britons [5].

2 AIMS

The aim of this paper is to develop an aetiological network model based on findings of environmental stress from the Romano-British and Poundbury Camp studies. Following that, we will consider the multiple independent, summative, synergistic and reciprocative interactions of excessive lead ingestion, poor nutrition, and recurrent infections on the immune and endocrine systems and the epigenome.

3 MATERIALS AND METHODS

This paper works further to Brook *et al.* [5], which summarised the findings for four outcomes of dental development: tooth number; crown size and shape; root size and shape; and enamel structure scored using standardised measurements [16–18].

Here, a thorough literature search of all the interrelated effects of excessive lead ingestion, recurrent infections and poor nutrition on the immune and endocrine systems, as well as on the epigenome is reported.

4 RESULTS AND DISCUSSION

Our findings [5] show that in comparison with Modern Britons of European ancestry, the Romano-Britons exhibited the same types of dental development variations and sexual dimorphism but a much higher prevalence of hypodontia, smaller crown and root size, and enamel defects. These results provide evidence for environmental stress affecting phenotypic outcomes arising throughout the morphological, differentiation and calcification stages of tooth development. Different genes act during different developmental stages, so that the finding of substantial numbers and degrees of phenotypic variations in each stage is a strong indication that the aetiology of the higher prevalence of these variations is environmental rather than genetic mutations. Set out in Fig. 1 is our proposal, developed from evidence in the literature reviewed below for the complex adaptive system emerging from the interactions of these environmental agents.

4.1 Lead

No level of lead exposure is considered safe and its cumulative effects are harmful to body systems. Lead kitchenware and Roman recipes, which included lead for preparation of drinks, were major sources in the Romano-Briton population [11]. After ingestion, lead is stored in teeth and bones, and later bone remodelling becomes an endogenous source of lead exposure [19]. Endogenous and

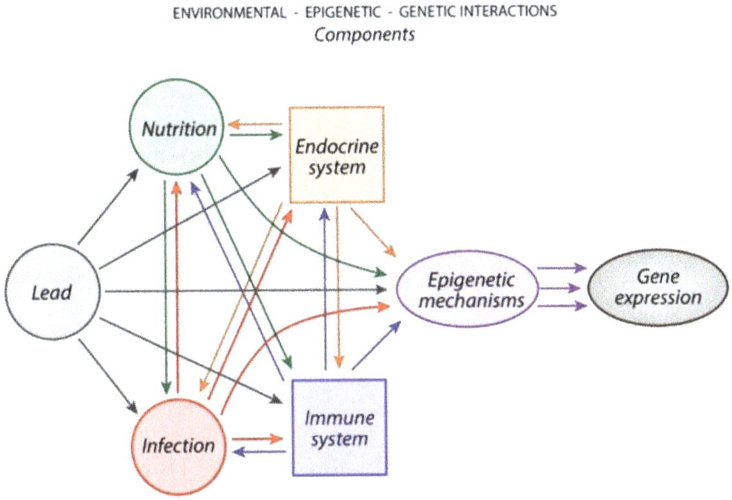

Figure 1: Environmental-epigenetic-genetic interactions. A complex adaptive system with multiple, often reciprocal interactions by which the environmental factors interact with body systems and epigenetic mechanisms to alter gene expression.

exogenous variables influencing bodily effects of lead include stress, dietary deficiencies and concomitant exposure to other chemical, biological and radiological hazards. These potentially modify, or even potentiate the toxic effects of lead [19]. In addition, lead can also cross the placenta, and the blood-brain barrier. Moreover, in-utero and lactational exposures further increase the risk of lead toxicity during the gestational and neonatal period [19].

4.2 Effect of lead on nutrition

Long-term effects of excessive lead ingestion include digestive disturbances, colic and diarrhoea [11]. When prolonged, these result in poor absorption of nutrients and digestive malnutrition. Besides that, environmental chemical exposure and nutrient deprivation can occur simultaneously and synergistically. For example, food chains are often contaminated by chemical agents found in soil and water. Lee *et al.* [20] hypothesise a unifying mechanism that attempts to link epigenetic alterations in relation to DNA hypomethylation due to chemical agents and nutrient deficiency. Both chemical exposure and nutrient deficiency may impair the synthesis of S-adenosylmethionine (SAM) via a common pathway, which results in DNA hypomethylation and modulation of the epigenome [20].

4.2.1 Effect of lead on the immune system, thus increasing the body's susceptibility to infection

Lead impairs immune function and alters host resistance. Lead exposure can result in impaired immune function of T-lymphocytes and erythrocytes, and a delayed immune response [21,22]. There was a decrease in the percentage of CD4+ and significant increase in CD8+ cells in children exposed to high lead levels [23]. Lead exposure in adults produces an impaired host resistance to bacterial and viral infections [24,25]. In addition, lead exposure produces a shift in the functional capacity of the immune system, which may lead to an increased risk of atopic and autoimmune diseases, while also impairing host defenses against infections and cancer [26].

4.2.2 Effect of lead on the endocrine system and epigenome

Lead has been found to be an endocrine disruptor in the Drosophila fruit fly [27]. The ability of lead to act directly on the epigenome is probably one of the most important, having consequences long after this environmental stressor has been removed. Perinatal lead exposure in mice during development and lactation leads to persistent life-course changes in energy expenditure, food intake, body weight, total body fat, glucose tolerance and insulin response [28].

4.3 Poor nutrition

4.3.1 Effect of poor nutrition on lead toxicity

Dietary deficiencies modify the toxicodynamics of lead: diets deficient in iron, zinc and calcium have been shown to increase the gastrointestinal absorption of lead [19]. Furthermore, maternal dietary protein levels modify the developmental immunotoxicity of lead in rat offsprings [29]. Individuals living in developing countries are more likely to have nutritional deficiencies that increase lead absorption [30]. Children who are engaged in child labour tend to belong to the same population that is at risk of nutritional deficiencies. This enhances the adverse health effects of lead on top of increasing its absorption into the body [31].

4.3.2 Effect of poor nutrition on the immune system

The immune system may be compromised by nutritional deficiencies during prenatal and early postnatal life. In rural Gambia, infection-related mortality was much higher for adults born in the annual climatically determined, nutrition-deficient ‘hungry’ season, which suggests immune function, may be compromised by this early-life event [32]. Furthermore, thymus development during infancy is critically sensitive to environmental exposures, with smaller thymuses observed in those born in the hungry season. The compromised thymus and hence T-lymphocyte development result in defective adaptive immunity later in life.

In contrast, the combined effects of some hormones may positively affect on the immune system. Davis [33] found that nutritional deprivation, physical and emotional stress and environmental temperature are mediated by hormones and the endocrine system, which alter the immune system. He proposed that Growth Hormone is involved in homeostasis of the immune system during nutrient deprivation. Partitioning of nutrients away from muscle growth makes it more available for use by the immune system. Also, Growth Hormone exerts a direct positive effect on the thymus, lymphocytes and other components of the immune system.

4.3.3 Effect of poor nutrition on recurrent infections

The relationship between infection and malnutrition is synergistic [34]. There is evidence for both the adverse effect of infection on nutritional status and the increased susceptibility of malnourished individuals to infection. Each may worsen the other and that the potentiated effects of malnutrition and infection combined are thus greater than the sum of the two.

Specific micronutrients, for example, vitamin A, beta-carotene, folic acid, vitamin B_{12}, vitamin C, riboflavin, iron, zinc, and selenium have immunomodulating functions. They influence the susceptibility of a host to infectious diseases, including their course and outcome. The antioxidant properties of the micronutrients may not only regulate immune homeostasis of the host, but also alter the genome of microbes particularly viruses [35].

4.3.4 Effect of poor nutrition on the endocrine system

Early exposure to undernutrition has been strongly associated with low birth weight, epigenetic programming and functional alteration of the hypothalamo-pituitary-adrenal (HPA) axis throughout the lifespan [36]. Studies using animal models have shown that pituitary corticotrophs and the adrenal cortex are affected. In humans, recent evidence shows that there is indeed HPA-axis dysregulation [36].

4.3.5 Effect of poor nutrition causing epigenetic changes

There is now compelling evidence to show that poor nutrition causes epigenetic changes in animals and humans. In the Dutch hunger winter study, individuals who were prenatally exposed to famine in 1944–1945 and had poor early-life nutritional status showed DNA methylation changes at the IGF2 locus compared with their unexposed, same-sex siblings [37]. Maternal diet consistently demonstrates alteration in the offspring's epigenome [38]. These findings support previous epidemiologic studies which suggested that adult disease risk is associated with early-life environmental conditions and is greatly influenced by changing pattern of the epigenome and gene expression. The association was specific for periconceptional exposure, reinforcing that very early mammalian development is a crucial period for establishing and maintaining epigenetic marks.

4.4 Recurrent infections

4.4.1 Effect of recurrent infections on nutrition

Infections have been found to aggravate micronutrient deficiencies by reducing nutrition intake, increasing nutritional losses and interfering with utilisation via alteration of metabolic pathways [35]. Infection is a common precipitating and aggravating factor of malnutrition, while malnutrition also tends to be a major factor in the occurrence of infection [39]. This is reflected in the reciprocal interactions in Fig. 1.

4.4.2 Effect of recurrent infections on the immune system

Prenatal infection may further reduce the capacity of the innate neonatal immune system to respond to endotoxin; neonatal rat pups responded to challenge with significantly reduced corticosterone while maternal infection suppresses the offspring's inflammatory response to lipopolysaccharide [40].

4.4.3 Effect of recurrent infections on the endocrine system

In a guinea pig model, offspring of dams exposed to endotoxin during pregnancy exhibited an attenuated cortisol response to environment stress in the weaning period. In adulthood, the female offsprings had diminished cortisol response to immune challenges while both male and female offsprings exhibited altered febrile response [41].

4.4.4 Effect of recurrent infections causing epigenetic changes

Recurrent infections can lead to epigenetic modification of an imprinted gene, such as IGF2. Bacteria translocate to the placenta of intrauterine growth-restricted mice foetuses, and cause structural alteration within the challenged placental layer, DNA hypermethylation and down regulation of IGF2 [42]. In addition, there is increased DNA methylation at several CpG islands in the gastric mucosae of individuals infected with *H. pylori* compared to uninfected individuals.

4.5 Endocrine system

4.5.1 Effect of the endocrine system on the immune system and epigenetics

A number of hormones, for example, growth hormones (GH), prolactin (PRL), Insulin-like Growth Factor-1 (IGF-1), glucocorticoids (GC) and thyroid hormones, have receptors in and exert biologic actions on tissues of the immune system [33]. PRL secretion in response to stress has an immunomodulatory role: it stimulates the immune system directly and reduces the degree to which GCs are secreted in response to stress. GH, PRL and IGFs all play an integral role in growth, maintenance, repair and function of the immune system [33]. These further interactions are also included in Fig. 1. Moreover, the effects of endocrine disruptors are heritable or transgenerational [44, 45].

5 CONCLUSION AND FUTURE WORK

Environmental stress is a significant health problem globally. Environmental factors such as heavy metals, malnutrition and infection affect developing countries disproportionately more. These facts emphasise the value of studying the dentition as a complex adaptive system and this Romano-British population as a well-documented, accessible source of material.

This paper builds upon previous work exploring of the dentition as a self-adaptive complex system. It demonstrated that the dentition is a valuable model for investigating the control of general development and the causes of developmental abnormalities and systemic illnesses. Many of the same genes, epigenetic and environmental influences and similar mechanisms are present during development of the dentition and of other body tissues and systems. A promising area of future research is exploring the possibility of reversing the long-term adverse outcomes of the epigenetic change.

ACKNOWLEDGEMENTS

This investigation was supported by the Australian Dental Research Foundation. We are grateful to Theya Molleson of the Natural History Museum, London, who provided not only access to material but also valuable advice and support. Contributions to the database used were made by a series of postgraduate students: L.K. Foo, C.A. Johns, M. Scheers, J.M. Smith and C.A. Underhill.

REFERENCES

[1] Honavar, V., Complex Adaptive Systems Group Iowa State University, available at http://www.cs.iastate.edu/~honavar/alife.isu.html

[2] Brook, A.H. & O'Donnell, M.B., The dentition: a complex system demonstrating self-principles. *Self-Adaptive and Self-Organizing Systems (SASO), Fifth IEEE International Conference*, pp. 208–209, 2011.
http://dx.doi.org/10.1109/saso.2011.41

[3] Brook, A.H., O'Donnell, M.B., Hone, A., Hart, E., Hughes, T.E., Smith, R.N. & Townsend, G.C., General and craniofacial development are complex adaptive processes influenced by diversity. *Australian Dental Journal*, **59**(s1), pp. 13–22, 2014.
http://dx.doi.org/10.1111/adj.12158

[4] Bei, M., Molecular genetics of tooth development. *Current Opinion in Genetics & Development*, **19**(5), pp. 504–510, 2009.
http://dx.doi.org/10.1016/j.gde.2009.09.002

[5] Brook, A.H., Koh, K.S.B. & Toh, V.K.L., Influences in a biologically complex adaptive system: environmental stress affects dental development in a group of Romano-Britons. *International Journal of Design & Nature and Ecodynamics*, **11**(1), pp. 33–40, 2016.
http://dx.doi.org/10.2495/DNE-V11-N1-33-40

[6] Kaidonis, J.A., Ranjitkar, S., Lekkas, D., Brook, A.H. & Townsend, G.C., Functional dental occlusion: an anthropological perspective and implications for practice. *Australian Dental Journal*, **59**(s1), pp. 162–173, 2014.
http://dx.doi.org/10.1111/adj.12133

[7] Painter, R.C., Roseboom, T.J. & Bleker, O.P., Prenatal exposure to the Dutch famine and disease in later life: an overview. *Reproductive Toxicology*, **20**(3), pp. 345–352, 2005.
http://dx.doi.org/10.1016/j.reprotox.2005.04.005

[8] Painter, R.C., Osmond, C., Gluckman, P., Hanson, M., Phillips, D.I.W. & Roseboom, T.J., Transgenerational effects of prenatal exposure to the Dutch famine on neonatal adiposity and health in later life. *An International Journal of Obstetrics & Gynaecology*, **115**(10), pp. 1243–1249, 2008.
http://dx.doi.org/10.1111/j.1471-0528.2008.01822.x

[9] Jirtle, R.L. & Tyson, F.L. (eds), *Enviromental Epigenomics in Health and Disease*, Springer-Verlag: Berlin Heidelberg, 2013.

[10] Boekelheide, K., Blumberg, B., Chapin, R.E., Cote, I., Graziano, J.H., Janesick, A., Lane, R., Lillycrop, K., Myatt, L., Thayer, K.A. & Waalkes, M.P., Predicting later-life outcomes of early-life exposures. *Environmental Health Perspectives*, **120**(10), pp. 1353–1361, 2012.
http://dx.doi.org/10.1289/ehp.1204934

[11] Farwell, D.E. & Molleson, T., *Poundbury Vol 2-The Cemeteries,* Friary Press: Dorchester, UK, 1993.

[12] Waldron, H.A., Mackie, A. & Townshend, A., The lead content of some Romano-British bones. *Archaeometry*, **18**(2), pp. 221–227, 1976.
http://dx.doi.org/10.1111/j.1475-4754.1976.tb00165.x

[13] Whittaker, D.K. & Stack, M.V., The lead, cadmium and zinc content of some Romano-British teeth. *Archaeometry*, **26**(1), pp. 37–42, 1984.
http://dx.doi.org/10.1111/j.1475-4754.1984.tb00315.x

[14] Lewis, M.E., Life and death in a civitas capital: metabolic disease and trauma in the children from late Roman Dorchester, Dorset. *American Journal of Physical Anthropology,* **142**(3), pp. 405–416, 2010.
http://dx.doi.org/10.1002/ajpa.21239

[15] Walker, P.L., Bathurst, R.R., Richman, R., Gjerdrum, T. & Andrushko, V.A., The causes of porotic hyperostosis and cribra orbitalia: a reappraisal of the iron-deficiency-anemia hypothesis. *American Journal of Physical Anthropology*, **139**(2), pp. 109–125, 2009.
http://dx.doi.org/10.1002/ajpa.21031

[16] Brook, A.H. & Smith, J.M., The aetiology of developmental defects of enamel: a prevalence and family study in East London, UK. *Connective Tissue Research*, **39**(1–3), pp. 151–156, 1998.
http://dx.doi.org/10.3109/03008209809023921

[17] Brook, A.H., Variables and criteria in prevalence studies of dental anomalies of number, form and size. *Community Dentistry and Oral Epidemiology*, **3**(6), pp. 288–293, 1975.
http://dx.doi.org/10.1111/j.1600-0528.1975.tb00326.x

[18] Moorrees, C.F., Thomsen, S.O., Jensen, E. & Yen, P.K.J., Mesiodistal crown diameters of deciduous and permanent teeth. *Journal of Dental Research*, **36**(1), pp. 39–47, 1957.
http://dx.doi.org/10.1177/00220345570360011501

[19] Healey, N., Lead toxicity, vulnerable subpopulations and emergency preparedness. *Radiation Protection Dosimetry*, **134**(3–4), pp. 143–151, 2009.
http://dx.doi.org/10.1093/rpd/ncp068

[20] Lee, D.H., Jacobs Jr, D.R. & Porta, M., Hypothesis: a unifying mechanism for nutrition and chemicals as lifelong modulators of DNA hypomethylation. *Environmental Health Perspectives*, **117**, pp. 1799–1802, 2009.
http://dx.doi.org/10.1289/ehp.0900741

[21] Zhao, Z., Li, R., Sun, L., Li, Z. & Yang, R., Effect of lead exposure on the immune function of lymphocytes and erythrocytes in preschool children. *Journal of Zhejiang University Science*, **5**(8), pp. 1001–1004, 2004.
http://dx.doi.org/10.1631/jzus.2004.1001

[22] Mishra, K.P., Singh, V.K., Rani, R., Yadav, V.S., Chandran, V., Srivastava, S.P. & Seth, P.K., Effect of lead exposure on the immune response of some occupationally exposed individuals. *Toxicology*, **188**(2), pp. 251–259, 2003.
http://dx.doi.org/10.1016/S0300-483X(03)00091-X

[23] Li, S., Zhengyan, Z., Rong, L. & Hanyun, C., Decrease of CD4+ T-lymphocytes in children exposed to environmental lead. *Biological Trace Element Research*, **105**(1–3), pp. 19–25, 2005.
http://dx.doi.org/10.1385/BTER:105:1-3:019

[24] Selye, H., Tuchweber, B. & Bertok, L., Effect of lead acetate on the susceptibility of rats to bacterial endotoxins. *Journal of Bacteriology*, **91**(2), pp. 884–890, 1966.

[25] Lee, J.E., Naqi, S.A., Kao, E. & Dietert, R.R., 2002. Embryonic exposure to lead: comparison of immune and cellular responses in unchallenged and virally stressed chickens. *Archives of Toxicology*, **75**(11–12), pp. 717–724, 2002.
http://dx.doi.org/10.1007/s00204-001-0283-9

[26] Dietert, R.R. & Piepenbrink, M.S., Lead and immune function. *Critical Reviews in Toxicology*, **36**(4), pp. 359–385, 2006.
http://dx.doi.org/10.1080/10408440500534297

[27] Hirsch, H.V., Possidente, D. & Possidente, B., Pb 2+: an endocrine disruptor in Drosophila? *Physiology & Behavior*, **99**(2), pp. 254–259, 2010.
http://dx.doi.org/10.1016/j.physbeh.2009.09.014

[28] Faulk, C., Liu, K., Barks, A., Goodrich, J.M. & Dolinoy, D.C., Longitudinal epigenetic drift in mice perinatally exposed to lead. *Epigenetics*, **9**(7), pp. 934–941, 2014.
http://dx.doi.org/10.4161/epi.29024

[29] Chen, S., Golemboski, K.A., Piepenbrink, M. & Dietert, R.R., Developmental immunotoxicity of lead in the rat: influence of maternal diet. *Journal of Toxicology and Environmental Health, Part A*, **67**(6), pp. 495–511, 2004.
http://dx.doi.org/10.1080/15287390490276520

[30] Mahaffey, K.R., Nutrition and lead: strategies for public health. *Environmental Health Perspectives*, **103**(Suppl 6), pp. 191, 1995.
http://dx.doi.org/10.1289/ehp.95103s6191

[31] World Health Organisation (WHO), Childhood lead poisoning, available at http://www.who.int/ceh/publications/childhoodpoisoning/en/index.html

[32] Moore, S.E., Collinson, A.C., N'Gom, P.T., Aspinall, R. & Prentice, A.M., Early immunological development and mortality from infectious disease in later life. *Proceedings of the Nutrition Society*, **65**(3), pp. 311–318, 2006.
http://dx.doi.org/10.1079/PNS2006503

[33] Davis, S.L., Environmental modulation of the immune system via the endocrine system. *Domestic Animal Endocrinology*, **15**(5), pp. 283–289, 1998.
http://dx.doi.org/10.1016/S0739-7240(98)00034-4

[34] Scrimshaw, N.S., Taylor, C.E. & Gordon, J.E., *Interaction of Nutrition and Infection. WHO Monograph Series 57*. World Health Organization: Geneva, 1968.
[35] Bhaskaram, P., Micronutrient malnutrition, infection, and immunity: an overview. *Nutrition Reviews*, **60**(suppl 5), pp. S40–S45, 2002.
[36] Lesage, J., Sebaai, N., Leonhardt, M., Dutriez-Casteloot, I., Breton, C., Deloof, S. & Vieau, D., Perinatal maternal undernutrition programs the offspring hypothalamo–pituitary–adrenal (HPA) axis. *Stress*, **9**(4), pp. 183–198, 2006. http://dx.doi.org/10.1080/10253890601056192
[37] Heijmans, B.T., Tobi, E.W., Stein, A.D., Putter, H., Blauw, G.J., Susser, E.S., Slagboom, P.E. & Lumey, L.H., Persistent epigenetic differences associated with prenatal exposure to famine in humans. *Proceedings of the National Academy of Sciences*, **105**(44), pp. 17046–17049, 2008. http://dx.doi.org/10.1073/pnas.0806560105
[38] Lillycrop, K.A. & Burdge, G.C., Maternal diet as a modifier of offspring epigenetics. *Journal of Developmental Origins of Health and Disease*, **6**(02), pp. 88–95, 2015. http://dx.doi.org/10.1017/S2040174415000124
[39] Scrimshaw, N.S. & SanGiovanni, J.P., Synergism of nutrition, infection, and immunity: an overview. *The American Journal of Clinical Nutrition*, **66**(2), pp. 464S–477S, 1997.
[40] Lasala, N. & Zhou, H., Effects of maternal exposure to LPS on the inflammatory response in the offspring. *Journal of Neuroimmunology*, **189**(1), pp. 95–101, 2007. http://dx.doi.org/10.1016/j.jneuroim.2007.07.010
[41] Hodyl, N.A., Walker, F.R., Krivanek, K.M., Clifton, V. & Hodgson, D.M., Modelling prenatal bacterial infection: functional consequences of altered hypothalamic pituitary adrenal axis development. *Behavioural Brain Research*, **178**(1), pp. 108–114, 2007. http://dx.doi.org/10.1016/j.bbr.2006.12.008
[42] Bobetsis, Y.A., Barros, S.P., Lin, D.M., Weidman, J.R., Dolinoy, D.C., Jirtle, R.L., Boggess, K.A., Beck, J.D. & Offenbacher, S., Bacterial infection promotes DNA hypermethylation. *Journal of Dental Research*, **86**(2), pp. 169–174, 2007. http://dx.doi.org/10.1177/154405910708600212
[43] Maekita, T., Nakazawa, K., Mihara, M., Nakajima, T., Yanaoka, K., Iguchi, M., Arii, K., Kaneda, A., Tsukamoto, T., Tatematsu, M. & Tamura, G., High levels of aberrant DNA methylation in Helicobacter pylori–infected gastric mucosae and its possible association with gastric cancer risk. *Clinical Cancer Research*, **12**(3), pp. 989–995, 2006. http://dx.doi.org/10.1158/1078-0432.CCR-05-2096
[44] Anway, M.D., Leathers, C. & Skinner, M.K., Endocrine disruptor vinclozolin induced epigenetic transgenerational adult-onset disease. *Endocrinology*, **147**(12), pp. 5515–5523, 2006. http://dx.doi.org/10.1210/en.2006-0640
[45] Jedeon, K., Marciano, C., Loiodice, S., Lavier, M.C.C., Berdal, A. & Babajko, S., Enamel defects due to exposure to combinations of endocrine disruptors (abstract). *IADR/ PER congress conference paper,* 2014. available at https://iadr.confex.com/iadr/per14/webprogram/Paper192149.html

AGENTS WITHIN A DEVELOPMENTAL COMPLEX ADAPTIVE SYSTEM: INTRAUTERINE MALE HORMONES INFLUENCE HUMAN TOOTH SIZE AND SHAPE

F. LAM[1], R. YONG[1], S. RANJITKAR[1], G.C. TOWNSEND[1] & A.H. BROOK[1, 2]
[1]School of Dentistry, University of Adelaide, Australia.
[2]Institute of Dentistry, Queen Mary University of London, England.

ABSTRACT

Dental development is a complex adaptive system (CAS) in which genetic, epigenetic and environmental factors interact during different developmental stages to influence final tooth size and shape. From these complex interactions, a self-regulating and self-organising system emerges with spatiotemporal, multidimensional, multilevel and multifactorial components. The aim of this study was to quantify the influence of intrauterine male hormones on the mesiodistal dimension of permanent teeth by studying females from opposite-sex (OS) and same-sex (SS) twin pairs to explore dental development as a CAS. A custom-designed 2D image analysis system was used to measure the mesiodistal dimension from the occlusal view in each permanent tooth of 30 dizygotic OS and 30 dizygotic SS female twin pairs. Linear mixed models were used to quantify the differences in tooth size and shape. The mean mesiodistal tooth dimension was greater in the OS cohort for the majority of the permanent dentition. These findings reflected the complex interactions between genetic, epigenetic and environmental factors and their outcomes during development. The results indicated the transmission of the male hormones to the female in utero influencing her endocrine system, leading to epigenetic changes and so affecting gene expression. This demonstrated prenatal sex hormones as an environmental agent influencing multiple interactions within a developmental CAS.

Keywords: complex adaptive systems, dental development, epigenetic, genetic, phenomics, prenatal sex hormones, sexual dimorphism, Y chromosome.

1 INTRODUCTION

1.1 The human dentition as a complex adaptive system

Dental development is a complex adaptive system (CAS) emerging from interactions between genetic, epigenetic and environmental factors during the initiation, morphogenesis, differentiation, calcification and eruption of teeth [1, 2]. Understanding the spatiotemporal, multidimensional, multilevel and multifactorial components of dental development has the potential to improve the knowledge of the role of CASs in craniofacial and general developments. Interactions between genetic, epigenetic and environmental factors on dental development can influence sexual dimorphism observed in the dentition, such as the final tooth size and shape [3].

1.2 Sexual dimorphism within the dentition

The phenotypic difference between males and females within the same species has been found in many human characteristics such as bone and teeth. Sexual dimorphism has been identified within

DOI: 10.2495/DNE-V11-N4-696-702

different tooth dimensions of the primary and permanent dentitions and is influenced by several factors, including the Y chromosome and intrauterine male hormones [3].

1.2.1 Influence of Y chromosome

The Y chromosome is present in males, and it contains the SRY gene, which is influential in the testicular development and testosterone release [4]. The influences of the Y chromosome and genes on these characteristics require further investigation.

1.2.2 Twin testosterone transfer hypothesis

Testosterone release in utero can transfer from a male twin to the female co-twin, causing more masculine traits in the female [5]. By measuring a phenotype that develops at the time of the testosterone surges in utero in females from different twin types, it is possible to investigate the influence of intrauterine testosterone on the development of that phenotype. In addition, the intrauterine testosterone may have a priming effect, possibly with an epigenetic change that will influence tooth development later on.

1.2.3 Epigenetics

Epigenetic mechanisms alter the production of different cell types, forming different tissues and performing different functions, whilst still possessing the same DNA sequence [6–10]. Epigenetics involves "the addition or removal of methyl groups to DNA or the attachment of acetyl groups of histones" [11]. MicroRNAs are also epigenetically regulated to control a variety of biological functions in plants and animals [11].

1.3 Dental phenomics

Dental phenotypes such as tooth size and shape can be analysed using state-of-the-art, enhanced dental phenomic techniques such as two-dimensional (2D) and three-dimensional (3D) image analysis systems. The outcomes of such analyses can help to elucidate the developmental factors in health and disease [12]. A clear understanding of the various dental developmental variations and diseases can assist operators in forming a patient-centred diagnosis and treatment plan.

2 AIMS

The aim of this study was to quantify the influence of intrauterine male hormones on the mesiodistal dimension of permanent teeth between females from the opposite-sex (OS) and the same-sex (SS) twin pairs to explore dental development as a CAS.

3 MATERIALS AND METHODS

Measurements were obtained from all teeth present in the permanent dentition of the 30 female OS and 30 female SS twins, excluding the second and third molars as they were not fully formed. The dental casts were placed on a platform under standardised illumination as shown in Fig. 1. Each tooth was imaged from the occlusal view using a digital camera (Canon EOS 50D digital SLR camera, Cannon, Australia) with a resolution of 15.1 megapixels and image array of 4,752 × 3,168 pixels. The camera was positioned horizontally above the dental casts on an adjustable rod. Illumination involved four multidirectional spotlights surrounding the cast (Fig. 1). The camera was connected to a personal computer, which utilised imaging software.

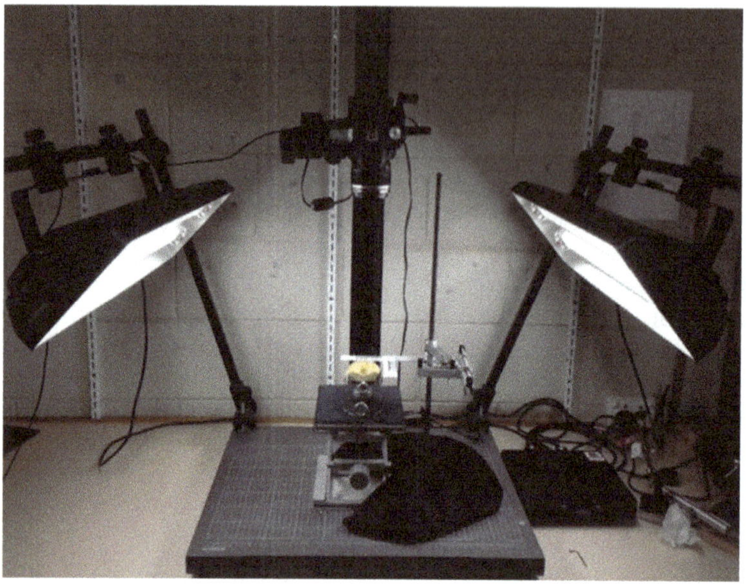

Figure 1: 2D image analysis system at the University of Adelaide Dental School.

Figure 2: Mesiodistal (MD) width (blue line) measured on a permanent mandibular second premolar from the occlusal view.

The mesiodistal width represented the maximum distance between mesial and distal proximal surfaces of the tooth crown and was calculated from the occlusal/incisal (MD) view (Fig. 2). For each individual tooth and tooth types (incisors, canines, premolars and molars) in the mandibular and maxillary arches, descriptive statistics such as mean values and standard deviations were calculated for each zygosity. Linear mixed models were used to compare the variables for OS and SS females for individual teeth and the tooth types.

Table 1: Comparison of mesiodistal widths (in millimetres) from occlusal view (MD) for every tooth between females from the OS and the SS cohorts.

			OS Females (n = 30)		SS Females (n = 30)	
MD		p (F)	Mean (mm)	SD (mm)	Mean (mm)	SD (mm)
			Maxillary			
Right	I1	0.967	8.480	0.519	8.474	0.519
	I2	0.901	6.556	0.738	6.581	0.811
	C	0.591	7.736	0.468	7.670	0.479
	P1	0.202	7.161	0.471	7.000	0.504
	P2	0.029	7.017	0.481	6.734	0.520
	M1	0.079	10.948	0.578	10.686	0.578
Left	I1	0.755	8.478	0.518	8.520	0.518
	I2	0.837	6.644	0.586	6.611	0.636
	C	0.67	7.574	0.490	7.519	0.512
	P1	0.121	7.158	0.437	6.979	0.455
	P2	0.209	6.989	0.448	6.837	0.483
	M1	0.577	10.729	0.548	10.650	0.548
			Mandibular			
Left	I1	0.805	5.377	0.406	5.351	0.411
	I2	0.489	5.924	0.402	5.852	0.402
	C	0.14	6.609	0.419	6.447	0.428
	P1	0.087	7.254	0.482	7.040	0.482
	P2	0.047	7.416	0.502	7.157	0.509
	M1	0.157	11.105	0.561	10.901	0.555
Right	I1	0.813	5.445	0.407	5.421	0.407
	I2	0.972	5.978	0.411	5.974	0.414
	C	0.295	6.646	0.429	6.528	0.440
	P1	0.133	7.173	0.505	6.976	0.511
	P2	0.053	7.470	0.525	7.206	0.532
	M1	0.076	11.079	0.588	10.812	0.582

I1 = permanent central incisor; I2 = permanent lateral incisor; C = permanent canine; P1 = permanent first premolar; P2 = permanent second premolar; M1 = permanent first molar; n = sample size; mean = mean values; SD = standard deviation; ($p < 0.05$); shading denotes direction of difference regardless of level of significance (bold) (red, OS>SS; green, OS < SS).

4 RESULTS

The mean values of MD widths showed a greater trend in the OS females compared with SS females for all teeth, except the maxillary right lateral incisors and left central incisors. The OS females

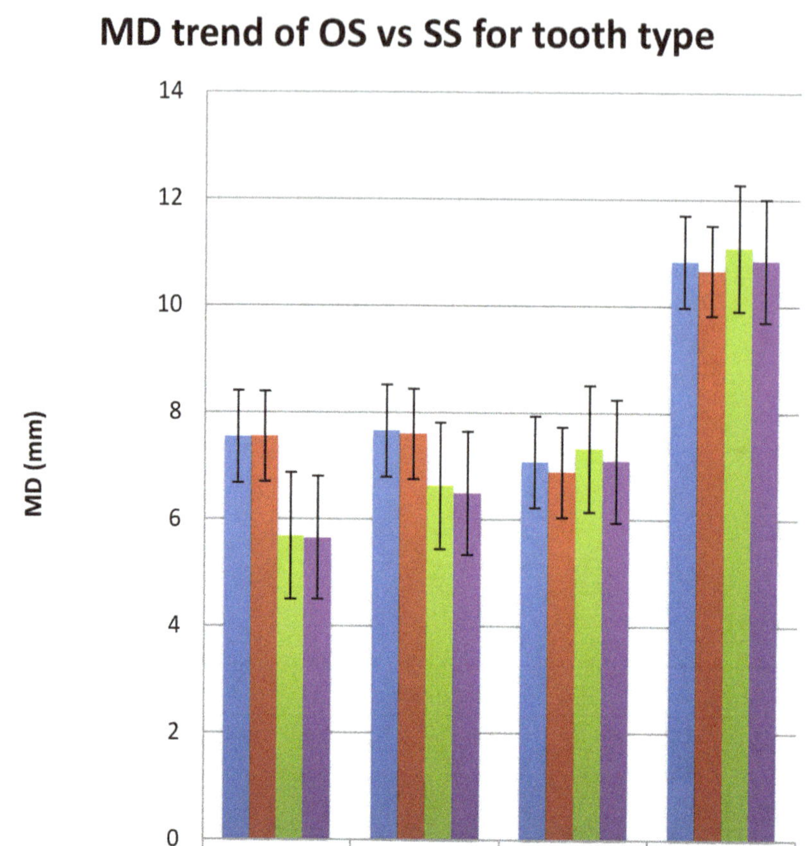

	Incisors	Canines	Premolars	Molars
Maxillary DZOS	7.54	7.655	7.081	10.838
Maxillary DZSS	7.547	7.595	6.888	10.668
Mandibular DZOS	5.681	6.627	7.328	11.092
Mandibular DZSS	5.65	6.488	7.095	10.856

Figure 3: The mesiodistal (MD) widths of OS Females vs. SS Females for tooth type.

Blue = maxillary OS females, red = maxillary SS females; blue vs red for maxillary arch; green = mandibular OS females, purple = mandibular SS females; green vs. purple for mandibular arch. Error bars represent standard errors.

showed significantly greater MD widths for the right second maxillary premolars and left mandibular second premolars in comparison to SS females ($p<0.05$) (Table 1).

The mean MD widths were significantly greater in the OS cohort than the SS cohort for maxillary and mandibular premolars when the teeth were grouped into four tooth types, excluding the maxillary incisors that displayed a difference of 0.007 mm ($p<0.05$) (Fig. 3). The mandibular premolars and molars showed the greatest difference between OS cohort and SS cohort for the MD widths (Fig. 3).

5 DISCUSSION

Using a twin model, this study further investigated the possible influence of prenatal sex hormones and the Y chromosome on tooth size and shape. Utilisation of tooth measurement data contributed to the understanding of the interactions of the genetic, epigenetic and environmental factors on tooth development. The tooth size and shape data made important contributions to the TTT hypothesis with regard to developments in research and clinical relevance.

Houle *et al.* [13] indicated that phenotypic characteristics like health, disease and evolutionary fitness have variations due to the complex interactions between genotype and environment. The authors suggested that genotype to phenotype association would be "inaccessible" without accurate phenotypic analysis such as the acquisition of crown dimensions using a 2D image analysis system. The genotype-phenotype map discussed was a metaphor for the multifactorial development of phenotypes due to the genetic, epigenetic and environmental influences. In essence, they were discussing the genotype-phenotype map in the context of the CAS in health and disease [13].

Dental development is a complex network within which multiple factors, including the intrauterine male hormones, influence the endocrine system and the epigenetic mechanisms controlling gene expression and development. A CAS with multiple reciprocating interactions was proposed to explore the environmental factors interaction with body systems and epigenetic mechanisms. The prenatal sex hormones and Y chromosome could influence the endocrine system and immune system leading to epigenetic changes, which result in enhanced gene expression. We propose that altered gene expression from this process has led to the increase in tooth size between the OS and SS females for the mesiodistal dimension.

6 CONCLUSION

The aim was fulfilled by measuring the MD dimension within the permanent dentition of OS and SS samples using an enhanced 2D analysis method to quantify the influence of intrauterine male hormones on tooth size and shape. The results further explore the relation of genotype to phenotype and the proposed network of interactions between male hormones and gene expression. The results indicated larger tooth crown dimensions in females from the OS cohort in comparison to the SS cohort, supporting the validity and value of using the human dentition to explore how genetic, epigenetic and environmental interactions influence in a biological CAS. The proposed modelling of these interactions explores the SRY gene on the Y chromosome influencing the development of male hormones, which transmits to the female in utero influencing her endocrine system, leading to epigenetic changes and so affecting gene expression. This demonstrates an environmental agent influencing multiple interactions in a developmental CAS. The difference in tooth size between OS and SS females provides further information of the pattern of sexual dimorphism, specifically the direct effects of the Y chromosome in males and supports the Twin Testosterone Transfer hypothesis. Further research is exploring more linear dimensions, the perimeter and surface area of the permanent teeth in the twin cohorts to investigate the dimensions most influenced by the Y chromosome and intrauterine male hormones.

ACKNOWLEDGEMENTS

The support of the Australian Dental Research Foundation to the first author in the form of the Undergraduate Research Grant is gratefully acknowledged.

REFERENCES

[1] Brook, A.H., The dentition: the outcomes of morphogenesis leading to variations of tooth number, size and shape. *Australian Dental Journal*, **59**, pp. 131–142, 2014.
http://dx.doi.org/10.1111/adj.12160

[2] Brook, A., Brook O'Donnell, M., Hone, A., Hart, E., Hughes, T., Smith R. & Townsend, G., General and craniofacial development are complex adaptive processes influenced by diversity. *Australian Dental Journal,* **59S**, pp. 13–22, 2014.
http://dx.doi.org/10.1111/adj.12158

[3] Ribeiro, D.C., Brook, A.H., Hughes, T.E., Sampson, W.J. & Townsend, G.C., Intrauterine hormone effects on tooth dimensions. *Journal of Dental Research,* **92**, pp. 425–431, 2013.
http://dx.doi.org/10.1177/0022034513484934

[4] Gilbert, S.F., *Chromosomal Sex Determination in Mammals. Developmental Biology*, 6th edn., Sinauer Associates: Sunderland, MA, 2001.

[5] Miller, E.M., Prenatal sex hormone transfer: a reason to study opposite sex twins. *Personality and Individual Differences,* **17**, pp. 511–529, 1994.
http://dx.doi.org/10.1016/0191-8869(94)90088-4

[6] Townsend, G. & Brook, A., Genetic, epigenetic and environmental influences on dental development. *Ortho Tribune,* **3**, pp. 4–6, 2008.

[7] Barros, S.P. & Offenbacher, S., Epigenetics: connecting environment and genotype to phenotype and disease. *Journal of Dental Research,* **88**, pp. 400–408, 2009.
http://dx.doi.org/10.1177/0022034509335868

[8] Brook, A., Multilevel complex interactions between genetic, epigenetic and environmental factors in the aetiology of anomalies of dental development. *Archives of Oral Biology,* **54S**, pp. 3–17, 2009.
http://dx.doi.org/10.1016/j.archoralbio.2009.09.005

[9] Townsend, G., Hughes, T., Bockmann, M., Smith, R. & Brook, A., How studies of twins can inform our understanding of dental morphology. *Frontiers of Oral Biology,* **13**, pp. 136–141, 2009.
http://dx.doi.org/10.1159/000242406

[10] Bell, J.T. & Spector, T.D., A twin approach to unraveling epigenetics. *Trends in Genetics,* **27**, pp. 116–125, 2011.
http://dx.doi.org/10.1016/j.tig.2010.12.005

[11] Bernal, J.E., Duran, C. & Papiha, S.S., Transcriptional and epigenetic regulation of human microRNAs. *Cancer Letters,* **331**(1), pp. 1–10, 2012.

[12] Yong, R., Ranjitkar, S., Townsend, G.C., Smith, R.N., Evans, A.R., Hughes, T.E., Lekkas, D. & Brook, A.H., Dental phenomics: advancing genotype to phenotype correlations in craniofacial research. *Australian Dental Journal,* **59**, pp. 34–47, 2014.
http://dx.doi.org/10.1111/adj.12156

[13] Houle, D., Govindaraju, D.R. & Omholt, S., Phenomics: the next challenge. *Nature Reviews Genetics,* **11**, pp. 855–866, 2010.
http://dx.doi.org/10.1038/nrg2897

AGENTS WITHIN A DEVELOPMENTAL COMPLEX ADAPTIVE SYSTEM: INTRAUTERINE MALE HORMONES AND DENTAL ARCH SIZE IN HUMANS

P. PATEL,[1] R. YONG,[1] S. RANJITKAR,[1] G. TOWNSEND,[1] & A. BROOK[1, 2]
[1]School of Dentistry, University of Adelaide, Australia.
[2]School of Dentistry, Queen Mary University of London, UK.

ABSTRACT

Oral development is a complex adaptive system influenced by genetic, epigenetic and environmental factors. The dental arch develops from 6 weeks in utero until adult life, forming an accessible record to study general growth and development. Increased tooth size in female dizygotic opposite-sex (DZOS) twins compared with female dizygotic same-sex (DZSS) twins provides evidence for the masculinisation of females gestated with a male co-twin, possibly due to the intrauterine influence of male sex hormones: the "Twin Testosterone Transfer" (TTT) hypothesis. This study aimed to investigate the potential influence of intrauterine male hormones on dental arch size of female DZOS twins. Serial dental models of the primary and permanent dentitions of 69 female DZOS and DZSS twins were examined. Intercanine width, intermolar width, arch length and arch circumference were measured using a customised 2D image analysis system. Unpaired t-tests showed significant differences for mandibular intercanine width ($p = 0.03$; effect size = 0.6) and borderline differences for mandibular intermolar width and arch circumference ($p = 0.05$; effect size = 0.5). No significant differences were found in the permanent dentition. These findings provide support for the TTT hypothesis with some arch dimensions being larger in female DZOS twins. We have developed a model of assessing the effects of intrauterine male hormones on the epigenetic changes that last into postnatal life. Our evidence suggests that this is a moderate effect possibly interacting with numerous other environmental factors that may influence arch size.
Keywords: complex adaptive system, epigenetics, sex chromosomes, sexual dimorphism, twins.

1 INTRODUCTION

The development of the human dentition is a complex adaptive system influenced by genetic, epigenetic and environmental factors [1]. Sexual dimorphism in the human craniofacial complex has also been related to the growth promoting effects of the Y-chromosome [2,3] and sex hormones [4,5]. In the dentition, sexual dimorphism has been described in tooth size [6], arch dimensions [7,8] and underlying bone size [3], suggesting common influencing factors on dental and craniofacial development. However, the specific mechanisms by which sex chromosomes and prenatal sex hormones interact and influence craniofacial development require further investigation.

The first surge of intrauterine testosterone occurs around week 6 of human gestation and peaks around week 14. The SRY gene on the Y chromosome is responsible for initiating testicular differentiation in a male foetus [9,10]. A second surge occurs soon after birth, and the hormonal levels remain high during the first year of life. A third surge occurs during the adolescent growth spurt and is responsible for the development of male secondary sexual characteristics. Both the primary and permanent teeth, along with the supporting alveolar structures, develop during the first two surges and therefore provide a good model to study the effects of sex hormones on craniofacial development.

DOI: 10.2495/DNE-V11-N4-703-711

Currently, foetal testosterone level is most accurately measured via amniocentesis [11], but the invasive nature of this procedure raises ethical issues and prevents its widespread application [10]. Alternatively, the Twin Testosterone Transfer (TTT) hypothesis, premised on hormonal transfer between foetuses from the same pregnancy [12–14], allows non-invasive and cost effective assessment of the influence of prenatal testosterone on human development. Previous twin studies supporting the TTT hypothesis have demonstrated increased masculine traits in human females from opposite-sex twin pairs as opposed to same-sex twin pairs. Examples include increased total brain volume [14], altered craniofacial growth and dental asymmetries [15], increased symptoms of Alcohol Use Disorder (AUD) [16], lower disordered eating attitudes [17] and increased tooth crown size [4,5]. These findings not only support the TTT hypothesis but also indicate that hormonal transfer may influence sensitivity of androgen receptors later in life (i.e. during pubertal hormone surges) [11].

2 AIMS

The aim of this study was to investigate the possible influence of intrauterine male hormones on dental arch dimensions (intercanine width, intermolar width, arch length and arch circumference) in both the primary and permanent dentitions of female dizygotic opposite-sex (DZOS) twins and female dizygotic same-sex (DZSS) twins. It was hypothesised that female DZOS twins would have larger dental arch dimensions than female DZSS twins, reflecting intrauterine male hormone diffusion.

3 MATERIALS AND METHODS

3.1 Materials

A convenience sample of 69 females from 33 DZOS and 36 DZSS twin pairs was selected for this study. Serial dental casts of the primary and permanent dentitions of the same individuals were obtained from the collection of records of twins enrolled in an ongoing study of dental and facial development at the School of Dentistry, University of Adelaide [18]. One co-twin from each pair of DZOS and DZSS twins was randomly selected by generating and assigning random numbers to avoid bias from inclusion of the same arch data from both co-twins. All participants were of European ancestry with no relevant medical or dental history. Twin zygosities were confirmed by analysing up to six highly variable genetic loci (*FES, vWA3l, F13A1, TH01, D21S11, FGA*) on six different chromosomes, using DNA obtained from buccal cells. The probability of dizygosity, given concordance for all systems, was <1% [19]. Dental impressions were taken when the twins were aged 4 to 7 years (mean age = 5.5 years) in the primary dentition and 12–17 years (mean age = 14.1 years) in the permanent dentition. The inclusion criteria included presence of all primary and permanent teeth, except second and third permanent molars. Participants with missing teeth, significant wear or partially erupted teeth were excluded.

Dental arches were examined using a 2D image analysis system, previously validated for tooth size and dental arch determination [20,21]. Each dental cast was clamped on a cast surveyor table and a tripod levelling device was used to level it accurately into a horizontal plane. The plane was defined by three reference points, including the central fossae of the second primary molars in the primary dentition, the first permanent molars in the permanent dentition and the contact area of the central incisors. An American Board of Forensic Odontology (ABFO) ruler No.2 (Lightening Powder Co Inc., Oregon, USA) was used as a reference scale (Fig. 1a–d). An occlusal photograph was then obtained for each dental cast for analysis.

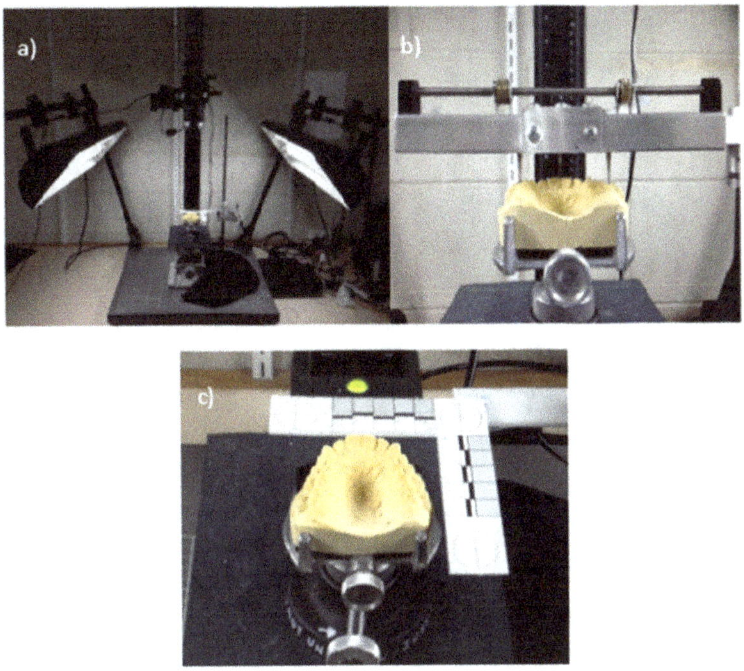

Figure 1: (a) 2D Image system (b) Tripod levelling device (c) Reference scale.

3.2 Measurement of arch dimensions

Based on previous studies [7,8,22], the four most repeatable landmarks (intercanine width, intermolar width, arch length and arch circumference) were chosen. Intercanine and intermolar widths were measured between the cuspal tips of canines and molars for both primary and permanent dentitions. Arch length was defined as a perpendicular distance between the contact area of the central incisors and the line passing through the most distal surface of the second primary molar in primary dentition, and the most mesial surface of the first permanent molar in permanent dentition. Arch circumference was measured as the line passing through the mesial and distal contact areas of each tooth from the first permanent molar/second primary molar of one side to the first permanent molar/second primary molar of the opposite side of the arch (Fig. 2a–d).

The images were analysed using ImageJ v. 1.48 (National Institute of Health, USA). Every image was first calibrated and landmarks digitised to identify the x, y coordinates which were then saved in an excel spreadsheet. A customised macro was used to calculate distances between landmarks corresponding to the parameters of interest.

Intra- and inter-operator repeatability tests were performed for all arch dimensions of primary and permanent dentitions in a randomly selected sub-group of 25% individuals. Mean differences between double determinations of intercanine width (0.15–0.27 mm), intermolar width (0.15–0.25 mm), arch length (0.02–0.10 mm) and arch circumference (0.15–0.25 mm) were non-significant, indicating that systematic errors were unlikely to introduce any bias to the analysis. In addition, the Dahlberg statistic was calculated and the error variance (Se^2) was expressed as a percentage of the observed variance for each variable to determine the extent to which variability due to experimental error affected the observed variance. Error percentages for all arch dimensions were less than 2%. Overall, random errors of method were small for all variables and unlikely to bias the results of the study.

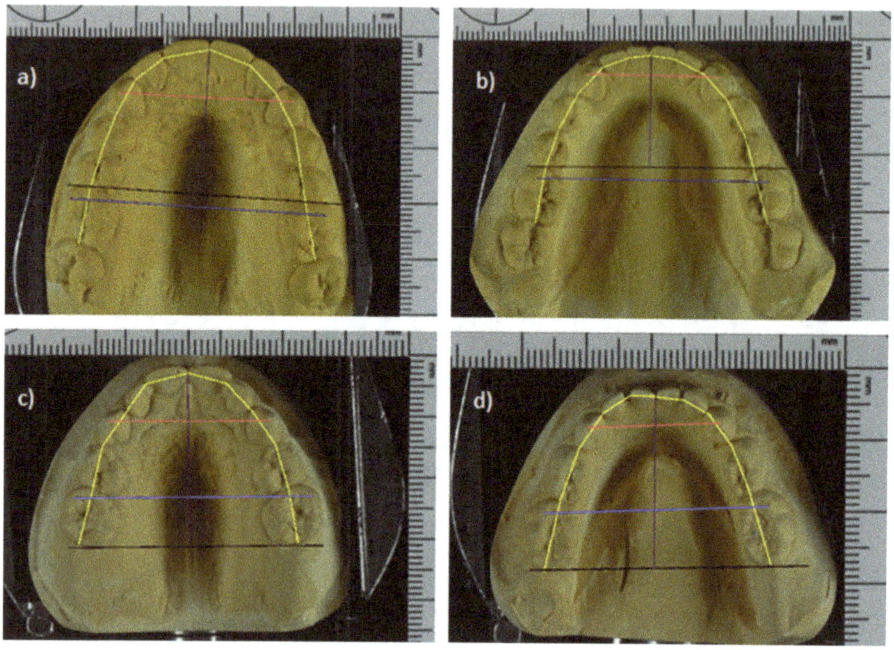

Figure 2: Landmarks on study models for arch dimensions: (a) permanent maxillary arch, (b) permanent mandibular arch, (c) primary maxillary arch and d) primary mandibular arch. Red - Intercanine width. Blue - Intermolar width. Purple - Arch length. Yellow – Arch circumference. Black – Reference line.

Analysis of histograms showed that all variables were approximately normally distributed and parameter descriptive statistics including mean and standard deviation (SD) were calculated. Unpaired t-tests were conducted to determine whether there were significant differences in arch dimensions among the two zygosity groups. The analysis was performed using SPSS version 20 (IBM 2011, USA). The effect size was also calculated to assist in interpreting the intensity of the male hormone effect (i.e. small, medium or large), given that p values alone are not adequate to fully understand the results [23]. Adjustments for multiple t-tests were not made to avoid Type II (β) error in our sample.

4 RESULTS

In the primary dentition, there were significant differences in the mandibular intercanine width ($p = 0.03$; effect size = 0.6) and borderline significance in mandibular intermolar width ($p = 0.05$; effect size = 0.5) and arch circumference ($p = 0.05$; effect size = 0.5) (Table 1). No significant differences were noted in any of the four measurements in the maxillary arch ($p>0.05$) (Table 1). In the permanent dentition, there were no significant differences in the arch dimensions between the two zygosity groups ($p>0.05$) (Table 2).

5 DISCUSSION

This study is the first to test the TTT hypothesis by comparing dental arch dimensions of both human primary and permanent dentitions in female twins. DZOS twin pairs provide a unique model to investigate the influence of prenatal sex hormones on dental development, since female DZOS twins

Table 1: Comparison of dental arch dimensions in the primary dentition of DZOS and DZSS female twins.

	DZOS females			DZSS females			Effect size
	n	mean	SD	n	mean	SD	
Maxillary arch							
Intercanine width	33	27.5	1.89	36	27.5	2.07	0.0
Intermolar width	33	40.5	1.92	36	40.7	2.13	0.1
Arch length	33	27.3	2.00	36	27.1	2.03	0.1
Arch circumference	33	71.5	3.59	36	71.3	4.03	0.0
Mandibular arch							
Intercanine width	33	22.9*	1.36	36	22.0*	1.87	0.6
Intermolar width	33	35.5†	1.76	36	34.6†	1.99	0.5
Arch length	27	24.4	1.53	34	23.8	1.56	0.4
Arch circumference	33	66.8†	2.93	36	65.4†	3.09	0.5

*Mean values in DZOS females were significantly greater than DZSS females at $p<0.05$

†Mean values in DZOS females were greater than DZSS females at borderline significance at $p = 0.05$

Table 2: Comparison of dental arch dimensions in the permanent dentition of DZOS and DZSS female twins.

	DZOS females			DZSS females			Effect size
	n	mean	SD	n	mean	SD	
Maxillary arch							
Intercanine width	29	33.6	2.26	36	27.5	2.07	0.0
Intermolar width	32	50.6	3.05	24	50.8	3.03	0.2
Arch length	32	25.0	2.49	24	25.4	3.02	0.1
Arch circumference	29	93.3	4.68	23	93.7	6.32	0.1
Mandibular arch							
Intercanine width	32	25.6	1.65	23	26.1	1.53	0.3
Intermolar width	33	43.1	2.66	24	43.5	2.67	0.2
Arch length	33	21.4	1.96	24	21.5	2.55	0.0
Arch circumference	31	84.2	4.27	23	85.0	5.08	0.2

No significant differences were found in the permanent arch dimensions between DZOS and DZSS females ($p>0.05$).

develop under different concentrations of sex hormones compared to female DZSS twins [12]. By studying arch size independently, it is possible to consider whether they vary together and are affected by the same factors or whether they vary independently and are influenced by different factors.

Some mandibular arch dimensions of the primary dentition were larger in DZOS female twins than DZSS female twins, providing support for the TTT hypothesis. It appears that the effect of intrauterine male hormones produced by the male counterpart on the female DZOS twin lasts until the age of 5–7 years, with limited effects on arch dimensions later in the permanent dentition. In the permanent dentition, postnatal environmental factors, such as tooth wear from diet and tooth grinding, play a major role in influencing the size and shape of the dental arch [24]. Comparison of data for the primary and permanent dentition in our study imply the occurrence of a priming effect (possibly an epigenetic change) in DZOS females in utero. In addition, our findings indicate greater effects of testosterone on the growth and development of the mandible than the maxilla, which is consistent with the notion that mandible displays greater plasticity to growth and adaptation than the cranium [25].

A proposed mechanism to explain the increased arch size in DZOS females is that testosterone binds to the androgen receptor in the cytoplasm of the cells, resulting in ligand-receptor activation and dimerisation. The activated complex then localises to the nucleus, where it bonds to DNA and affects transcription [26]. Once primed, the effect of testosterone can continue even at low levels. Immunohistochemical studies have also shown the presence of oestrogen, progesterone and testosterone receptors in human foetal cartilaginous tissue [27], indicating a possible direct effect of sex hormones on prenatal dentoalveolar skeletal development. Thus, part of the action of intrauterine testosterone on the dental arches of female DZOS twins may be on the supporting dentoalveolar bone.

Our findings of dental arch dimensions conforming to the TTT hypothesis may have broader applications in craniofacial biology research and clinical practice, given that dental arch development underpins craniofacial morphometry [24]. Interestingly, our findings are consistent with reports of accelerated craniofacial and statural growth in boys with delayed puberty [28,29]. So far, little attention has been given to the dental arch development in children undergoing testosterone treatment, or in children with sex hormone imbalance, but this is an evolving field of research involving multi-disciplinary management.

6 CONCLUSIONS

This is the first study to provide evidence of the effect of intrauterine testosterone on dental arch dimensions of female DZOS twins. In addition to earlier findings of the influence of testosterone on tooth size, this study highlights intrauterine testosterone as an important developmental factor contributing to sexual dimorphism in the human dental arches. Further investigations into the organisational effects of testosterone on other cranial structures will provide additional insights into how genetic, epigenetic and environmental factors influence craniofacial growth and development [30].

ACKNOWLEDGEMENTS

The support of the National Health and Medical Centre of Australia and the Australian Dental Research Foundation towards supporting the twin research is gratefully acknowledged.

REFERENCES

[1] Brook, A., O'Donnell, M.B., Hone, A., Hart, E., Hughes, T., Smith, R. & Townsend, G.C., General and craniofacial development are complex adaptive processes influenced by diversity. *Australian Dental Journal*, **59** (1 Suppl), pp. 13–22, 2014. http://dx.doi.org/10.1111/adj.12158

[2] Guatelli-Steinberg, D., Sciulli, P.W. & Betsinger, T.K., Dental crown size and sex hormone concentrations: another look at the development of sexual dimorphism. *American Journal of Physical Anthropology*, **137**, pp. 324–333, 2008.

http://dx.doi.org/10.1002/ajpa.20878

[3] Alvesalo, L., Human sex chromosomes in oral and craniofacial growth. *Archives of Oral Biology*, **54**, pp. 18–24, 2009. http://dx.doi.org/10.1016/j.archoralbio.2008.06.004

[4] Dempsey, P.J., Townsend, G.C. & Richards, L.C., Increased tooth crown size in females with twin brothers: evidence for hormonal diffusion between human twins in utero. *American Journal of Human Biology*, **11**, pp. 577–586, 1999. http://dx.doi.org/10.1002/(SICI)1520-6300(199909/10)11:5%3C577::AID-AJHB1%3E3.0.CO;2-Y

[5] Ribeiro, D.C., Brook, A.H., Hughes, T.E., Sampson, W.J. & Townsend, G.C., Intrauterine hormone effects on tooth dimensions. *Journal of Dental Research*, **92**, pp. 425–431, 2013. http://dx.doi.org/10.1177/0022034513484934

[6] Buschang, P.H., Demirjian, A., Tanguay, R. & Charbonneau, C., Multivariate patterns of variation in dental arch morphology of French-Canadians. *Human Biology*, **59**, pp. 911–920, 1987.

[7] Bishara, S.E., Ortho, D., Jakobsen, J.R., Treder, J. & Nowak, A., Arch width changes from 6 weeks to 45 years of age. *Amercian Journal of Orthodontics and Dentofacial Orthopaedics*, **111**, pp. 401–409, 1997. http://dx.doi.org/10.1016/S0889-5406(97)80022-4

[8] Thilander, B., Dentoalveolar development in subjects with normal occlusion. A longitudinal study between the ages of 5 and 31 years. *European Journal of Orthodontics*, **31**, pp. 109–120, 2009. http://dx.doi.org/10.1093/ejo/cjn124

[9] Christine Knickmeyer, R. & Baron-Cohen, S., Fetal testosterone and sex differences. *Early Human Development*, **82**, pp. 755–760, 2006. http://dx.doi.org/10.1016/j.earlhumdev.2006.09.014

[10] Tapp, A.L., Maybery, M.T. & Whitehouse, A.J., Evaluating the twin testosterone transfer hypothesis: a review of the empirical evidence. *Hormonal Behaviour*, **60**, pp. 713–722, 2011. http://dx.doi.org/10.1016/j.yhbeh.2011.08.011

[11] Cohen-Bendahan, C.C., Buitelaar, J.K., Van Goozen, S.H., Orlebeke, J.F. & Cohen-Kettenis, P.T., Is there an effect of prenatal testosterone on aggression and other behavioral traits? A study comparing same-sex and opposite-sex twin girls. *Hormonal Behaviour*, **47**, pp. 230–237, 2005. http://dx.doi.org/10.1016/j.yhbeh.2004.10.006

[12] Miller, E.M., Prenatal sex hormone transfer: a reason to study opposite-sex twins. *Personality and Individual Differences*, **17**, pp. 511–529, 1994. http://dx.doi.org/10.1016/0191-8869(94)90088-4

[13] Miller, E.M. & Martin, N., Analysis of the effects of hormones on opposite-sex twin attitudes. *Acta Geneticae Medicae Gemellologiae*, **44**, pp. 41–44, 1995.

[14] Peper, J.S., Brouwer, R.M., van Baal, G.C., Schnack, H.G., van Leeuwen, M., Boomsma, D.I., Kahn, R.S. & Hulshoff Pol, H.E., Does having a twin brother make for a bigger brain? *European Journal of Endocrinology*, **160**, pp. 739–746, 2009. http://dx.doi.org/10.1530/EJE-08-0915

[15] Boklage, C.E., Interactions between opposite-sex dizygotic fetuses and the assumptions of Weinberg difference method epidemiology. *American Journal of Human Genetics*, **37**, pp. 591–605, 1985.

[16] Ellingson, J.M., Slutske, W.S., Richmond-Rakerd, L.S. & Martin, N.G., Investigating the influence of prenatal androgen exposure and sibling effects on alcohol use and alcohol use disorder in females from opposite-sex twin pairs. *Alcoholism: Clinical and Experimental Research*, **37**, pp. 868–876, 2013.
http://dx.doi.org/10.1111/acer.12035

[17] Culbert, K.M., Breedlove, S.M., Sisk, C.L., Burt, S.A. & Klump, K.L., The emergence of sex differences in risk for disordered eating attitudes during puberty: a role for prenatal testosterone exposure. *Journal of Abnormal Psychology*, **122**, pp. 420–432, 2013.
http://dx.doi.org/10.1037/a0031791

[18] Townsend, G., Richards, L., Messer, L.B., Hughes, T., Pinkerton, S., Seow, K., Gotjamanous, T., Gully, N. & Bockmann, M., Genetic and environmental influences on dentofacial structures and oral health: studies of Australian twins and their families. *Twin Research and Human Genetics*, **9**, pp. 727–732, 2006.
http://dx.doi.org/10.1375/twin.9.6.727

[19] Hughes, T.E., Bockmann, M.R., Seow, K., Gotjamanos, T., Gully, N., Richards, L.C. & Townsend, G.C., Strong genetic control of emergence of human primary incisors. *Journal of Dental Research*, **86**, pp. 1160–1165, 2007.
http://dx.doi.org/10.1177/154405910708601204

[20] Brook, A., Smith, R., Elcock, C., Al-Sharood, M., Shah, A. & Karmo, M., The measurement of tooth morphology: development and validation of a new image analysis system. *Dental Morphology*, pp. 380–387, 1998.

[21] Khalaf, K., Robinson, D., Elcock, C., Smith, R. & Brook, A., Tooth size in patients with supernumerary teeth and a control group measured by image analysis system. *Archives of Oral Biology*, **50**, pp. 243–248, 2005.
http://dx.doi.org/10.1016/j.archoralbio.2004.09.013

[22] Eguchi, S., Townsend, G.C., Richards, L.C., Hughes, T. & Kasai, K., Genetic contribution to dental arch size variation in Australian twins. *Archives of Oral Biology*, **49**, pp. 1015–1024, 2004.
http://dx.doi.org/10.1016/j.archoralbio.2004.07.006

[23] Sullivan, G. & Feinn, R., Using effect size-or why the P value is not enough. *Journal of Graduate Medical Education*, **4**, pp. 279–282, 2012.
http://dx.doi.org/10.4300/JGME-D-12-00156.1

[24] von Cramon-Taubade, N., Global human mandibular variation reflects differences in agricultural and hunter-gatherer subsistence strategies. *Proceeding of the National Academy of Sciences of the United States America*, **108**, pp. 19546–19551, 2011.
http://dx.doi.org/10.1073/pnas.1113050108

[25] Brown, T., Townsend, G.C., Pinkerton, S.K. & Rogers, J.R., Facial growth patterns in the Warlpiri. *In Yuendumu, Legacy of a Longitudinal Growth Study in Central Australia*, University of Adelaide Press: Adelaide, pp. 135–170, 2011.

[26] Whitehouse, A.J., Mattes, E., Maybery, M.T., Sawyer, M.G., Jacoby, P., Keelan, J.A. & Hickey, M., Sex-specific associations between umbilical cord blood testosterone levels and language delay in early childhood. *Journal of Child Psychology and Psychiatry,* **53**, pp. 726–734, 2012.
http://dx.doi.org/10.1111/j.1469-7610.2011.02523.x

[27] Ben-Hur, H., Thole, H., Mashiah, A., Insler, V., Berman, V., Shezen, E., Elias, D., Zuckerman, A. & Ornoy, A., Estrogen, progesterone and testosterone receptors in human fetal cartilaginous tissue: immunohistochemical studies. *Calcified Tissue International*, **60**, pp. 520–526, 1997.
http://dx.doi.org/10.1007/s002239900274

[28] Bergada, I. & Bergada, C., Long term treatment with low dose testosterone in constitutional delay of growth and puberty: effect on bone age maturation and pubertal progression. *Journal of Paediatric Endocrinology and Metabolism*, **8**, pp. 117–122, 1995.
http://dx.doi.org/10.1515/JPEM.1995.8.2.117

[29] Verdonck, A., Gaethofs, M., Carels, C. & Zegher, F., Effect of low-dose testosterone on craniofacial growth in boys with delayed puberty. *European Journal of Orthodontics*, **21,** pp. 137–143, 1999.
http://dx.doi.org/10.1093/ejo/21.2.137

[30] Yong, R., Ranjitkar, S., Townsend, G., Smith, R., Evans, A., Hughes, T.E., Lekkas, D. & Brook, A.H., Dental phenomics: advancing genotype to phenotype correlations in craniofacial research. *Australian Dental Journal*, **59** (1 Suppl), pp. 34–47, 2014.
http://dx.doi.org/10.1111/adj.12156

TRANSIENT INDUCED GLOBAL RESPONSE SYNCHRONIZATION

WILLIAM SULIS
Collective Intelligence Laboratory, McMaster University, Canada.

ABSTRACT

Synchronization has a long history in physics where it refers to the phase locking of identical oscillators. This notion has been applied in biology to such widely varying phenomena as the flashing of fireflies and the binding problem in the brain. The relationship between neural activity and the behaviour of the organism is complex and still poorly understood. There have been attempts to explain this using the notion of synchronization, but the participating neurons are fungible, their activity transient and stochastic, and their dynamics highly variable. In spite of this, the behaviour of the organism may be quite robust. The phenomenon of transient induced global response synchronization (TIGoRS) has been used to explain the emergence of stable responses at the global level in spite of marked variability at the local level. TIGoRS is present when an external stimulus to a complex system causes the system's responses to cluster closely in state space. In some models, a 10% input sample can result in a concordance of outputs of more than 90%. This occurs even though the underlying system dynamics is time varying and inhomogeneous across the system. Previous work has shown that TIGoRS is a ubiquitous phenomenon among complex systems. The ability of complex systems exhibiting TIGoRS to stably parse environmental transients into salient units to which they stably respond led to the notion of Sulis machines which emergently generate a primitive linguistic structure through their dynamics. This paper reviews the notion of TIGoRS and its expression in several complex systems models including driven cellular automata, cocktail party and dispositional cellular automata.

Keywords: cellular automata, cooperation, emergence, synchronization, TIGoRS.

1 INTRODUCTION

Conflict and competition are dominant themes in the social and economic sciences. The vast majority of models in economics are predicated on the assumption that the agents involved in economic transactions are all competing for a limited set of resources and capital. Likewise, many models of group behaviour in the social sciences also make a presumption of competition between the group members, particularly when the members are themselves large social groups such as ethnic populations or nations. Nevertheless there is much in the life of individuals that involves at least some degree of cooperation. While conflict may form a prominent aspect of human designs, cooperation abounds in the natural world. Organisms provide quintessential examples of cooperative systems. Symbiosis between organisms provides another set of examples of cooperation. Social insect colonies provide examples of cooperation both within colonies and sometimes with other species [1].

The most extreme form of cooperative behaviour, yet paradoxically the simplest dynamically, can be seen in the synchronization of the activities of different agents engaged in some task. The actions of two men sawing down a tree using a long saw, or driving spikes into a railway tie, exhibit a form of phase locked synchronization. There is little or no freedom in the actions of each member of the dyad. This type of extreme cooperation resembles the traditional concept of synchronization in classical physics where the prototypical example is of two pendulums sitting on the same mantle piece,

DOI: 10.2495/DNE-V11-N4-712-721

set in motion separately and whose oscillations gradually adjust until they come to move in lock step (possibly with a difference in phase). Synchronization does not always involve external behaviour. Many studies have been carried out looking for synchronization between unconscious physiological processes such as heart rate and galvanic skin responses [2].

Phase locked synchronized behaviour is much easier to study than cooperative behaviour on the whole since the behaviour of the synchronized system is overtly constrained. This is not true of synchronization phenomena in general and certainly not true of cooperative behaviour where the only constraint may rest with the goal of the behaviour and not the specifics of the behaviour itself.

Phase locked synchronization provides too severe a constraint to be useful for understanding cooperative behaviour among humans. A more general form of synchronization may be observed in the action of a symphony orchestra. This is certainly an example of cooperation among a group of individuals. No single member of the orchestra can perform an orchestral work. Only the ensemble has this capability. Performers play different instruments and produce different musical lines, yet for the work to be musically coherent and consistent they must follow the score with near perfect timing in the coordination of their parts. In that sense their activities are synchronized even though they are not identical.

Synchronization has been suggested as one means by which the brain is able to integrate information from different sensory domains into a coherent experience [3,4]. Unfortunately synaptic activity is noisy and unreliable [5] and neurons generate rather noisy and unreliable rate or frequency modulated signals [6,7]. Their responses appear to be meta-stable (context dependent) [8–10].

Stigmergy (the incitement to work by the products of work) in collective intelligence systems provides another example in which there is context-dependent cooperation at the colony level capable of producing complex nest constructions, yet the behaviour of individual members of the systems appears to be random and uncorrelated [11]. Dynamic networks, in which the network describing the interactions among participating agents is in continual flux, provide another challenge to the modelling of cooperative behaviour, requiring concepts such as compatibility and sociability [12].

These examples suggest that in a complex hierarchical system there may be patterned behaviour appearing at one level, generated by the activity of system components at a lower level, in the absence of any fixed relationship between these two levels of activity. The global patterns are emergent from the interactions of the system's agents with their environment and among themselves, and these interactions are spontaneous, contextual, emergent, and not pre-programmed. Although humans tend to program patterned responses into the behaviour of their constructions, this need not be true of naturally occurring computational systems [13].

The search for a form of cooperative behaviour which possessed the characteristics of being spontaneous, contextual and emergent, while at the same time constrained enough to analyze and general enough to be salient for complex systems led to the discovery of transient induced global response synchronization (TIGoRS).

2 TIGORS

TIGoRS was discovered while studying the behaviour of a variety of formal complex systems models (tempered neural networks, cellular automata) interacting with a dynamic environment [14–18]. The behaviour of these systems in the presence of external stimulation proved to be quite distinct from that under autonomous conditions. TIGoRS refers to the phenomenon in which a transient stimulus applied to a dynamical system produces a set of responses which cluster more closely in pattern space than would be expected based upon random chance alone. In many models, these responses clustered in spite of variations in the initial states of the system, initial rule configurations, final rule configurations, variability in the agents receiving external stimulation, variability in the sampling of the external transient. Nevertheless a stable correspondence appeared between an external transient (stimulus) and the subsequent global behaviour of the system (responses). This correspondence depended upon symmetry and complexity of the input stimulus and the strength of the stimulus measured as the percentage of agents of the system receiving direct stimulation by the external transient.

Most of the research into TIGoRS has been carried out using cellular automata, which are discrete space, discrete time complex systems composed of individual cells capable of being in any of a set of states, each of which is assigned a fixed neighbourhood of cells with which they interact, and where each subsequent state is determined by a function of its current state and the states of all cells within its local neighbourhood. For example, a 2 state, 3 neighbour cellular automata consists of a set of cells arranged around a circular lattice. Each cell can exist in one of 2 states, 0 or 1. At time step n, the cell *i* determines its state a_i^{n+1} at time step n+1 by entering its state value and those of its adjacent neighbours into a function *f*, hence $a_i^{n+1} = f\left(a_{i-1}^n, a_i^n, a_{i+1}^n\right)$.

Each neighbourhood is described by a triple of states (*x, y, z*) where the value of *x,y,z* may be 0 or 1. We may encode each neighbourhood as a number from 0 to 7 using the formula: $N_{(x,y,z)} = x + 2*y + 2^2*z$. Each vector $\left(a_0, a_1, \ldots a_7\right)$ corresponds to a rule as follows: if the neighbourhood of a cell *k* is (*x, y, z*) having the value *n*, then the next state of *k* will be the value a_n. Each rule is given as a number: $a_0 + 2*a_1 + 2^2*a_2 + \ldots 2^7*a_7$.

The Hamming distance *d* (the total number of sites discordant between two patterns) is a common metric used to study cellular automata patterns. The greater the number of differences between two patterns, the greater their Hamming distance. Let the transients induced in a system Σ under a transient stimulus η, starting from two distinct initial conditions, be denoted Ψ_η, Ψ'_η. TIGoRS was originally defined to be present whenever $d\left(\Psi_\eta, \Psi'_\eta\right) < \frac{1}{2} d\left(rand\left(\Psi_\eta\right), rand\left(\Psi'_\eta\right)\right)$, where $rand\left(\Psi\right)$ denotes a random permutation in space and time of the states of a pattern Ψ.

This definition proved to be inadequate in general since in situations in which the input stimulus directly altered the states of individual agents, it did not distinguish between clustering which was a direct result of the stimulus and clustering which was due to an effect of the dynamics of interaction among the agents. The use of the term TIGoRS has been refined to refer to only those situations in which there is an enhanced clustering above and beyond the input effect and due to the system dynamics. Figure 1 provides an example in which clustering occurs as the sampling rate increases in the absence of TIGoRS. The near linear relationship between stimulus rate and Hamming distance indicates that the correspondence is driven entirely by the stimulus. Figure 2 on the other hand provides an example of TIGoRS. The Hamming distance response is bimodal and non-linear, consisting of an initial region of near linear and steep decline, followed by a transition region, followed a region with a much slower, also near linear decline. The initial response reflects the effect of the internal dynamic of the system, the later response that of the stimulus alone.

Three classes of cellular automata have been studied. *Driven cellular automata* [17] are standard finite deterministic cellular automata in which a transient temporal pattern of states is randomly sampled and input into the cellular automaton, which is then allowed to evolve under its natural dynamic. Many classes of driven cellular automata exhibited pseudo TIGoRS (pure input driven correspondence) but cases of true TIGoRS were observed. Driven cellular automata have a rigid dynamic. *Cocktail party automata* [18] are adaptive, allowing for changes in the rule space of the automaton. Briefly, at time *t*, a cell *A* with neighbourhood *N* updates its state to *s* and then checks throughout to lattice to observe which states were chosen by those cells whose neighbourhoods had the same values as *N*. If the majority of cells chose state *s*', then the cell updates

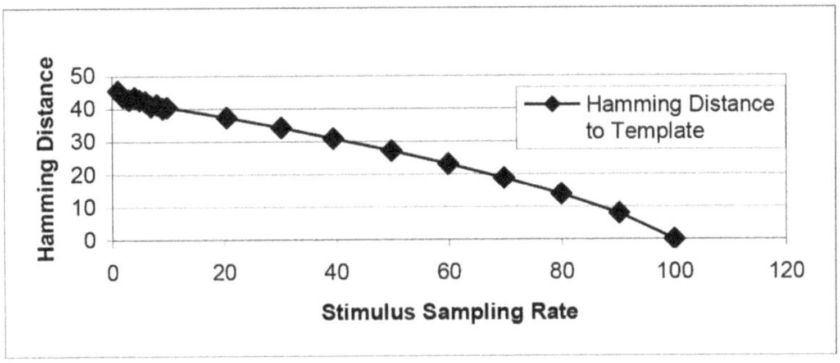

Figure 1: Absence of TIGoRS (Driven C.A. 41, Input 70).

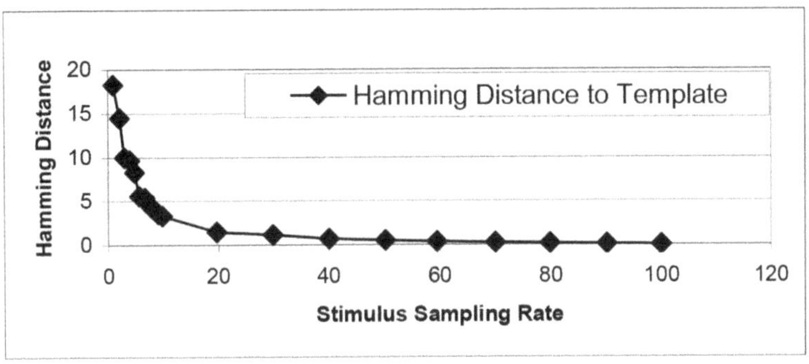

Figure 2: Presence of TIGoRS (Dispositional C.A. 41, Input 70).

its rule, assigning the state *s*' to the neighbourhood value *N*. *Dispositional cellular automata* [19] lie somewhat between driven and cocktail parry automata. The rule space is again dynamic but local. If a given cell at time *t* receives an input, *s*, then it updates its rule so that a state *s* will be generated whenever the neighbourhood matches the current neighbourhood of the cell. For dispositional and cocktail party automata, both the state space and the rule space change as the automaton evolves.

3 COMPUTATIONAL COMPETENCE AND TRANSIENT LANGUAGE

An understanding of TIGoRS requires some familiarity with the formal notions of computational competence and computational performance [13]. Computational competence refers to the ability or capability of the system under scrutiny to carry out the computation under consideration. Computational performance refers to the actual mechanism by which the computation is carried out.

Computational competence requires

1. a specification, in advance, of the environmental context in which the computation is to take place,
2. a specification of the features of the environment which will be manipulated by the observer,
3. a specification of the behaviours to be exhibited which indicate that the required computation did indeed take place, and
4. a specification of the duration within which this task is to be performed.

The determination of computational competence consists first of specifying some set of initial conditions together with the environmental constraints to be imposed by the experimenter. Next, the system-environment pair is allowed to evolve over the allotted time. Finally, the experimenter observes whether or not the subsequent behaviours of both the environment and the organism lie within the previously assigned sets. If so, then the requisite computation took place, otherwise not.

Computational performance requires

1. that the system satisfy the conditions of computational competence for the computation in question and
2. that the system be physically capable of implementing this computational computation in reality using whatever means are appropriate to the organism.

There is no assumption as to the form of the process underlying this performance.

The transient language formulation of a dynamical system was introduced to address questions related to computational competence and performance as well as TIGoRS [15–17]. Classically, a dynamical system is described by a set of instantaneous states which forms the state space S together with a dynamics, defined by a mapping $f_t : T \times S \to S$ where $t \in T$ is a time parameter (which may be discrete or continuous) such that for any initial state x, $f_t(x) = x_t$ and $f_t f_{t'} = f_{t+t'}$. A trajectory in S from an initial state x is a path of the form $\{f_t(x) \mid t \in T\}$

In the transient language formulation, the set P of all short duration transients forms the state space and the trajectory of a dynamical system is represented as a sequence of finite duration transients. Each transient is thus a trajectory of the form $\Pi_k(x) = \{f_t(x) \mid t \in (0,k)\}$. The dynamic is constructed on p by defining a map $g : T \times P \to P$ such that for any transient Π, $g_t(\Pi) = \Pi_t$ where $\pi_t \in P$ and has duration t. These short duration transients can be thought of as letters in an alphabet so that a trajectory becomes a sequence of letters, much like a word. The complex system acts as a transducer, converting stimulus 'words' into response 'words' according to some emergent 'grammar'.

In a typical experiment there will be an open set of stimuli Q representing the actions of the experimenter and the effects of the environment, an open set R of responses deemed representative of the appropriate response or correct computation, and some set Σ of possible initial histories. The system Δ is competent to carry out the required behaviour or computation if the set of responses to the stimulus Q, $\Delta(\Sigma, Q)$, satisfies $\Delta(\Sigma, Q) \subset R$.

TIGoRS provides a mechanism which can create such a pairing for at least a subset of all possible stimuli available to a complex system. Stimuli that are capable of inducing TIGoRS at acceptable levels of input (which must be specified according to the particular circumstances of the situation) are termed salient. A discussion of salience and its implications appears elsewhere [19]. The stimulus-response connection using TIGoRS is an example of a 'good enough' information processing paradigm typical of naturally occurring computational systems [13].

Automata possessing architectures and dynamics consistent with these ideas have been called Sulis machines (Dufort & Lumsden [21]).

4 RESULTS

The combinatorics of cellular automata provides a daunting challenge to formal and experimental analysis. A simple 2-state 3 neighbour automaton having 100 cells evolved for just 450 time steps has a pattern space of some 2^{45000} patterns. Each cell takes one of 256 possible rules so at any time the rule configuration of the automaton is but one of 256^{100} possibilities. In spite of this, classifications of cellular automata based on the symmetry properties of the patterns generated under autonomous evolution have been made. One such classification is: uniform, fixed, vertical, diagonal, complex and chaotic [16]. It is impossible to exhaustively search the entire states space. Instead, a smaller sample must be chosen, usual derived from the symmetry structure of the automata and of the stimulus patterns.

As an illustration of the essential qualitative features of TIGoRS, a series of simulations of the three types of cellular automata (driven, dispositional and cocktail party) were carried out for 2 state 3 neighbour systems having 100 cells and observed for 450 time steps. In previous studies TIGoRS generally occurred within this duration. Each automaton was initialized with a uniform rule configuration and a random initial state with 10% of the cells having state 1. The initial rules were 7 (diagonal), 14 (diagonal), 37 (vertical), 40 (uniform), 41 (complex), 43 (diagonal), 44 (fixed), 45 (chaotic). The input patterns were taken as outputs from autonomous cellular automata having rules 70 (vertical), 72 (fixed), 73 (complex), 74 (diagonal), 75 (chaotic).

The sampling template used to generate the input was randomly generated and distinct for each automaton simulated. A fixed input pattern was sampled according to the sampling template and whenever the pattern was sampled, the corresponding cell had its state changed to that of the input pattern at that site. After sampling, the automaton was allowed to update under its own dynamic. Simulations were carried out for durations from 50–450 times steps in 50 step increments but only results for the 450 time step are reported here.

In order to avoid problems of pseudo-TIGoRS, a new measure of TIGoRS was developed in order to detect only genuine TIGoRS. Using a fixed sampling template corresponding to a fraction p of pattern sites, a fraction p of automaton cells will automatically match the pattern and the responses of other identically stimulated automata. We are interested, however, in the excess synchronization provided by the dynamics of the system itself, and so need to eliminate the input contribution. Thus, the excess synchronization (efficacy) of TIGoRS was defined as follows: $E(p)=-P+H(0)-H(p)$, where P is the percentage of pattern cells sampled, $H(0)$ is the percentage Hamming distance between the current automaton response and the template response in the absence of input and $H(p)$ is the percentage Hamming distance between the template response and the current automaton response under input by pattern p with sampling P.

Three different cellular automata were studied: driven cellular automata, dispositional cellular automata and cocktail party automata. Each class of automata was simulated for the 40 possible input-rule pairings. Of the 40 combinations studied, genuine TIGoRS appeared in 3 cases for driven cellular automata, 28 cases for dispositional cellular automata and 21 cases for cocktail party automata. It should be noted that when the same sampling template is used for all of the simulations, the cocktail party automaton showed TIGoRS in 36 cases. The cocktail party automaton appears to be more discriminating, and there are fewer patterns capable of forcing a global binding of automaton output. It is clear that dispositional and adaptive systems are better able to support emergent processes such as TIGoRS than fixed systems although at the expense of a loss of simple causal relationships between low- and high-level processes.

In the case of driven automata, in two cases (14–70,37–72) TIGoRS applied only to the matching of automaton responses, not of automaton response and pattern. This also occurred in two cases (14–74, 40–73) for dispositional cellular automata, and not at all for the cocktail party automata. The converse situation occurred three times for the dispositional automata (14–70, 14–72, 44–72) and eleven times for the cocktail party automaton (7–72, 14–72, 37–72, 37–73, 40–72, 41–73, 41–74, 43–72, 43–73, 44–72, 45–72). The cocktail party automaton behaved more as a pattern transducer and less like a pattern replicator.

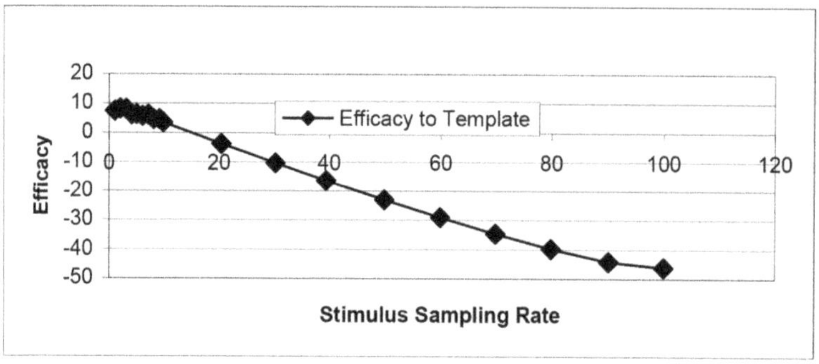

Figure 3: Absence of TIGoRS (Driven C.A. 41, Input 70).

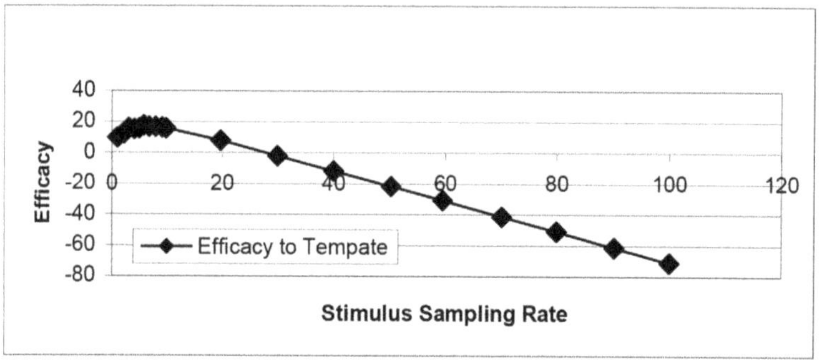

Figure 4: Presence of TIGoRS (Dispositional C.A. 41, Input 70).

Figures 3 and 4 depict the change in efficacy between different automata runs as a function of sampling frequency in the presence of TIGoRS.

The efficacy curve is approximately linear in the case of pseudo-TIGoRS, steadily declining as the sampling rate increases, while it is definitely non-linear in the case of genuine TIGoRS. The efficacy rises rapidly for the first few sampling frequencies, achieving a positive maximum value and then steadily declines as the stimulating frequency increases further. This suggests the presence of two dynamical regimes: an initial regime in which the automaton dynamics augments the effect of the pattern stimulus resulting in rapid synchronizing effect, followed by a second regime in which the bulk of the synchronization is passive arising as a direct result of pattern sampling. These figures provide convincing evidence that TIGoRS is a real dynamical response of the system and is not merely an artefact of the stimulus paradigm.

In both cases the same complex rule (71) was used for the initial rule configuration, while a diagonal pattern (70) was used to provide the input. This pattern has a complicated multi-frequency structure which propagates around the automaton cells. The difference lies in the dynamics – standard cellular automaton in the first case and dispositional in the second. This shows one benefit of the dispositional dynamic. A cocktail party automaton with the same initial rule configuration also exhibits TIGoRS under this input pattern, and the maximal excess synchronization occurs two time steps earlier than for the dispositional automaton.

A direct comparison of the three automata types can be seen in Figs 5 and 6 which depict the changes in Hamming distance and excess synchronization to the pattern for the same initial rule (41) and input pattern (70).

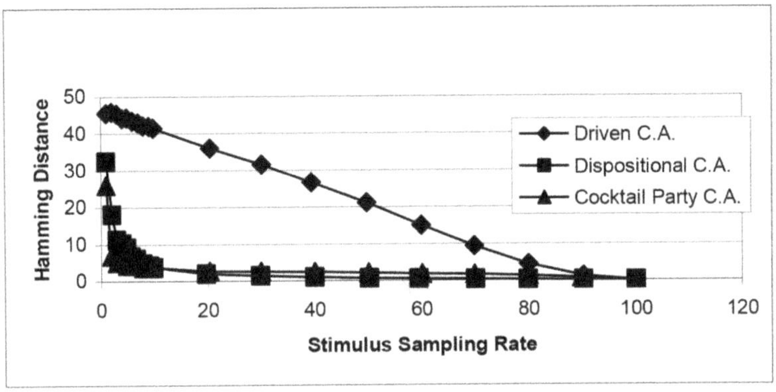

Figure 5: Hamming Distance to Pattern. Comparisons of Three Automata (41–70).

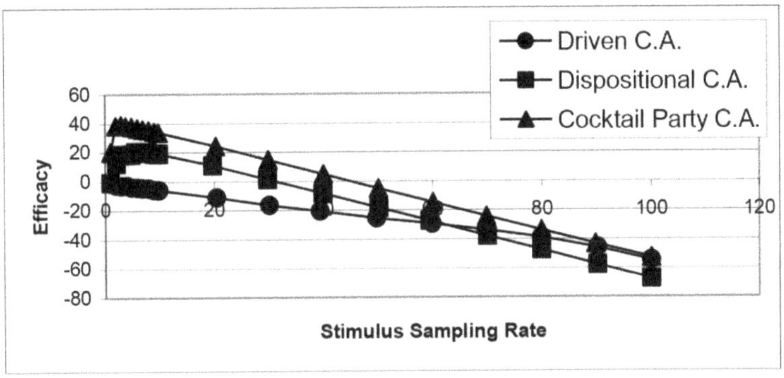

Figure 6: Efficacy to Pattern. Comparisons of Three Automata (41–70).

5 CONCLUSIONS

The phenomenon of TIGoRS is ubiquitous across a wide range of complex systems models and this may be true of naturally occurring computational systems as well [20]. TIGoRS appears to be a emergent non-linear reaction to the presence of an external stimulus which depends upon the particular underlying dynamic of the system. TIGoRS enables a complex system to produce an emergent response to an environmental stimulus that is stable across a broad range of initial, rule and stimulus conditions. TIGoRS provides a means through which information is stored, not as an invariant property of the system, but as a dynamical response. The system extracts salient information out of the environment to construct its response rather than storing the necessary information within its own structure. This is particularly true of systems evolving according to a dispositional dynamic in which the presence of an external input stimulus alters not just the state of the system but its internal underlying dynamic as well.

TIGoRS is a form of synchronization that may have wide spread expression throughout the biological and social realms and is deserving of further scrutiny.

REFERENCES

[1] Wilson, E.O. & Holldobler, B., *The Ants,* Belknap Press: Cambridge, 1990.

[2] Guastello, S.J., Pincus, D. & Gunderson, P.R., Electrodermal arousal between participants in a conversation: nonlinear dynamics for linkage effects. *Nonlinear Dynamics Psychology and Life Sciences*, **10**, pp. 365–399, 2006.

[3] Aertsen, A.D. & Arndt, M., Response synchronization in the visual cortex. *Current Opinion in Neurobiology*, **3**(4), pp. 586–594, 1993.
http://dx.doi.org/10.1016/0959-4388(93)90060-C

[4] Eckhorn, R., Bauer, R., Jordan, W., Kruse, W., Munk, M. & Reitboeck, H., Coherent oscillations: a mechanism of feature linking in the visual cortex. Multiple electrode and correlation analyses in the cat. *Biological Cybernetics*, **60**(2), pp. 121–130, 1988.
http://dx.doi.org/10.1007/BF00202899

[5] Mason, A., Nicoll, A. & Stratford, K., Synaptic transmission between individual pyramidal neurons of the rat visual cortex in vivo. *The Journal of Neuroscience*, **11**(1), pp. 72–84, 1991.

[6] Mainen, Z.F. & Sejnowski, T.J., Reliability of spike timing in neocortical neurons. *Science*, **268**(5216), pp. 1503–1506, 1995.
http://dx.doi.org/10.1126/science.7770778

[7] Shadlen, M.N. & Newsome, W.T., Noise, neural codes and cortical organization. *Current Opinion in Neurobiology*, **4**, pp. 569–579, 1994.
http://dx.doi.org/10.1016/0959-4388(94)90059-0

[8] Barnes, C.A., Suster, M.S., Shen, J. & McNaughton, B.L., Multistability of cognitive maps in the hippocampus of old rats. *Nature*, **388**, pp. 272–275, 1997.
http://dx.doi.org/10.1038/40859

[9] Jung, M.W. & McNaughton, B.L., Spatial selectivity of unit activity in the hippocampal granular layer. *Hippocampus*, **3**(2), pp. 165–182, 1993.
http://dx.doi.org/10.1002/hipo.450030209

[10] Quirk, G.J., Mueller, R.U. & Kubie, J.L., The firing of hippocampal place cells in the dark depends on the rat's recent experience. *The Journal of Neuroscience*, **10**(6), pp. 2008–2017, 1990.

[11] Sulis, W., Collective intelligence: observations and models. In *Chaos and Complexity in Psychology,* eds. S. Guastello, M. Koopmans & D. Pincus, Cambridge University Press: Cambridge, 2009.
http://dx.doi.org/10.1109/ijcnn.1992.227138

[12] Trofimova, I.N., Sociability, diversity, and compatibility in developing systems: EVS approach. In *Formal Descriptions of Developing System*, eds. J. Nation, I. Trofimova, J. Rand & W. Sulis, Boston: Kluwer, 2003.

[13] Sulis, W., Naturally occurring computational systems. *World Futures,* **39**(4), pp. 229–241, 1993.

[14] Sulis, W., Tempered neural networks. In *Proceedings of the International Joint Conference on Neural Networks '92, Vol. III,* IEEE Press: Baltimore, 1992.

[15] Sulis, W., Emergent computation in tempered neural networks: dynamical automata. In *Proceedings of the World Congress on Neural Networks '93 Vol. IV 449*, Lawrence Erlbaum: New York, 1993.

[16] Sulis, W., Emergent computation in tempered neural networks: computation theory. In *Proceedings of the World Congress on Neural Networks '93,* Vol. IV 452, Lawrence Erlbaum: New York, 1993.

[17] Sulis, W., Driven cellular automata. In *1993 Lectures in Complex Systems*, eds. D. Stein & L. Nadel, Addison-Wesley: New York, 1993.

[18] Sulis, W., Driven cellular automata, adaptation, and the binding problem. In *Advances in Artificial Life, Lectures Notes in Artificial Intelligence 929,* eds. F. Moran, A. Moreno, J. Merelo & P. Chacon, Springer-Verlag: New York, 1995.
http://dx.doi.org/10.1007/3-540-59496-5_346

[19] Sulis, W., Synchronization, TIGoRS, and information flow in complex systems. *Nonlinear Dynamics Psychology and Life Science April*, In Press, 2016.

[20] Sulis, W., Archetypal dynamics, emergent situations and the reality game. *Nonlinear Dynamics Psychology and Life Science,* **14**(3), pp. 209–238, 2010.

[21] Dufort, P.A. & Lumsden, C.J., Dynamics, complexity and computation. In *Physical Theory in Biology: Foundations and Exploration,* eds. C.J. Lumsden, W.A. Brandts & L.E.H. Trainor, World Scientific Press: Singapore, 1997.
http://dx.doi.org/10.1142/9789812819598_0004

MECHANICAL CHARACTERIZATION OF BONE ALLOGRAFTS, ENRICHED WITH MESENCHYMAL CELLS

JOSÉ LUIS DÍAZ LEÓN[1], RAÚL LESSO ARROYO[1], RAMÓN RODRÍGUEZ CASTRO[1] & ALFONSO LÓPEZ VÁZQUEZ[2]
[1]Departamento de Ingeniería Mécanica y Bió mecanica. Instituto Técnologico de Celaya, Celaya, Guanajuato, México.
[2]Investigació n y Desarrollo de Top Health. Zapopan, Jalisco, México.

ABSTRACT
Bone allografts, which are used as bone regeneration material, must satisfy two functions: a biologic function and a mechanical function. The biologic task is satisfied by enrichment of the osseous reconstructive material with mesenchymal cells, radio-sterilized and lyophilized, which helps to a more efficient formation of new bone. Regarding the mechanical function, the allografts must be as rigid and strong as intact bone for immediate load-bearing capability. Consequently, a good mechanical characterization is needed to guarantee the structural integrity of the allografts in the host tissue. Thus, in this work results are presented from compression testing of cancellous and cortical/cancellous allograft tissue chips, as well as results on flexure and pure shear testing of cortical allograft strips. The test specimens were fabricated according to standard procedures. For the cancellous graft chips, different elastic moduli were obtained a long the three perpendicular directions, 32.2, 98.2, and 162.4 MPa, showing orthotropic behavior. For the cortical/cancellous chips, compression loads were only applied along the longitudinal and transversal directions of the cortical phase; the corresponding elastic moduli displayed were 518.6 and 384.8 MPa. On the other hand, for the cortical graft strips, the flexure elastic modulus obtained was 38.9 GPa; reported flexure elastic modulus in the literature for fresh human bone are between 1.525 and 31.5 GPa. Finally the shear strength exhibited by the cortical graft was 43.6 MPa.
Keywords: biomechanics, complex material, mechanical bone, tissue engineering.

1 INTRODUCTION

Bones are rigid organs that consist of osseous tissue, bone marrow, endosteum, periosteum, cartilage, nerves, and vascular channels. Osseous tissue, which realizes mechanical functions, is formed by connective tissue cells in an extracellular matrix composed mainly of minerals, proteins, and water. The bone composition and configuration will vary according to factors such as the anatomical location, supported load, age and gender of the person, and the possible diseases that such individual could suffer. In regard to bone composition, mineral phase comprises about 65 wt.% and water around 10 wt.%, whereas the rest part is an organic matrix of collagen. According to its structure, osseous tissue may be cancellous (trabecular) or cortical (lamellar). Cancellous bone is a network of interconnected porosities, ranging between 50% to 90% of empty space, with a solid portion that is formed of struts and plates that can adopt different configurations. It is located at the epiphysis of long bones and the interior of cuboid bones. Cortical tissue is located at the bone surface, and it has a homogeneous and compact macrostructure. It is found mainly at the bone diaphysis and its thickness varies according to the bone anatomical location [1].

Bones have mechanical, synthetic, and metabolic functions. The mechanical functions are protection of internal organs, body support, and interaction with muscles and tendons to generate body movement. The synthetic function is conducted by the bone marrow, where both bone and blood

DOI: 10.2495/DNE-V11-N4-722-730

cells are synthesized. Metabolic functions are related to act as a reservoir of calcium, phosphorus, growth factors and fat.

Regarding the mechanical function, bones are the structural elements of the human body. Skeletal system supports loads due to the different activities of an individual as holding things, walking, running, and so forth. These loads induce tensile, compressive, or shear stresses on the bone tissue. More complex stress states, such as those caused by bending or twisting of bone, can be decomposed into the three basic above-mentioned stresses. To study these stresses, bone mechanical properties such as elasticity modulus, compressive, and tensile strength are important. These properties are highly dependent on the position of the bone and the condition of the individual. Besides, mechanical properties of bone vary depending on the load orientation with respect to the orientation of the tissue (anisotropy) and the speed to which the load is applied (viscoelasticity). Some important mechanical properties are described in Table 1.

Bone tissue as an engineering material can fail because mechanical loads originate stresses over the limits a healthy bone can bear or because the mechanical properties of bone are decreased by various pathologies making the bone weak and prone to be damaged. Usually, small bone defects heal spontaneously but large defects cannot regenerate without intervention. There are several options for the reconstruction of large bone defects that include bone grafts (autograft, allograft and xenograft) as well as bone constructs created by bone tissue engineering principles [2].

Autografts are osseous tissue taken from the same individual and are considered the "gold standard" because they are osteoinductive, osteoconductive, and osteogenic. This material is normally taken from a site that is not under mechanical load such as the iliac crest. Autografts contain cells and growth factors that support the process of bone regeneration and do not exhibit risk of rejection and disease transmission. Some drawbacks of autografts are the necessity of additional surgeries, possible infections, morbidity of the bone, pain, and its limited availability. Depending on the source of the osseous tissue there may also be allografts (tissue from individuals of the same species) or xenografts (tissue from individual of different species, animals in particular). Xenografts, could be osteoinductive and osteoconductive and with low cost with high availability but have the disadvantages of immune response and risk of transmission of animal diseases.

Allografts presented benefits as ready availability and easy handling but require treatments such as freeze drying, irradiation, and washing with acid, among others, to prevent rejection by the receptor and remove any possible infections from the tissue to be implanted; these processes can affect their mechanical and biological properties. Thus, allografts retain osteoconductivity and

Table 1: Property of bone.

Property	Cortical bone	Cancellous bone
Tensile strength (MPa)	50–150	10–100
Compressive strength (MPa)	130–230	2–12
Young's modulus (GPa)	7–30	0.02–0.5
Strain to failure (%)	1–3	5–7
Shear strength (MPa)	53–70	
Shear modulus (GPa)	3	

osteoinductivity but they do not possess osteogenic cellular components [3]. This lack of osteogenic cells leads to the creation of allogenic bone grafts added with mesenchymal stem cells (MSCs), from bone marrow aspirates, to provide all three aspects of bone healing. Then, it remains to know how the mechanical properties of the allografts enhanced with MSCs are affected. Consequently, in this work, results are pre- sented from compression testing of allograft tissue chips, as well as results on flexure and pure shear testing of cortical allograft strips.

2 MATERIALS AND METHODS

Human bones are subjected to different loading conditions: tensile, compression, flexure, shear, torsion and combinations of all of them. Fractures produced by tensile loading generally appear in bones with a considerable portion of cancellous osseous tissue. Fractures due to compression loading frequently emerge in vertebras and in some articulations of elderly individuals. Shear failures are present in almost all bones, mainly those where torsion or combined loads are present. Flexure fractures become visible in long bones where the uniform stress distribution is altered due to unsymmetrical cross sections distinctive in such bones [4, 5].

The bone allografts characterized in the present work, which derive from long and flat human-donated bones, are used for filling of bone defects, prosthetic or bone graft augmentation for general orthopaedic reconstruction [6, 7]. Once implanted, the grafts will be under a combination of tension, compression and shear stresses [8]. Thus, the mechanical behavior of the bone allografts is experimentally studied by performing compression, shear, and bending tests to get their corresponding elastic and strength properties. It is worth noting that the mechanical properties obtained under dry conditions are different from those obtained in hydrated bones (Fig. 1a), which occurs when the bone grafts are being osteo-integrated after implantation [910–11].

The raw material to be tested consisted of cancellous and cortical/cancellous allograft chips (average size 10 mm), as well as small cortical allograft strips. Thus, particular fixtures and specimens were designed according to standard test procedures. In addition, because the mechanical response of bone is loading rate dependent, Fig. 1b [12], the different tests were performed at a travelling head speed of 1 mm/min in the testing machines, which corresponds to the lower limit of the recommended speed range (1–5 mm/min) established in the standards [13, 14, 15].

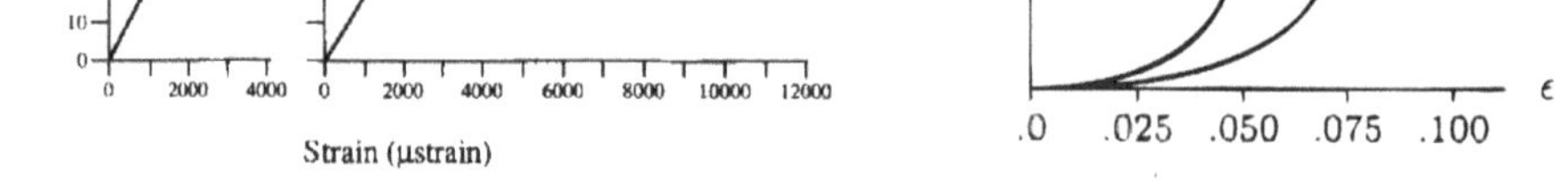

(a) (b)

Figure 1: Graphics. (a) Stress-strain behavior of dry and hydrated bone. (b) Effect of loading rate on stress-strain behavior.

2.1 Compression testing

Several researchers have studied the orthotropic behavior of bone by testing full and partial specimens conformed by cancellous, cortical or cancellous/cortical osseous tissue [12, 16]. Fantner *et al.* [17] tested cancellous chips of 5 × 5 × 4 mm in compression. In the present investigation, cubic chips of 5 × 5 × 5 mm were carefully prepared from irregular-shaped cancellous allograft chips added with MSCs. These specimens were tested in compression along the three perpendicular directions to study their orthotropic behavior (Fig. 2a) [18]. In addition, cubic chips of 5 × 5 × 5 mm were also prepared from cancellous/cortical allograft material; however, in this case compression testing was done along the longitudinal and transversal directions, respect to cortical fibers direction (Fig. 2b) [19]. All tests were performed according to ASTM D1621 standard [13], in a Instron 1011 testing machine.

Turner and Burr [9] proposed a modified compression test fixture to circumvent possible misalignment respect to the loading axis and thus prevent bending effects (Fig. 3a). This misalignment can be due to non-parallel specimen loading faces or possible premature failure of struts or plates of the cancellous structure. The designed fixture consisted of replacing one of the compression plates for an articulated plate, which is free to rotate around the three perpendicular axes such that no bending moments are generated on the loading faces (Fig. 3b).

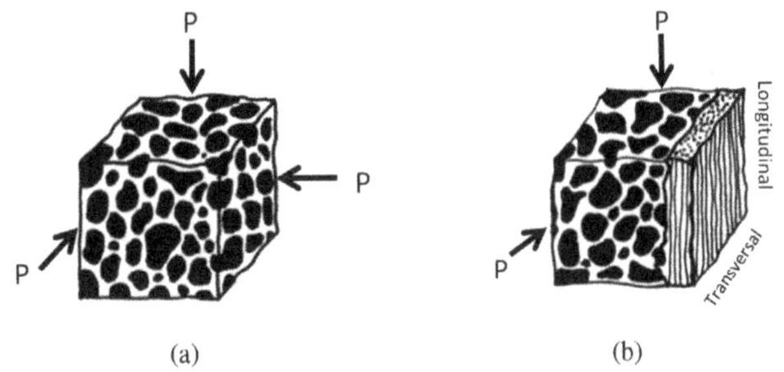

Figure 2: (a) Cancellous tissue chip. (b) Cancellous/cortical chip.

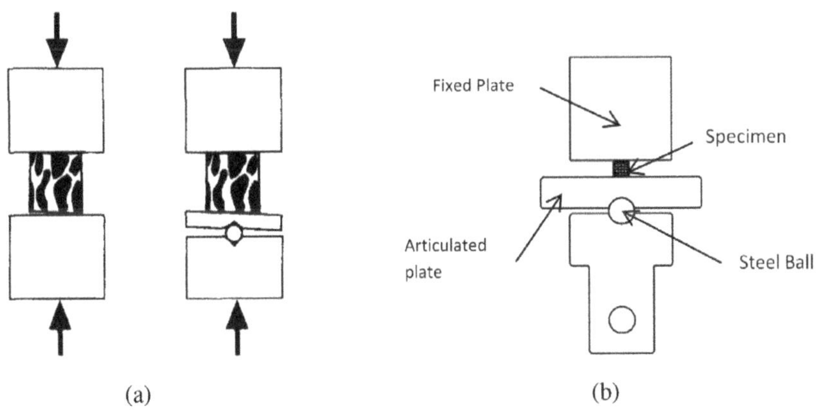

Figure 3: (a) Compression fixture, Turner and Burr [9]. (b) Designed fixture.

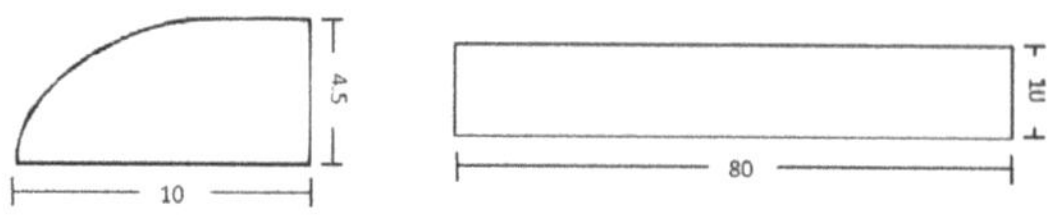

Figure 4: Bending specimen, dimensions in mm.

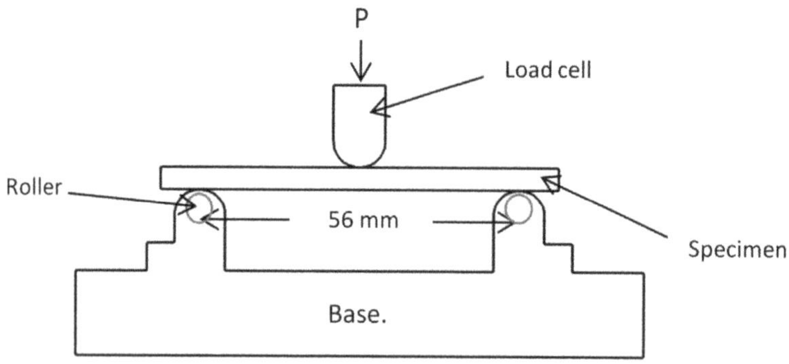

Figure 5: Three point bending fixture.

2.2 Flexure testing

Draper and Goodship [20] used four-point bending for flexure testing of hydrated cortical bone samples 10 mm width, 5 mm thickness, and 45 mm length. In the present case, three point bending tests were performed according to ISO 178 standard [14] on lyophilized cortical bone strips. The samples were cut along the bone fiber direction and the dimensions of the bending specimens are 80 mm length, 4.5 mm thickness, 10 mm width (Fig. 4). It is important to mention that the cross sectional area resembles an elliptical segment, which was maintained during the testing to fulfill the thickness/width aspect ratio established in the standard (Fig. 4). The three-point bending test arrangement is shown in Fig. 5, where the distance between supports (rollers) is 56 mm.

2.3 Shear testing

Shear tests were performed in accordance with the ASTM 5379 standard [15], which is based on the Iosipescu method. Due to material availability, the specimen size was reduced to one quarter of the standard size, following the guidelines for resizing established in the ASTM standard. One of the most important criteria to fulfill is a minimum thickness of 2.5 mm, although some favorable results have been reported when using 0.5 mm thickness [21]. Shear specimen length must be four times the width, with a 90 degree angle length centered notch. The notch depth is 20% of specimen width [9, 21]. The final shear specimen size was 20 mm length, 4 mm thickness, 5 mm width (Fig. 6).

As mentioned above, the Iosipescu method was used for shear testing. The Iosipescu fixture size was reduced to one quarter the size proposed in ASTM D5379 standard [15]. Stiffness of the fixture was a concern, as well as the alignment of the lower and upper grips with the specimen to provide pure shear loading (Fig. 7).

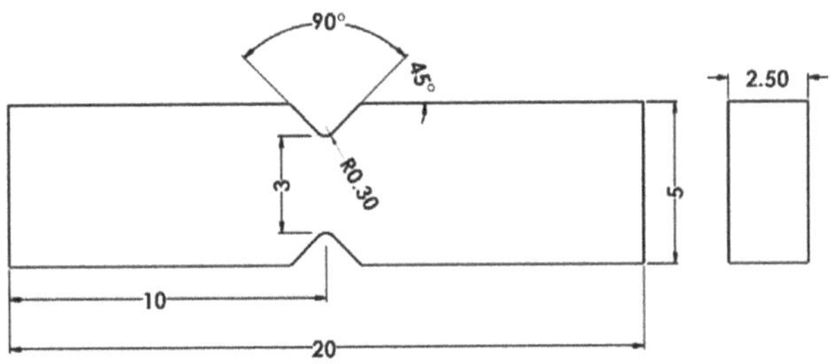

Figure 6: Shear specimen, dimensions in mm.

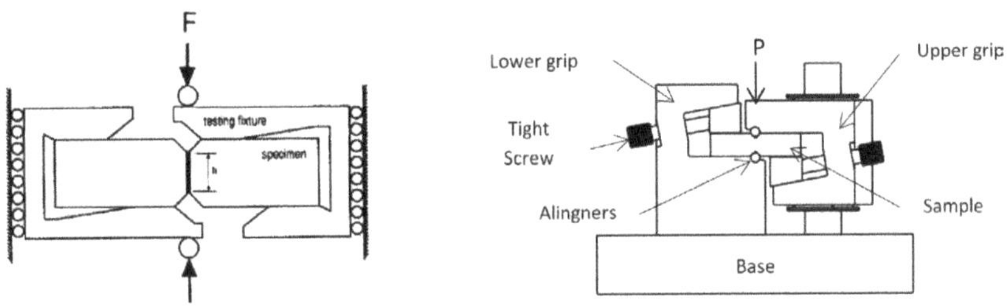

(a) (b)

Figure 7: (a) Iosipescu fixture, Turner and Burr [9], (b) Designed fixture.

3 RESULTS

Elastic compression moduli along the three orthogonal directions were obtained from compression testing cancellous allograft chips. Figure 8a shows the initial linear part of stress–strain plots for the three loading directions and chip 8b. Each line represents the average of 5 tests, and the corresponding moduli obtained are 162.4 MPa, 98.2 MPa, and 32.2 MPa. Fig. 8c displays initial stress–strain plots for cancellous/cortical allograft chips tested in compression. Load was applied along the longitudinal (cortical fiber) and transverse (perpendicular to cortical fiber 8.d) direction. The moduli obtained were 518.6 MPa and 384.6 MPa, respectively.

Shear stiffness modulus was obtained from shear testing allograft cortical strips. Stress–strain plot shown in Fig. 9 represents the average of five tests. The average shear stiffness modulus obtained was 132.14 MPa.

Considering flexure testing of allograft cortical specimens, three tests were performed and Fig. 10a, shows the average stress-strain plot and shear specimen tested 10.b. From such plot, yield stress in bending was 204.15 MPa while the flexural elastic modulus was 24.411 GPa.

4 CONCLUSIONS

Mechanical properties were obtained from compression testing of cancellous and cortical/cancellous allograft chips, and from flexure and pure shear testing of cortical allograft strips. Design of scaled fixtures and specimens was crucial to carry out the mechanical characterization, due to restrictions in the availability of sampling materials as well as to specimen size that can be obtained from different bones. Orthotropic behavior was observed for the cancellous and cortical/cancellous allograft chips.

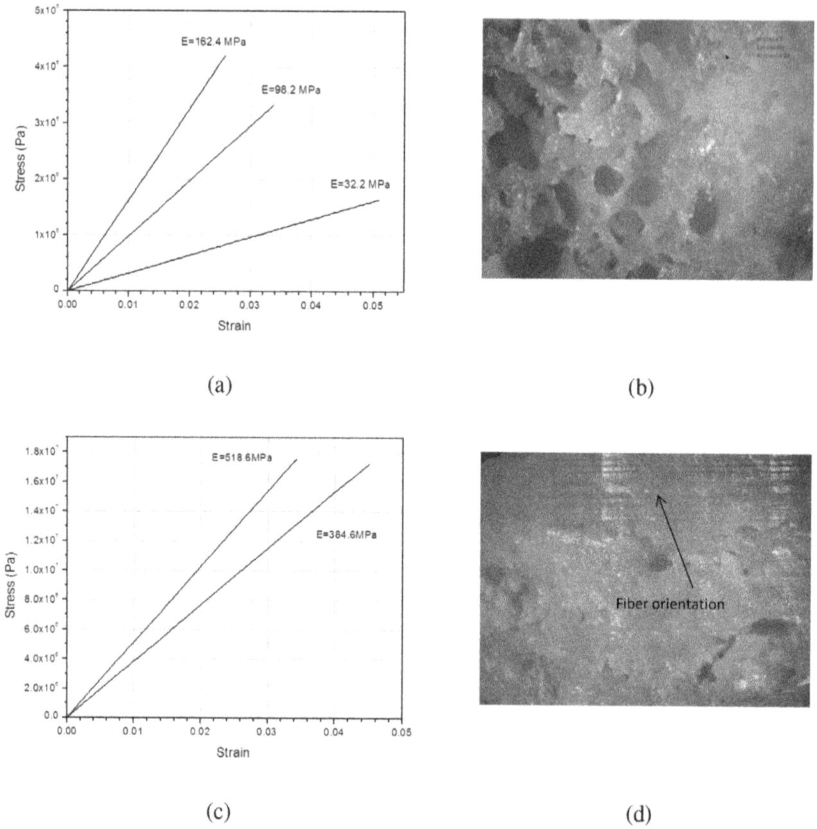

Figure 8: Stress–strain behavior. (a) Compression in cancellous chips along three orthogonal directions, (b) cancellous chip mcirostructure, (c) compression in cancellous/cortical chips along two orthogonal directions, (d) cancellous/cortical microstructure.

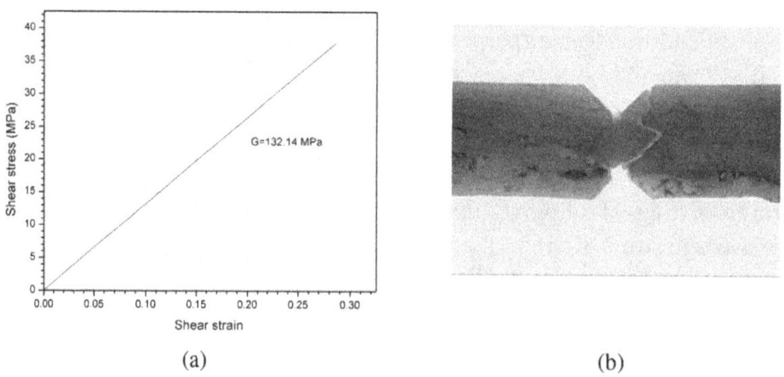

Figure 9: (a) Shear stress–strain behavior in cortical strips, (b) fractured specimen.

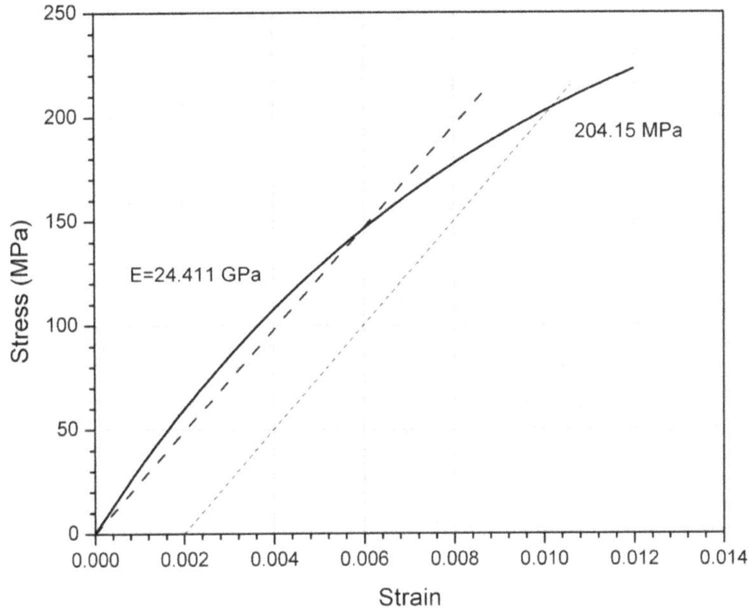

Figure 10: Bending stress–strain behavior in cortical strips.

By comparing the elastic moduli obtained in the present study with those reported in the literature (Table 1), it is observed that elastic moduli values (162.4 MPa, 98.2 MPa, and 32.2 MPa) for the cancellous allograft bones added with MSCs are in the range reported in the literature (20–500 MPa). On the other hand, elastic moduli values obtained for cortical/cancellous allografts added with MSCs (518.6 MPa and 384.6 MPa) are well below of those reported in the literature (7–30 GPa). Likewise, shear stiffness modulus (132.14 MPa) obtained in the present work is very low compared to the corresponding modulus reported (3 GPa). Finally, flexure elastic modulus (24.41 GPa) for cortical strips fall in the range reported in the literature.

REFERENCES

[1] Velasco, M.A., Narváez-Tovar, C.A. & Garzó n-Alvarado, D.A., Design, materials, and mechanobiology of biodegradable scaffolds for bone tissue engineering. *BioMed Research International*, **2015**, p. 21, 2015, Article ID 729076, Hidawi Publishing Corporation.

[2] Eslaminejad, M.B. & Faghihi, F., *Mesenchymal* stem cell-based bone engineering for bone regeneration. *Regenerative Medicine and Tissue Engineering - Cells and Biomaterials,* eds. D. Eberli, InTech, pp. 57–82, 2011.

[3] Grabowski, G. & Robertson, R.N., Bone allograft with mesenchymal stem cells: a critical review of the literature. *Hard Tissue,* **2**(2), p. 20, 2013.

[4] Nordin, M. & Frankel, V., *Biomechanics of bone. Basic Biomechanics of the Musculoskeletal System*, Lippincott Williams & wilkins: Philadelphia, Pennsylvania USA, (pp. 26-–58)., 2001.

[5] Joon, P. & Lakes, R.S., *Biomaterials: An Introduction. Composites as Biomaterials,* Springer: New York USA, (pp. 208-–222)., 2007.

[6] Arévalo, R., Jenny, A., Páez-Guerrero, D.M. & Rodríguez-Pardo, V.M., Celulas madre mesenquimales: características bioló gicas y aplicaciones clínicas. *NOVA*, **5**, 2007.

www.ingramcontent.com/pod-product-compliance
Lightning Source LLC
LaVergne TN
LVHW062230200926
848484LV00006B/1509